Spring +Spring MVC +MyBatis
框架技术精讲与整合案例

缪勇 施俊◎编著

清华大学出版社
北京

内 容 简 介

本书全面地讲解了使用最新流行轻量级框架 SSM 进行 Java EE Web 开发的技术，重点介绍了 Eclipse 开发平台、Spring 框架、Spring MVC 和 MyBatis 框架等基础知识，并用三个 SSM 框架整合案例演示框架应用技巧和连接技术，内容由浅入深，引人入胜。

本书共分 21 章，各基础章节在知识点讲解中，均结合了小案例的精讲，以帮助读者更好地理解和掌握。综合实例部分涉及三个 SSM 整合案例，均按功能分类，采用三层架构(数据访问层、业务逻辑层和视图层)进行精讲，各层之间分层清晰，层与层之间耦合方法简单，读者可以全面理解实现过程，同时三个案例分别使用了三个流行前端 UI：Easy UI、Bootstrap 和 Vue，可以进一步拓展读者的知识面。为方便读者学习和教学开展，本书提供了全程真实课程录像。

本书不仅适合初学者按部就班地学习，也适合网络开发人员作为技术参考，同时，也可作为高等院校计算机相关专业学生的课堂教材。

本书封面贴有清华大学出版社防伪标签，无标签者不得销售。
版权所有，侵权必究。举报：010-62782989，beiqinquan@tup.tsinghua.edu.cn。

图书在版编目(CIP)数据

Spring+Spring MVC+MyBatis 框架技术精讲与整合案例/缪勇，施俊编著. —北京：清华大学出版社，2019（2022.1重印）
 ISBN 978-7-302-52899-9

Ⅰ. ①S… Ⅱ. ①缪… ②施… Ⅲ. ①JAVA 语言—程序设计 Ⅳ. ①TP312.8

中国版本图书馆 CIP 数据核字(2019)第 083521 号

责任编辑：杨作梅
装帧设计：杨玉兰
责任校对：王明明
责任印制：杨 艳

出版发行：清华大学出版社
网　　址：http://www.tup.com.cn, http://www.wqbook.com
地　　址：北京清华大学学研大厦 A 座　　邮　编：100084
社 总 机：010-62770175　　邮　购：010-62786544
投稿与读者服务：010-62776969, c-service@tup.tsinghua.edu.cn
质量反馈：010-62772015, zhiliang@tup.tsinghua.edu.cn

印 装 者：三河市铭诚印务有限公司
经　　销：全国新华书店
开　　本：190mm×260mm　　印　张：30.75　　字　数：750 千字
版　　次：2019 年 6 月第 1 版　　印　次：2022 年 1 月第 6 次印刷
定　　价：99.00 元

产品编号：079817-01

前　言

SSM 框架是继 SSH 之后，目前比较主流的 Java EE 企业级框架，适用于搭建各种大型的企业级应用系统。SSM 框架，是 Spring + Spring MVC + MyBatis 的缩写，Spring 通过依赖注入来管理各层的组件，使用面向方面编程 AOP 管理事务、日志、权限等。Spring MVC 代表了 Model(模型)、View(视图)、Controller(控制)，接收外部请求，进行分发和处理。MyBatis 基于 JDBC 的框架，主要用来操作数据库，并将业务实体和数据表联系起来。

1. 本书内容结构

本书全面介绍了 Eclipse 开发平台、Spring 框架、Spring MVC 框架和 MyBatis 框架等基础知识，最后通过三个具体实例详细讲解了 SSM 框架的整合和运用。全书共分 21 章，具体内容如下。

第 1 章　搭建 Java Web 开发环境，主要介绍 Java 开发包(Java Development Kit)、应用服务器 Tomcat、MySQL 数据库和集成开发环境 Eclipse。

第 2 章　Spring 的基本应用，主要介绍 Spring 框架入门的一些基础知识，重点讲解 Spring 的核心机制：依赖注入/控制反转。

第 3 章　Spring Bean 的装配模式，主要介绍 Bean 工厂 ApplicationContext、Bean 的配置、Bean 的作用域和 Bean 的装配方式。

第 4 章　Spring AOP(面向方面编程)，主要介绍 Spring AOP 的相关概念，并以日志通知为例先后讲解基于 XML 配置文件的 AOP 实现和基于@AspectJ 注解的 AOP 实现。

第 5 章　Spring 的数据库编程，主要介绍 Spring 中的 JDBC 编程。

第 6 章　Spring MVC 简介，主要介绍 Spring MVC 的模式、基础知识和工作流程。

第 7 章　Spring MVC 常用注解，介绍 Spring MVC 的常用注解和 3 种请求映射方式，参数绑定注解和转换 JSON 格式。

第 8 章　Spring MVC 标签库，介绍 Spring MVC 的表单标签和如何使用表单标签绑定数据。

第 9 章　Spring MVC 类型转换、数据格式化和数据校验，介绍 Spring MVC 的数据处理。

第 10 章　Spring MVC 的文件上传和下载，介绍 MultipartResolver 接口和 ResponseEntity 类型。

第 11 章　Spring MVC 的国际化和拦截器，介绍 messageSource、LocaleResolver 国际化语言区域解析器接口以及拦截器的配置。

第 12 章　MyBatis 入门，介绍 MyBatis 框架的概念、下载与安装和工作原理，并详细讲解 MyBatis 框架的基本用法。

第 13 章　MyBatis 的关联映射，介绍使用 MyBatis 框架处理三种关联关系的具体过程。

第 14 章　动态 SQL，介绍 MyBatis 框架的动态 SQL 及动态 SQL 的主要元素。

第 15 章　MyBatis 的注解配置，介绍 MyBatis 框架基于注解的单表增删改查、多表关联映射和动态 SQL 等。

第 16 章　MyBatis 缓存，介绍 MyBatis 框架的缓存概念和一级缓存、二级缓存的用法。

第 17 章　Spring 整合 MyBatis，介绍 SSM 框架，并以登录功能为例，采用注解方式实现 Spring 与 MyBatis 框架的整合。

第 18 章　前端 UI 框架，介绍 jQuery Easy UI、Bootstrap 和 Vue 三种前端框架。

第 19 章　电商平台后台管理系统，结合前端 Easy UI 框架，详细讲解典型的电商平台后台管理系统的具体实现过程。

第 20 章　校园通讯管理系统，结合前端 Bootstrap 的 H+框架，详细讲解校园通讯管理系统的具体实现过程。

第 21 章　电商网站，结合前端 Vue 框架，详细讲解简单的电商网站的具体实现过程。

2. 本书的特点和优势

本书作者在 Java EE Web 领域具有多年的开发和教学讲解经验，熟悉 Java 开发理论知识体系，凭着娴熟的笔法和渊博的理论知识，采取精雕细琢的写作方式，将 SSM 开发技术展现得淋漓尽致，能使读者很快进入实际开发角色。本书与市场上其他类似书籍相比，具有以下与众不同的特色。

(1) 细致全面：本书内容的编排从开发环境搭建开始，从基本知识入手，由浅入深地逐渐转入到高级部分，所讲解的内容囊括了 SSM 框架的重要知识点。注重介绍如何在实际工作中活用基础知识，做到高质量地进行程序开发。

(2) 结合示例：本书在各章知识点的讲解中，都结合了小示例的精讲加以验证。对特别难懂的知识点，通过恰当的示例帮助读者进行分析、加以理解。

(3) 讲解透彻：本书在项目案例讲解的过程中，均按功能分类，采用三层架构(模型、视图、控制)进行相关组件的讲解，各层之间分层清晰，层与层之间以松耦合的方法组织在一起，便于读者理解每个功能的实现过程。

(4) 实用性强：本书的实用性较强，以经验为后盾、以实践为导向、以实用为目标，深入浅出地讲解 Java Web 开发中的各种问题。

(5) 课堂实录：采用知识讲解+课堂实录的方式，提供一套全过程课程录像，更利于读者跟进学习，既可以直接用于学校教学，又方便读者自学，是很多初学者和教学老师的选择。

3. 本书读者对象

- 有一定 Java 基础，但是没有 Java EE 系统开发经验的初学者。
- 有其他 Web 编程语言(如 ASP、ASP.NET)开发经验，欲快速转向 Java EE 开发的程序员。
- 对 JSP 有一定了解，但是缺乏 Java EE 框架开发经验，并希望了解流行开源框架 Spring、Spring MVC 和 MyBatis 以及欲对这些框架进行整合的程序员。
- 有一定 Java Web 框架开发基础，需要对 Java EE 主流框架技术核心进一步了解和掌握的程序员。
- 大中专院校正在学习编程开发的计算机及相关专业的学生。

- 公司管理人员或人力资源管理人员。

4. 本书配套资源

本书附赠完整的学习资源,包括同步教学录像、教学 PPT、源代码、素材文件等内容,可供学习者使用,请从清华大学出版社官网(http://www.tup.tsinghua.edu.cn)下载。

5. 本书作者及致谢

本书由扬州职业大学的缪勇和施俊编写。其中,施俊编写第 1~11 章,主要内容是开发环境搭建和 Spring、Spring MVC 基础知识;缪勇编写第 12~21 章,主要内容是 MyBatis 基础知识和三个整合案例。李新锋对全书进行了审核和统筹,其他参与编写的人员还有王梅、陈亚辉、李艳会、刘娇、王晶晶、游名扬、李云霞、王永庆、蒋梅芳、谢伟、纪航、沈勇等,同时扬州国脉通信发展有限责任公司、江苏智途科技股份有限公司也为本书的编写提供了帮助,在此一一向他们致谢。

由于作者水平有限,书中难免存在一些不足和疏漏之处,敬请读者批评指正。

目 录

第 1 章 搭建 Java Web 开发环境 1
1.1 建立 JDK 的环境 1
- 1.1.1 下载与安装 JDK 1
- 1.1.2 配置 JDK 环境变量 3
- 1.1.3 验证 JDK 是否配置 4

1.2 建立 Tomcat 的环境 4
- 1.2.1 下载与安装 Tomcat 5
- 1.2.2 配置 Tomcat 环境变量 5
- 1.2.3 启动与停止 Tomcat 6
- 1.2.4 Tomcat 的目录结构 6

1.3 创建 MySQL 数据库环境 7
- 1.3.1 MySQL 概述 7
- 1.3.2 下载 MySQL 8
- 1.3.3 安装与配置 MySQL 10
- 1.3.4 使用 MySQL 数据库 12

1.4 搭建 Java Web 开发环境 14
- 1.4.1 下载与安装 Eclipse 14
- 1.4.2 在 Eclipse 中配置 JDK 14
- 1.4.3 在 Eclipse 中配置 Tomcat 15

1.5 创建和发布 Java Web 工程 16
- 1.5.1 创建 Web 项目、设计项目目录结构 16
- 1.5.2 编写页面代码，部署和运行 Web 项目 18

1.6 小结 19

第 2 章 Spring 的基本应用 20
2.1 Spring 概述 20
- 2.1.1 Spring 的概念 20
- 2.1.2 Spring 的优点 21
- 2.1.3 Spring 的体系结构 21
- 2.1.4 Spring 的下载 23

2.2 搭建 Spring 的入门程序 24
2.3 Spring 的核心机制：依赖注入/控制反转 26
- 2.3.1 依赖注入的概念 26
- 2.3.2 依赖注入的类型 27
- 2.3.3 依赖注入的示例 29

2.4 小结 32

第 3 章 Spring Bean 的装配模式 33
3.1 Spring IoC 容器 33
- 3.1.1 Bean 工厂 BeanFactory 33
- 3.1.2 Bean 工厂 ApplicationContext 34

3.2 Bean 的配置 35
3.3 Bean 的作用域 37
3.4 Bean 的装配方式 38
- 3.4.1 基于 XML 的 Bean 装配 38
- 3.4.2 基于 Annotation 的 Bean 装配 39
- 3.4.3 自动装配 41

3.5 小结 42

第 4 章 Spring AOP(面向方面编程) 43
4.1 AOP 概述 43
- 4.1.1 认识 AOP 43
- 4.1.2 AOP 术语 45

4.2 基于 XML 配置文件的 AOP 实现 46
- 4.2.1 前置通知 46
- 4.2.2 返回通知 49
- 4.2.3 异常通知 50
- 4.2.4 环绕通知 51

4.3 基于@AspectJ 注解的 AOP 实现 52
4.4 小结 56

第 5 章 Spring 的数据库编程 57
5.1 Spring JDBC 57
- 5.1.1 Spring JdbcTemplate 类 57
- 5.1.2 Spring JDBC 的配置 58

5.2 JdbcTemplate 的常用方法 59
    5.2.1 execute()方法 59
    5.2.2 update()方法 61
    5.2.3 query()方法 67
5.3 小结 70

第 6 章 Spring MVC 简介 71
6.1 MVC 模式概述 71
    6.1.1 Model I 和 Model II 71
    6.1.2 MVC 模式及其优势 72
6.2 Spring MVC 概述 73
6.3 Spring MVC 环境搭建 74
6.4 Spring MVC 请求流程 78
6.5 小结 79

第 7 章 Spring MVC 常用注解 80
7.1 基于注解的控制器 80
    7.1.1 @Controller 注解 80
    7.1.2 @RequestMapping 注解 83
7.2 请求映射方式 84
    7.2.1 根据请求方式进行映射 84
    7.2.2 Ant 风格的 URL 路径映射 85
    7.2.3 REST 风格的 URL 路径
         映射 86
7.3 绑定控制器类处理方法入参 88
7.4 控制器类处理方法的返回值类型 92
7.5 保存模型属性到 HttpSession 92
7.6 在控制器类的处理方法执行前执行
    指定的方法 93
7.7 直接页面转发、自定义视图与页面
    重定向 94
7.8 Spring MVC 返回 JSON 数据 96
7.9 小结 100

第 8 章 Spring MVC 标签库 101
8.1 Spring MVC 表单标签库概述 101
8.2 Spring MVC 表单标签库 102
    8.2.1 form 标签 102
    8.2.2 input 标签 103
    8.2.3 password 标签 105

    8.2.4 hidden 标签 105
    8.2.5 textarea 标签 105
    8.2.6 checkbox 标签 106
    8.2.7 radiobutton 标签 107
    8.2.8 select 标签 108
    8.2.9 option 标签 109
    8.2.10 options 标签 109
    8.2.11 errors 标签 113
8.3 小结 114

第 9 章 Spring MVC 类型转换、数据
格式化和数据校验 115
9.1 数据绑定简介 115
9.2 数据类型转换 116
    9.2.1 使用 ConversionService 进行
         类型转换 116
    9.2.2 使用@InitBinder 注解进行
         类型转换 120
9.3 数据格式化 121
9.4 数据校验 121
9.5 小结 124

第 10 章 Spring MVC 文件上传和
下载 125
10.1 文件上传 125
    10.1.1 单文件上传 126
    10.1.2 多文件上传 128
10.2 文件下载 130
10.3 小结 132

第 11 章 Spring MVC 的国际化和
拦截器 133
11.1 Spring MVC 国际化 133
    11.1.1 Spring MVC 国际化概述 133
    11.1.2 基于浏览器请求的国际化
          实现 135
    11.1.3 基于 HttpSession 的国际化
          实现 139
    11.1.4 基于 Cookie 的国际化
          实现 141

11.2	Spring MVC 拦截器 143	
	11.2.1 拦截器概述 143	
	11.2.2 拦截器执行流程 146	
	11.2.3 使用拦截器实现用户登录权限验证 149	
11.3	小结 153	

第 12 章 MyBatis 入门 154

12.1	MyBatis 概述 154
12.2	MyBatis 的下载与安装 155
12.3	MyBatis 的工作原理 155
12.4	MyBatis 的增删改查 157
	12.4.1 查询用户 157
	12.4.2 添加用户 162
	12.4.3 修改用户 163
	12.4.4 删除用户 164
12.5	使用 resultMap 属性映射查询结果 165
12.6	使用 Mapper 接口执行 SQL 166
12.7	小结 167

第 13 章 MyBatis 的关联映射 168

13.1	一对一关联映射 168
13.2	一对多关联映射 172
13.3	多对多关联映射 180
13.4	小结 183

第 14 章 动态 SQL 184

14.1	<if>元素 184
14.2	<where>、<if>元素 186
14.3	<set>、<if>元素 187
14.4	<trim>元素 189
14.5	<choose>、<when>和<otherwise>元素 191
14.6	<foreach>元素 193
14.7	小结 195

第 15 章 MyBatis 的注解配置 196

15.1	基于注解的单表增删改查 196
15.2	基于注解的一对一关联映射 199
15.3	基于注解的一对多关联映射 201
15.4	基于注解的多对多关联映射 204
15.5	基于注解的动态 SQL 206
	15.5.1 @SelectProvider 注解 206
	15.5.2 @InsertProvider 注解 208
	15.5.3 @UpdateProvider 注解 209
	15.5.4 @DeleteProvider 注解 211
15.6	小结 212

第 16 章 MyBatis 缓存 213

16.1	一级缓存 213
16.2	二级缓存 215
16.3	小结 216

第 17 章 Spring 整合 MyBatis 217

17.1	环境搭建 217
17.2	编写 SSM 整合的相关配置文件 222
17.3	创建实体类 227
17.4	数据访问层开发 227
17.5	业务逻辑层开发 228
17.6	控制器开发 228
17.7	表示层开发 229
17.8	小结 230

第 18 章 前端 UI 框架 231

18.1	Easy UI 框架 231
	18.1.1 Layout 控件 232
	18.1.2 Tabs 控件 233
	18.1.3 Tree 控件 234
	18.1.4 DataGrid 控件 235
18.2	Bootstrap 框架 236
	18.2.1 Bootstrap 简介 236
	18.2.2 环境安装 237
	18.2.3 Bootstrap 按钮 237
	18.2.4 Bootstrap 表格 239
	18.2.5 Bootstrap 网格系统 240
	18.2.6 Bootstrap 下拉菜单 242
	18.2.7 Bootstrap 面板 243
	18.2.8 Bootstrap 模态框 245
	18.2.9 Bootstrap 标签页 247

- 18.3 Vue 框架 248
  - 18.3.1 Vue 简介 248
  - 18.3.2 第一个 Vue 应用 249
  - 18.3.3 生命周期 250
  - 18.3.4 模板语法 251
  - 18.3.5 计算属性 256
  - 18.3.6 条件渲染 256
  - 18.3.7 列表渲染 257
  - 18.3.8 方法和事件 259
  - 18.3.9 Vue 组件 260
  - 18.3.10 Vue 脚手架 262
  - 18.3.11 Vue 路由 264
  - 18.3.12 Vuex 状态管理 266
- 18.4 小结 269

第 19 章 电商平台后台管理系统 270

- 19.1 需求与系统分析 270
- 19.2 数据库设计 271
- 19.3 环境搭建与配置文件 274
- 19.4 创建实体类 274
- 19.5 创建几个 Dao 接口及动态提供类 ... 278
- 19.6 创建 Service 接口及实现类 287
- 19.7 后台登录与管理首页面 293
- 19.8 商品管理 301
  - 19.8.1 商品列表显示 301
  - 19.8.2 查询商品 306
  - 19.8.3 添加商品 308
  - 19.8.4 商品下架 311
  - 19.8.5 修改商品 313
- 19.9 订单管理 314
  - 19.9.1 创建订单 314
  - 19.9.2 查询订单 324
  - 19.9.3 删除订单 328
  - 19.9.4 查看订单明细 330
- 19.10 客户管理 333
  - 19.10.1 客户列表显示 333
  - 19.10.2 查询客户 336
  - 19.10.3 启用和禁用客户 337
- 19.11 小结 339

第 20 章 校园通讯管理系统 340

- 20.1 需求与系统分析 340
- 20.2 数据库设计 342
- 20.3 环境搭建与配置文件 345
- 20.4 创建实体类 346
- 20.5 后台登录 349
- 20.6 平台管理员功能 359
  - 20.6.1 院校管理员管理 359
  - 20.6.2 院校管理 382
- 20.7 院校管理员功能 398
  - 20.7.1 单位管理 399
  - 20.7.2 角色管理 410
  - 20.7.3 用户管理 420
- 20.8 单位用户功能 435
  - 20.8.1 发送消息 435
  - 20.8.2 接收消息 450
- 20.9 小结 457

第 21 章 电商网站 458

- 21.1 需求与系统分析 458
- 21.2 数据库设计 458
- 21.3 环境搭建与配置文件 459
- 21.4 创建实体类 461
- 21.5 创建几个 Dao 接口 462
- 21.6 创建 Service 接口及实现类 463
- 21.7 商品列表页 465
- 21.8 商品详情页 473
- 21.9 购物车页 475
- 21.10 订单提交 480
- 21.11 小结 482

第 1 章 搭建 Java Web 开发环境

搭建软件开发环境是开发软件的第一步,优秀的开发环境能帮助程序员提高开发速度。本章将讲述如何搭建 Java Web 的开发环境,包括如下内容:Java 开发包(Java Development Kit)、应用服务器 Tomcat、MySQL 数据库和集成开发环境 Eclipse。

1.1 建立 JDK 的环境

JDK 是 Java Development Kit 的缩写,是整个 Java 的核心,包括 Java 运行环境、大量的 Java 工具和 Java 基础类库。主流的集成开发环境(IDE),比如 Eclipse、NetBeans、IntelliJ IDEA 等,都基于 JDK 环境,有些 IDE 在安装时内置了 JDK,有些则需要单独安装。JDK 由 Sun 公司开发,现已被 Oracle 公司收购,它为 Java 程序提供编译和运行环境,不管是做 Java 开发还是安卓开发都需要在计算机上安装 JDK。

1.1.1 下载与安装 JDK

JDK 可从 Oracle 官网下载,目前 JDK 的最新版本是 jdk-9.0.4,下载网址是 http://www.oracle.com/technetwork/java/javase/downloads/index.html,如图 1-1 所示。

单击 DOWNLOAD 按钮后,进入如图 1-2 所示页面,选择 Accept License Agreement 单选按钮,根据自己的系统类型选择下载相应的版本。

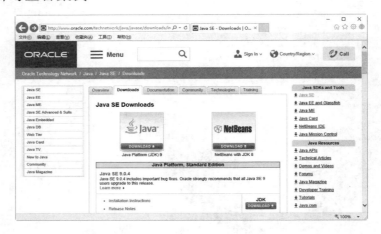

图 1-1　JDK 下载页面

Java SE Development Kit 9.0.4

You must accept the Oracle Binary Code License Agreement for Java SE to download this software.

○ Accept License Agreement　◉ Decline License Agreement

Product / File Description	File Size	Download
Linux	305.07 MB	jdk-9.0.4_linux-x64_bin.rpm
Linux	338.21 MB	jdk-9.0.4_linux-x64_bin.tar.gz
macOS	382.11 MB	jdk-9.0.4_osx-x64_bin.dmg
Windows	375.56 MB	jdk-9.0.4_windows-x64_bin.exe
Solaris SPARC	206.97 MB	jdk-9.0.4_solaris-sparcv9_bin.tar.gz

图 1-2　JDK 下载版本选择

提示：笔者的系统为 64 位的 Windows 10，下载的是 jdk-9.0.4_windows-x64_bin.exe 文件。

安装 JDK 9 的步骤如下。

(1) 双击下载的 exe 程序，进入安装向导界面，单击“下一步”按钮，如图 1-3 所示。

(2) 进入定制安装界面，选择相应的功能，这里我们设置为默认路径，也可单击“更改”按钮，修改为其他路径，然后单击“下一步”按钮，如图 1-4 所示。

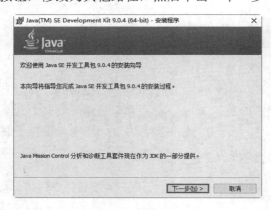

图 1-3　安装向导界面

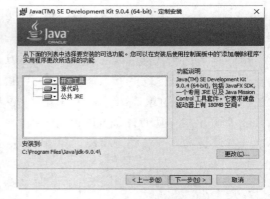

图 1-4　自定义安装界面

(3) JDK 安装完成之后，安装向导还会自动进入 JRE 安装界面。用户可以选择继续安装或取消安装，若要安装也可以更改 JRE 的安装目录，如图 1-5 所示。

(4) 单击“下一步”按钮，安装 JRE，直到最后的完成界面，如图 1-6 所示。

图 1-5　安装 JRE 界面

图 1-6　完成界面

1.1.2　配置 JDK 环境变量

JDK 安装后，如果要在 DOS 控制台窗口编译执行 Java 程序，需要对 JDK 进行环境变量配置，配置过程如下。

(1) 右击"我的电脑"，在弹出的快捷菜单中选择"属性"命令(或进入控制面板，选择"系统")，单击左侧的"高级系统设置"按钮，在弹出的"系统属性"对话框中的"高级"选项卡中单击"环境变量"按钮，弹出"环境变量"对话框，如图 1-7 所示。

图 1-7　系统属性和环境变量

(2) 在"系统变量"选项组中，单击"新建"按钮，弹出"新建系统变量"对话框，输入变量名"JAVA_HOME"，变量值为"C:\Program Files\Java\jdk-9.0.4"(这里是默认的安装路径，可根据自己安装的路径填写)，如图 1-8 所示。

(3) 再次新建系统变量，变量名为"CLASSPATH"，变量值为".;%JAVA_HOME%\lib\;"(注意，前面的"."表示当前路径，此处不可少)，如图 1-9 所示。

图 1-8　新建 JAVA_HOME 变量

图 1-9　新建 CLASSPATH 变量

(4) 在图 1-7 所示的"环境变量"对话框中,选择系统变量 Path,单击下方的"编辑"按钮,在弹出的"编辑环境变量"对话框中单击"编辑文本"按钮,弹出"编辑系统变量"对话框,新增变量值:"%JAVA_HOME%;%JAVA_HOME%\bin;",如图 1-10 所示。

图 1-10 修改 Path

1.1.3 验证 JDK 是否配置

JDK 环境变量配置完成后,在"开始"菜单的"搜索程序和文件"文本框或"运行"对话框中,输入"cmd",打开 cmd.exe 程序界面,在命令提示符后输入"java –version"命令,屏幕上会显示 JDK 的版本信息;再在命令提示符后输入"javac"命令,出现用法提示信息,表示 JDK 已经配置成功,如图 1-11 所示。

图 1-11 查看 Java 版本测试 JDK 是否已配置成功

1.2 建立 Tomcat 的环境

Tomcat 是 Apache 软件基金会(Apache Software Foundation)的 Jakarta 项目中的一个核心项目,是一个免费的开源 Web 容器,随着 Web 应用的发展,Tomcat 被越来越多地应用于商业用途,由 Apache、Sun 和其他一些公司及个人共同开发完成。最新的 Servlet 和 JSP 规范总是能在 Tomcat 中得到体现。目前,官网上的最新版本是 Tomcat 9.0.4。

1.2.1 下载与安装 Tomcat

从 Apache 官方网站可获取相应版本，Tomcat 提供了安装版本和解压缩版本的文件，可以根据需要进行下载。

(1) Tomcat 的官网地址为 http://tomcat.apache.org/，如图 1-12 所示。

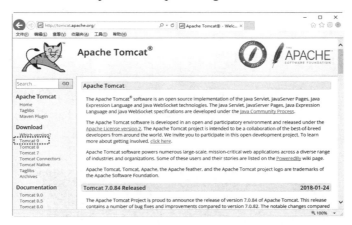

图 1-12 Tomcat 的官网首页

(2) 单击左侧 Download 下方的相应版本 Tomcat 9，进入下载页面，往下拖动滚动条，找到 Tomcat 9.0.4 版本的下载超链接，如图 1-13 所示。

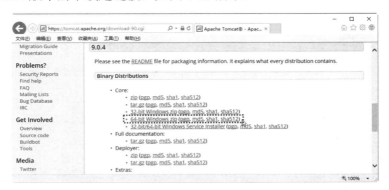

图 1-13 Tomcat 9.0.4 的下载页面

(3) Core 节点下包含 Tomcat 9.0.4 在不同平台的安装文件(根据自己的系统选择)，此处选择 "64-bit Windows zip(pgp,md5,sha1,sha512)"，单击该超链接，即可下载到本地计算机。

> 提示：这里下载的是 Tomcat 的免安装版本，在软件开发过程中，结合使用 IDE 开发工具时，建议使用免安装版，安装版一般在实际部署中使用。

1.2.2 配置 Tomcat 环境变量

Tomcat 的免安装版本配置比较简单，解压缩后需设置 Tomcat 的环境变量，配置的方法与配置 Java 环境变量类似，过程如下。

(1) 将下载的 apache-tomcat-9.0.4-windows-x64.zip 文件解压缩后，将其复制至 C:\Program Files\目录下，也可放在其他任何地方。

(2) 在"环境变量"对话框的"系统变量"中，新建系统变量 CATALINA_HOME，将值设置为 C:\Program Files\apache-tomcat-9.0.4。

(3) 修改系统变量 CLASSPATH，新增值"%CATALINA_HOME%\lib;"，单击"确定"按钮完成配置。

1.2.3 启动与停止 Tomcat

(1) 解压版 Tomcat 的启动方式为：进入 Tomcat 在本地目录下的 bin 子目录，笔者所用计算机为 C:\Program Files\apache-tomcat-9.0.4\bin，执行 startup.bat，就可启动服务，效果如图 1-14 所示。shutdown.bat 文件用于关闭 Tomcat 服务。

图 1-14 启动 Tomcat 服务成功

(2) 在浏览器地址栏中输入 http://localhost:8080/(这里 8080 为 Tomcat 的默认端口号，读者可以根据自己的实际配置修改)，进入 Tomcat 的 Web 管理页面，如图 1-15 所示。

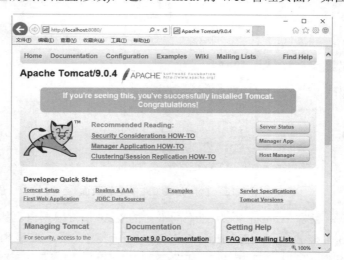

图 1-15 Tomcat 成功安装出现的管理页面

1.2.4 Tomcat 的目录结构

下面以 Tomcat 9.0.4 版本为例，介绍 Tomcat 的目录结构，如表 1-1 所示。

表 1-1　Tomcat 的目录结构

目录	说明
/bin	存放 Tomcat 命令，以.sh 结尾为 Linux 命令，以.bat 结尾为 Windows 命令
/conf	存放 Tomcat 服务器的各种配置文件，例如 server.xml
/lib	存放 Tomcat 服务器运行过程中需要加载的各种 JAR 文件包
/logs	存放 Tomcat 服务器运行过程中产生的日志文件
/temp	存放 Tomcat 服务器运行过程中产生的临时文件
/work	存放 Tomcat 在运行时的编译后文件，例如 JSP 编译后的文件
/webapps	发布 Web 应用，默认情况下将 Web 应用的文件存放在此目录中

提示：不同版本的 Tomcat，目录结构略有区别。

1.3　创建 MySQL 数据库环境

MySQL 是一个小型关系数据库管理系统，也是著名的开放源码的数据库管理系统。由于其体积小、速度快、总体运营成本低，许多中小型网站为了降低网站总体运营成本而选择 MySQL 作为网站数据库。

1.3.1　MySQL 概述

MySQL 由瑞典 MySQL AB 公司开发，后被 Sun 公司收购，现如今 Sun 公司又被 Oracle 公司收购。MySQL 针对不同的用户有不同的版本，分别为社区版和企业版。
- MySQL Community Server：社区版完全免费，但是官方不提供技术支持。
- MySQL Enterprise Server：企业版能为企业提供高性能数据库应用，高稳定性的数据库系统，以及完整的数据库提交、回滚以及锁机制等功能，但该版本收费。

注：MySQL Cluster 主要用于建立数据库集群服务器，需在以上两个版本的基础上使用。

MySQL 的命名机制由 3 个数字组成，例如，MySQL-5.7.21。
- 第 1 个数字 5 是主版本号，用于描述文件格式，表示版本 5 的所有发行版都有相同的文件格式。
- 第 2 个数字 7 是发行级别，它与主版本号组合在一起构成发行序列号。
- 第 3 个数字 21 是此发行系列的版本号，目前 MySQL 5.7.21 是最新版本。

由于其社区版的性能卓越，搭配 Linux、PHP 和 Apache 可组成良好的 LAMP 开发环境。与大型的关系型数据库(如 Oracle、DB2 和 SQL Server 等)相比，MySQL 的规模小，功能有限，但对于中小企业和个人学习使用来说，其提供的功能已经足够用，本书的后续程序，就是使用 MySQL 数据库作为后台数据库管理系统。

1.3.2　下载 MySQL

可以从官网下载 MySQL，其最新版本为 5.7.21，下面介绍如何从官网下载。

（1）进入官网主页 http://www.mysql.com/，在官网下载需要注册，单击右上角的 Register 链接，如图 1-16 所示。

图 1-16　MySQL 官网首页

（2）跳转进入注册页面，因现在都属于 Oracle 公司，所以会跳转到 Oracle 的注册页面，填写信息，如图 1-17 所示。

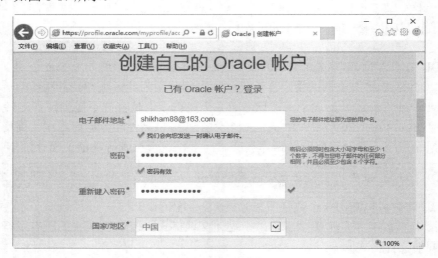

图 1-17　注册页面

（3）注册成功后，在 MySQL 官网上登录，进入网页 http://dev.mysql.com/downloads/，单击 MySQL Community Server 社区版本，如图 1-18 所示。

（4）进入下载页面，在 Select Operating System:下拉列表中选择 Microsoft Windows 选项，在 Select OS Version:下拉列表中选择 Windows (x86,64-bit)选项，然后可以选择安装版，也可选择压缩配置版，这里单击 Windows(x86,32 & 64-bit),MySQL Installer MSI 右侧的 Go to Download Page 按钮，如图 1-19 所示。

图 1-18　版本选择页面

图 1-19　选择系统版本

（5）进入类型选择页面，单击 Windows(x86，32-bit)，MSI Installer (mysql- install-community-5.7.21.0.msi)安装版右侧的 Download 按钮，即可下载，如图 1-20 所示。

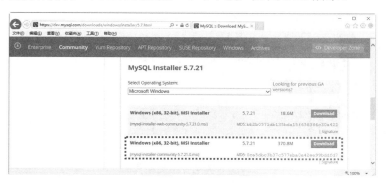

图 1-20　MySQL 下载页面

提示：社区安装版没有 64 位的安装程序，32 位的安装程序也可安装在 64 位的系统上。

1.3.3　安装与配置 MySQL

下载 MySQL 的安装程序后，进行 MySQL 的安装与配置过程。

(1) 双击 mysql-installer-community-5.7.21.0.msi 安装文件，进入 License Agreement(许可协议)界面，勾选下方的 I accept the license terms 复选框，如图 1-21 所示。

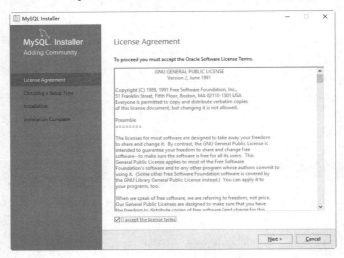

图 1-21　用户许可协议界面

(2) 单击 Next 按钮，进入 Choosing a Setup Type(选择安装类型)界面，根据需要选择，这里我们选择 Custom(自定义)类型，如图 1-22 所示。

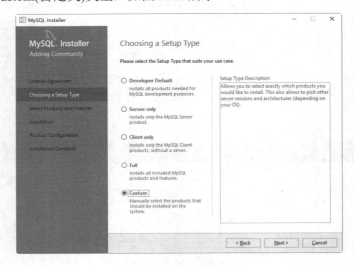

图 1-22　选择安装类型界面

(3) 单击 Next 按钮，进入 Select Products and Features(选择产品和功能)界面，在 Available Products 下方组件中，依次展开 MySQL Servers→MySQL Server→MySQL Server 5.7，选中 MySQL Server 5.7.21-X64，单击绿色的右向箭头，就会添加到右侧，选中右侧 MySQL Server

5.7.21-X64，单击 Advanced Options 链接，在弹出的对话框中可修改安装路径，如图 1-23 所示。

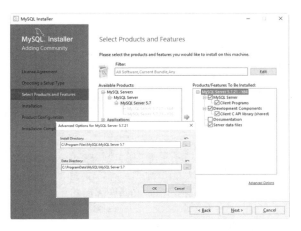

图 1-23　选择安装类型、修改安装路径界面

(4) 在选择产品和功能界面中单击 Next 按钮后，进入安装界面，单击 Execute 按钮，进行安装。

(5) 安装完成后，进入 Product Configuration(产品配置)界面，单击 Next 按钮，进入 Type and Networking(类型和网络配置)界面，对于学习用户来说，在 Config Type 下拉列表框中选择 Development Machine 选项，默认选中 TCP/IP 复选框，Port Number 为 3306，如图 1-24 所示。

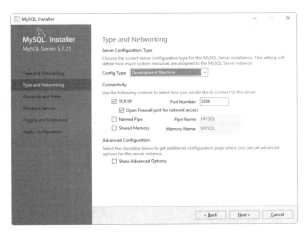

图 1-24　类型和网络配置界面

(6) 单击 Next 按钮，进入 Accounts and Roles(账户和角色)界面，设置 MySQL Root 用户的密码(123456)，可单击 Add User 按钮，添加用户并设置角色和密码，如图 1-25 所示。

(7) 单击 Next 按钮，进入 Windows Service(服务)界面，默认选中 Configure MySQL Server as a Windows Service 和 Start the MySQL Server at System Startup 复选框，Windows Service Name(服务名称)默认为 MySQL57，可以修改，选中 Standard System Account 单选按钮，如图 1-26 所示。

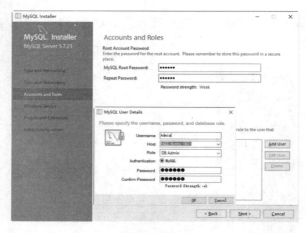

图 1-25 账户和角色界面

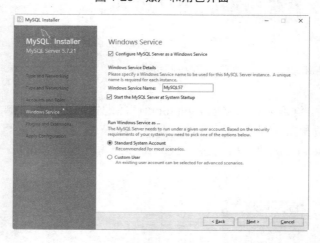

图 1-26 Windows 服务界面

(8) 单击 Next 按钮，进入 Plugins and Extensions(插件和扩展)界面，这里不做选择，单击 Next 按钮，进入 Apply Configuration(插件和扩展)界面，单击 Execute 按钮，进行安装。安装完成后，单击 Finish 按钮返回到 Product Configuration(产品配置)界面，在界面上状态显示为 Configuration Complete(配置结束)。单击 Next 按钮，进入 Installation Complete(安装完成)界面，单击 Finish 按钮，结束安装。

1.3.4 使用 MySQL 数据库

完成以上任务后，可以进入 MySQL 5.7 Command Line Client 进行测试，确保正常使用。操作方法：选择"开始"→"所有程序"→ MySQL → MySQL Server 5.7 → MySQL Command Line Client 命令，出现 DOS 窗口，在其中输入刚刚安装过程中设置的 Root 用户密码(123456)，按 Enter 键，出现"mysql>"提示界面，表示已经安装成功，如图 1-27 所示。

绝大多数的关系数据库都有两个部分：后端作为数据仓库，前端作为用于数据组件通信的用户界面。这种设计非常巧妙，它并行处理两层编程模型，将数据层从用户界面中分离出来，同时使数据库软件制造商专注于它们的产品强项：数据存储和管理，并为第三方创建大量的应用程序提供了便利，使各种数据库间的交互性更强。MySQL 数据库也不例外，常见的前端工具有 SQLyog、WorkBench、Navicat 等。

图 1-27　进入 MySQL

　　SQLyog 是业界著名的 Webyog 公司出品的一款简洁高效、功能强大的图形化 MySQL 数据库管理工具。SQLyog 官方网址为 https://www.webyog.com/，这里使用 SQLyog10.2(汉化版)图形化前端工具操作 MySQL 数据库。

　　启动 SQLyog 程序，第一次使用时会出现选择语言的界面，这里选择简体中文，显示试用信息，单击"继续"按钮，弹出"连接到我的 SQL 主机"对话框，如图 1-28 所示，这里单击"新建"按钮，设置一个名称，我们输入"My"作为名称，单击"确定"按钮，在"密码"文本框中输入密码，也可先测试连接，如图 1-29 所示。

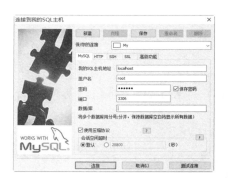

图 1-28　"连接到我的 SQL 主机"对话框(1)　　图 1-29　"连接到我的 SQL 主机"对话框(2)

　　单击"连接"按钮，进入 SQLyog 主窗口，SQLyog 的界面操作方式与 SQL Server 相似，如图 1-30 所示。

图 1-30　SQLyog 图形界面

1.4 搭建 Java Web 开发环境

1.4.1 下载与安装 Eclipse

Eclipse 开发工具的官网网址是 http://www.eclipse.org/，在首页上单击 IDE & Tools 选项；进入下一页面，单击 Java EE 选项；进入下一页面，单击右侧的 Windows 64-bit 链接；进入下一页面，单击 DOWNLOAD 按钮；进入下载页面，下载后的文件为 eclipse-jee-oxygen-2-win32-x86_64.zip 的压缩文件，安装步骤如下。

(1) 解压 eclipse-jee-oxygen-2-win32-x86_64.zip 文件，得到 eclipse 文件夹。

(2) 运行 eclipse 文件夹中的 eclipse.exe 文件，第一次启动 Eclipse 时，会弹出 Eclipse Launcher 对话框，要求设置工作空间以存放项目文档，可设置自己的工作空间，这里将工作空间设置为 F:\eclipse-workspace，如果同时选中 Use this as the default and do not ask again 复选框，下次启动时就不会再显示 Eclipse Launcher 对话框了，如图 1-31 所示。

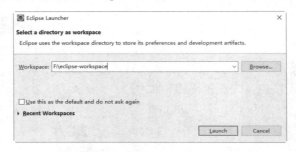

图 1-31 工作空间选择对话框

(3) 单击 Launch 按钮，进入 Eclipse 的初始界面，如图 1-32 所示。

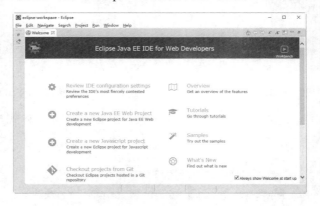

图 1-32 Eclipse 初始界面

1.4.2 在 Eclipse 中配置 JDK

在 Eclipse 中指定使用安装的 jdk-9.0.4 版本的 JRE，可进行如下设置。

(1) 从菜单栏选择 Window → Preferences(首选项)命令，在弹出的 Preferences 对话框的

左侧选择 Java → Installed JREs(已安装的 JRE)，在右侧单击 Add(添加)按钮，在弹出的 Add JRE 对话框中选择 Standard VM 选项，如图 1-33 所示。

(2) 单击 Next 按钮，进入 JRE Definition 设置界面，单击 Directory 按钮，在弹出的界面中指定 JRE 的安装路径，也可在 JRE home 文本框中输入 JRE 安装路径。此处，通过 Directory 按钮找到 jdk-9.0.4 版本的 JRE 的安装路径。确定后，JRE home、JRE name 和 JRE system libraries 会自动添加进来，单击 Finish 按钮完成添加，如图 1-34 所示。

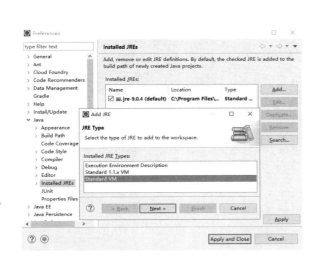

图 1-33　Preferences 对话框　　　　　图 1-34　JRE Definition 设置界面

提示：可根据自己所用计算机的情况，选择已安装的 JRE。

1.4.3　在 Eclipse 中配置 Tomcat

Eclipse 安装好后，你会发现系统里配置了相应的 Tomcat 服务器，若我们想要使用自己安装的 Tomcat，就需要重新进行如下配置。

(1) 在 Eclipse 的菜单栏中选择 Window → Preferences 命令，弹出 Preferences 对话框，在左侧选择 Server → Runtime Environment，单击右侧的 Add(添加)按钮，选择 Apache → Apache Tomcat v9.0，并选中 Create a new local server 复选框，如图 1-35 所示。

(2) 单击 Next 按钮，在 Tomcat Server 界面中，选择 Tomcat 的安装路径，单击 Browse 按钮选择 Tomcat 的安装路径，在 JRE 下拉列表中可用默认的 Workbench default JRE，也可选择之前添加的 JRE，单击 Finish 按钮完成，如图 1-36 所示。

(3) 配置完成后，在 Eclipse 主界面下方的 Servers 选项卡中，就可以看见添加的 Tomcat v9.0 Server at localhost 服务器了，如图 1-37 所示。

(4) 双击配置好的 Tomcat v9.0 Server at localhost 服务器，在 Overview 下的 Server Locations 中选中 Use Tomcat installation (takes control of Tomcat installation)单选按钮，并将 Deploy path 文本框中的内容修改为 webapps，修改相关内容后保存设置，如图 1-38 所示。

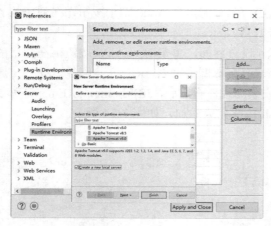 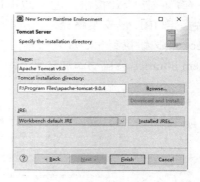

　　图 1-35　Tomcat 的设置界面　　　　　　图 1-36　Tomcat Server 界面

图 1-37　Tomcat 服务器添加完成

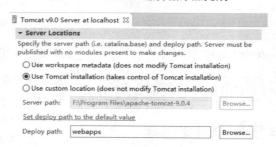

图 1-38　服务器设置修改界面

1.5　创建和发布 Java Web 工程

　　安装和配置 Eclipse 后，就可以通过在 Eclipse 中创建和发布一个 Web 应用程序来学习 Eclipse 的大致使用方法。下面的操作都是基于 Eclipse 进行的。

1.5.1　创建 Web 项目、设计项目目录结构

　　（1）在 Eclipse 菜单栏中选择 File → New → Dynamic Web Project 命令，弹出 New Dynamic Web Project 对话框。在 Dynamic Web Project 界面的 Project name 文本框中输入"myweb"，在 Target runtime 下拉列表框中选择 Apache Tomcat v9.0，在 Dynamic web module version 下拉列表框中选择 3.1 版本，在 Configuration 下拉列表框中选择 Default Configuration for Apache Tomcat v9.0，单击 Next 按钮，如图 1-39 所示。

(2) 进入 Java 设置界面，可以在 src 下添加文件夹，这里不用修改，单击 Next 按钮，如图 1-40 所示。

图 1-39　新建 Web 项目对话框　　　　　　图 1-40　Java 设置界面

(3) 进入 Web Module 界面，选中 Generate web.xml deployment descriptor 复选框，单击 Finish 按钮，如图 1-41 所示。

(4) 设置完成后，在窗体左侧的包资源管理器视图中，就可以看到 myweb 项目的目录结构，如图 1-42 所示。

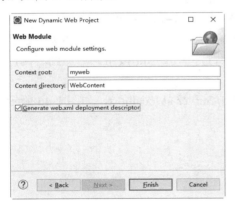

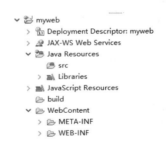

图 1-41　Web 模块设置界面　　　　　　图 1-42　项目的目录结构

> 提示：如果要使项目目录和 MyEclipse 尽量一致，可将图 1-41 所示界面中的 Content directory 文本框中的值修改为 WebRoot。

我们通常把 Java 类文件放在 Java Resources 的 src 目录下，可在 src 下定义包；把网页文件放在 WebContent 目录下，可在根路径下定义文件夹，这样方便管理。

1.5.2 编写页面代码，部署和运行 Web 项目

下面使用集成开发工具 Eclipse 来编写一个 JSP 页面，并部署运行。

(1) 创建一个 JSP 文件，选择 WebContent 并右击，在弹出的快捷菜单中选择 New → JSP File 命令，如图 1-43 所示。

(2) 在弹出的对话框中选择路径和输入文件名，这里为了方便，只输入一个 index.jsp 页面，直接放在 WebContent 的根路径下，如图 1-44 所示。

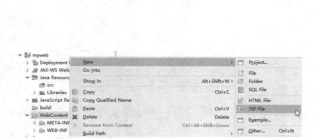

图 1-43 创建 JSP 文件　　　　　　　　图 1-44 输入 JSP 文件名称

(3) 单击 Finish 按钮，完成 JSP 页面的创建，当然页面内容需要我们自己编写。双击 index.jsp 页面，在主体部分编写提示"欢迎来到 Java Web 开发的世界！"，并且把字符编码设置为 contentType="text/html; charset=UTF-8"及 pageEncoding="utf-8"。

(4) 选择 Tomcat v9.0 Server at localhost 服务器并右击，在弹出的快捷菜单中选择 Add and Remove 命令，在对话框的左侧选择 myweb 项目，单击 Add 按钮，添加到右侧，单击 Finish 按钮完成部署，如图 1-45 所示。

(5) 启动 Tomcat，在工具栏中启动 Tomcat v9.0 Server at localhost，如图 1-46 所示，此时会在 Console(控制台)输出 Tomcat 的启动信息。

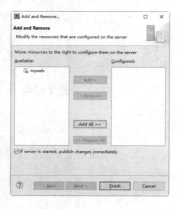
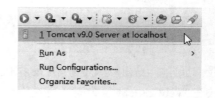

图 1-45 部署 myweb 项目　　　　　　　　图 1-46 启动 Tomcat

(6) 打开浏览器,输入"http://localhost:8080/myweb/index.jsp",按 Enter 键,运行结果如图 1-47 所示。

图 1-47　JSP 程序的运行效果

1.6　小　　结

本章详细讲述了搭建 Java Web 环境所需的各种软件的下载及安装方法,包括 JDK、Tomcat、MySQL 和 Eclipse IDE,以及在 Eclipse 中配置 JRE 和 Tomcat 的方法。本章所选择的软件,也是在开发过程中经常用到的组合。最后在 Eclipse 中创建和发布一个 Web 应用程序来学习 Eclipse 的大致使用方法。

第 2 章　Spring 的基本应用

本章开始学习 Spring 框架。Spring 框架可以说是 Java 世界最为成功的框架,已经发展为一个功能丰富并易用的轻量级集成框架,是当前主流的 Java Web 开发框架。Spring 是为解决企业级应用开发的复杂性而产生的,其核心是一个完整的基于控制反转(IoC)的轻量级容器,用户可以使用它建立自己的应用程序。在容器上,Spring 提供了大量使用的服务,将很多高质量的开源项目集成到统一的框架上。从某个程度上来看,Spring 框架充当了黏合剂和润滑剂的角色,它对 Hibernate、MyBatis 和 Struts 2 等框架提供了良好的支持,能够将相应的 Java Web 系统柔顺地整合起来,并让它们更易使用,同时其本身还提供了声明式事务等企业级开发不可或缺的功能。

2.1　Spring 概述

2.1.1　Spring 的概念

Spring 从 2004 年发布第一个版本至今已经十几年了。Spring 是由 Rod Johnson 组织和开发的一个分层的 Java SE/EE 一站式轻量级框架,它以 IoC(Inversion of Control,控制反转)和 AOP(Aspect Oriented Programming,面向方面编程)为内核。在 Spring 中,认为一切 Java 类都是资源,而资源都是类的实例对象(Bean),容纳并管理这些 Bean 的是 Spring 所提供的 IoC 容器,所以 Spring 是一种基于 Bean 的编程,它深刻地改变着 Java 开发世界,使用基本的 JavaBean 来完成以前只有 EJB 才能完成的工作,避免了 EJB 臃肿、低效的开发模式,因此迅速地取代 EJB 成为了实际的开发标准。

Spring 是一个轻量级框架,它大大地简化了 Java 企业级开发,提供了强大、稳定的功能,又没有带额外的负担,让使用 Spring 的人做每一件事情的时候都有得体和优雅的感觉。Spring 致力于 Java EE 应用各层的解决方案,而不是仅仅专注于某一层的方案。在表现层它提供了 Spring MVC 以及与 Struts 2 框架的整合功能;在业务逻辑层可以管理事务、记录日志等;在持久层可以整合 Hibernate、MyBatis、JdbcTemplate 等技术。这就充分体现出 Spring 是一个全面的解决方案,对于已经有较好解决方案的领域,Spring 绝不做重复的事情。

2.1.2　Spring 的优点

Spring 作为实现 JavaEE 的一个全方位应用程序框架,为开发企业级应用提供了一个健壮、高效的解决方案。它不仅可以应用于服务器端开发,也可应用于任何 Java 应用的开发。Spring 框架具有以下几个特点。

(1) 非侵入式:所谓非侵入式,是指 Spring 框架的 API 不会在业务逻辑上出现,也就是说业务逻辑应该是纯净的,不能出现与业务逻辑无关的代码。针对应用而言,这样才能将业务逻辑从当前应用中剥离出来,从而在其他的应用中实现复用;针对框架而言,由于业务逻辑中没有 Spring 的 API,所以业务逻辑也可以从 Spring 框架快速地移植到其他框架。

(2) 容器。Spring 提供了容器功能,容器可以管理对象的生命周期,以及对象与对象之间的依赖关系。可以写一个配置文件(通常是 xml 文件),在上面定义对象的名字,是否是单例,以及设置与其他对象的依赖关系。那么在容器启动之后,这些对象就被实例化好了,直接用就可以,而且依赖关系也建立好了。

(3) IoC:控制反转,即依赖关系的转移,如果以前都是依赖于实现,那么现在反转为依赖于抽象,其核心思想就是要面向接口编程。

(4) 依赖注入:对象与对象之间依赖关系的实现,包括接口注入、构造注入、属性 setter 方法注入,在 Spring 中支持后两种注入。

(5) AOP:面向方面编程,将日志、安全、事务管理等服务(或功能)理解成一个"方面",以前这些服务通常是直接写在业务逻辑的代码中,这有两个缺点:首先是业务逻辑不纯净,其次是这些服务被很多业务逻辑反复使用,不能做到复用。AOP 解决了上述问题,可以把这些服务剥离出来形成一个"方面",可以实现复用;然后将"方面"动态地插入到业务逻辑中,让业务逻辑能够方便地使用"方面"提供的服务。

其他还有一些特点但不是 Spring 的核心,例如对 JDBC 的封装与简化,提供事务管理功能,对 O/R mapping 工具(Hibernate、MyBatis)的整合,提供 MVC 解决方案;也可以与其他 Web 框架(Struts、JSF)进行整合;还有对 JNDI、mail 等服务进行封装。

2.1.3　Spring 的体系结构

Spring 框架(Spring Framework)在不断发展和完善,目前 Spring 框架由 20 个功能模块构成,这些模块被分组到 Core Container、Data Access/Integration、Web、AOP(Aspect Oriented

Programming)、Instrumentation、Messaging 和 Test 中，Spring Framework 包含的内容如图 2-1 所示。

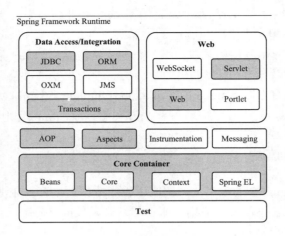

图 2-1　Spring Framework 结构

组成 Spring 框架的每个模块(或组件)都可以单独存在，或者与其他一个或多个模块联合实现。下面对体系结构中的模块作简单介绍，具体如下。

(1) Core Container，核心容器提供了 Spring 的基本功能，是其他模块建立的基础，它主要由 Beans 模块、Core 模块、Context 模块和 Spring EL 模块组成，介绍如下。

- Beans 模块：提供了 BeanFactory，是工厂模式实现的经典，Spring 将管理对象称为 Bean。
- Core 核心模块：提供了 Spring 框架的基本组成部分，包括 IoC 和 DI 功能。
- Context 上下文模块：构建于核心模块之上，它是访问定义配置的任何对象的媒介。扩展了 BeanFactory 的功能，其中 ApplicationContext 是 Context 模块的核心接口。
- Spring EL 模块：是 Spring 3.0 后新增的模块，提供了 Spring Expression Language 支持，是运行时查询和操作对象图的强大的表达式语言。

(2) Data Access/Integration，数据访问/集成层包括 JDBC、ORM、OXM、JMS 和 Transactions 模块，介绍如下。

- JDBC 模块：提供了一个 JDBC 的抽象层，大幅度地减少了在开发中对数据库的操作的编码。
- ORM 模块：提供了与多个第三方持久层框架的良好整合。
- OXM 模块。提供了一个支持对象/XML 映射的抽象层实现，如 JAXB、Castor、XMLBeans、JiBX 和 XStream。
- JMS 模块。指 Java 消息传递服务，包含使用和产生消息的特性，自 Spring 4.1 版本以后，提供了与 Spring-messaging 模块的集成。
- Transactions 模块：支持对实现特殊接口以及所有 POJO 类的编程和声明式的事务管理。

(3) Web，Web 层包括 WebSocket、Servlet、Web 和 Portlet 模块，介绍如下。

- Web 模块：提供了基础的针对 Web 开发的集成特性，例如多方文件上传，利用

Servlet 监听器进行 IoC 容器初始化以及 Web 应用上下文。
- Servlet 模块：也称做 Spring-webmvc 模块，包含 Spring 的模型-视图-控制器(MVC) 和 REST Web Services 实现的 Web 应用程序。
- WebSocket 模块：Spring 4.0 以后新增功能，提供了 WebSocket 和 SockJS 的实现，以及对 STOMP 的支持。
- Portlet 模块：类似 Servlet 模块的功能，提供了 Portlet 环境下的 MVC 实现。

(4) 其他模块。Spring 的其他模块还有 AOP、Aspects、Instrumentation、Messaging 以及 Test 模块，介绍如下。
- AOP 模块：提供了面向方面编程的支持，允许定义方法拦截器和切入点，将代码按照功能进行分离，以降低耦合性。
- Aspects 模块：提供了与 AspectJ 的集成功能，AspectJ 是一个功能强大且成熟的面向方面编程的框架。
- Instrumentation 框架：提供了类工具的支持和类加载器的实现，可以在特定的应用服务器中使用。
- Messaging 模块：Spring 4.0 以后新增的模块，提供了对消息传递体系结构和协议的支持。
- Test 模块：提供了对单元测试和集成测试的支持。

2.1.4　Spring 的下载

前面说过，Spring 的第一个版本在 2004 年发布，经过 10 多年的发展，版本也在不断升级优化。本书编写时，Spring 的最新版本是 5.0.4，本书的代码也是 Spring 5.0.4 版本测试通过，建议也下载该版本。

Spring 是一个独立的框架，它不需要依赖任何 Web 服务器或容器，既可在独立的 Java SE 项目中使用，当然也可在 Java Web 项目中使用。下载 Spring 框架可按如下步骤进行。

(1) 登录 https://repo.spring.io/webapp/#/artifacts/browse/tree/General/libs-release-local/ 或者登录 http://repo.springsource.org/libs-release-local/，依次进入 org→springframework→spring 路径，即可看到 Spring 框架各版本压缩包的下载链接，这里我们选择 RELEASE 5.0.4 版本，单击 spring-framework-5.0.4.RELEASE-dist.zip 下载该文件。

(2) 下载完成后，将压缩文件解压缩后得到一个名为 spring-framework-5.0.4.RELEASE 的文件夹，目录结构如图 2-2 所示。

图 2-2　Spring 框架包的目录结构

- docs 文件夹：该目录下存放 Spring 的相关文档，包括开发指南、API 参考文档。
- libs 文件夹：该目录下包含开发所需的 jar 包和源码。打开 libs 目录可以看到 63 个 jar 包文件。分为三类，其中以 RELEASE.jar 结尾的是 Spring 框架 class 文件的 jar 包；以 RELEASE-javadoc.jar 结尾的是 Spring 框架 API 文档的压缩包；以 RELEASE-sources.jar 结尾的是 Spring 框架源文件的压缩包。整个 Spring 框架由 21 个模块组成，该目录下 Spring 为每个模块都提供了三个压缩包。
- schema 文件夹：该目录包含 Spring 各种配置文件的 XML Schema 文档。
- readme.txt、notice.txt、license.txt 等说明性文档。

(3) 在 libs 目录中，有四个 Spring 的基础包，分别对应 Spring 核心容器的四个模块：spring-core-5.0.4.RELEASE.jar、spring-beans-5.0.4.RELEASE.jar、spring-context-5.0.4.RELEASE.jar、spring-expression-5.0.4.RELEASE.jar。

(4) 除此之外，使用 Spring 开发，除使用自带的 jar 包外，还要依赖于 commons-logging 的 jar 包文件，可通过 http://commons.apache.org/proper/commons-logging/download_logging.cgi 网址下载。下载完成得到一个 commons-logging-1.2-bin.zip 压缩包。将该压缩包解压后，即可找到 commons-logging-1.2.jar 文件。

2.2 搭建 Spring 的入门程序

下面通过示例程序来演示 Spring 框架的简单应用，其中只用到了 Spring 框架而没有使用其他技术，这样能使初学者更加容易理解。实现步骤如下。

(1) 在 Eclipse 中，创建一个名为 spring-1 的 Java 项目，在项目中新建文件夹 lib，用于存放项目所需的 jar 包。

(2) 将前面介绍的四个 Spring 的基础包，即 spring-core-5.0.4.RELEASE.jar、spring-beans-5.0.4.RELEASE.jar、spring-context-5.0.4.RELEASE.jar、spring-expression-5.0.4.RELEASE.jar 复制到 spring-1 项目的 lib 目录中。

(3) 将 Spring 依赖的日志包 commons-logging-1.2.jar 也复制到 lib 目录中。

(4) 选中该项目 lib 目录下的所有 jar 包，右击并选择 Build Path → Add to Build Path 命令，将这些 jar 包添加到项目的构建路径中。

(5) 在 spring-1 项目中创建 com.ssm 包，在包中新建一个名为 HelloSpring 的类。

```java
package com.ssm;
public class HelloSpring {
    private String userName;
    public void setUserName(String userName) {
        this.userName = userName;
    }
    public void show() {
        System.out.println(userName + "：欢迎您来学习Spring框架");
    }
}
```

第 2 章　Spring 的基本应用

(6) 在项目的 src 目录下创建 applicationContext.xml 文件，内容如下：

```xml
<?xml version="1.0" encoding="UTF-8"?>
<beans xmlns="http://www.springframework.org/schema/beans"
    xmlns:xsi="http://www.w3.org/2001/XMLSchema-instance"
    xsi:schemaLocation="http://www.springframework.org/schema/beans
     http://www.springframework.org/schema/beans/spring-beans.xsd">
    <!-- 配置一个bean，将指定类配置给Spring，让Spring创建其对象的实例 -->
    <bean id="helloSpring" class="com.ssm.HelloSpring">
        <!-- 为属性赋值 -->
        <property name="userName" value="张三"></property>
    </bean>
</beans>
```

在 applicationContext.xml 文件中，通过<bean>元素来实例化 HelloSpring 类，id 属性用来标识实例名 helloSpring，class 属性指定待实例化的全路径类名 com.ssm.HelloSpring。子元素<property>用来为类中的属性赋值，name 属性指定 HelloSpring 类中的属性 userName，value 属性给 userName 指定了值"张三"。

在 applicationContext.xml 文件中，第 2～5 行代码是 Spring 的约束配置，该配置信息不需要读者手写，可以在 Spring 的帮助文档中找到，方法如下：

打开 Spring 解压文件夹中的 docs 目录，在 spring-framework-reference 文件夹下打开 html5 文件夹，找到 index.html 文件，使用浏览器打开该 index.html 文件，单击 Core 链接进入，在 Table of Contents 下，找到 1.The IoC container→1.2.Container overview→1.2.1.Configuration metadata 目录，即可找到配置文件的约束信息，这里我们只需将信息复制到项目的配置文件中使用即可，如图 2-3 所示。

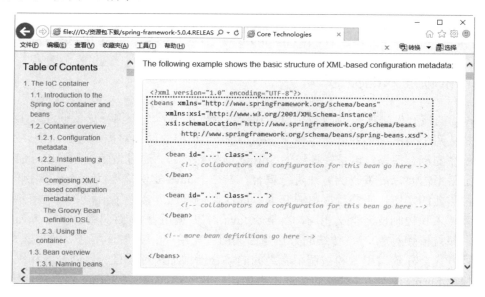

图 2-3　配置文件的约束信息

注意：在学习每一章时，如果涉及配置文件的约束信息，可以将相应章节源代码中的配置文件约束信息复制过来直接使用。

(7) 在 com.ssm 包中创建测试类 TestHelloSpring，在 main()方法中，需要初始化 Spring 容器，并加载 applicationContext.xml 配置文件，通过 Spring 容器获取 HelloSpring 类的 helloSpring 实例(即 Java 对象)，然后调用类中的 show()方法在控制台输出信息。

```
package com.ssm;
import org.springframework.context.ApplicationContext;
import org.springframework.context.support.ClassPathXmlApplicationContext;
public class TestHelloSpring {
    public static void main(String[] args) {
        // 初始化 spring 容器，加载 applicationContext.xml 配置
        ApplicationContext ctx = new ClassPathXmlApplicationContext("applicationContext.xml");
        // 通过容器获取配置中 helloSpring 的实例
        HelloSpring helloSpring = (HelloSpring)ctx.getBean("helloSpring");
        helloSpring.show();// 调用方法
    }
}
```

执行测试类 TestHelloSpring，控制台输出如下：

张三：欢迎您来学习 Spring 框架

从运行结果可以看出，控制台已成功输出了 HelloSpring 类中 show()方法的输出语句。在 main()方法中，并没有通过 new 关键字创建 HelloSpring 类的对象，而是通过 Spring 容器获取实现类对象，这就是 Spring IoC 容器的实现机制。

2.3 Spring 的核心机制：依赖注入/控制反转

2.3.1 依赖注入的概念

Spring 的核心机制就是 IoC(控制反转)容器，IoC 的另外一个称呼是依赖注入(DI)，这两个称呼是从两个角度描述的同一个概念。IoC 是一个重要的面向对象编程的法则，用来削减计算机程序的耦合问题，也是轻量级的 Spring 框架的核心。通过依赖注入，Java EE 应用中的各种组件不需要以硬编码的方法进行耦合，当一个 Java 实例需要其他 Java 实例时，系统自动提供需要的实例，无需程序显式获取。因此，依赖注入实现了组件之间的解耦。

依赖注入和控制反转含义相同，当某个 Java 对象(调用者)需要调用另一个 Java 对象(被调用者，即被依赖对象)时，传统的方法是由调用者采用"new 被调用者"的方式来创建对象，这种方式会导致调用者和被调用者之间的耦合性增加，对项目的后期升级和维护不利。

在使用 Spring 框架后，对象的实例不再由调用者创建，而是由 Spring 容器来创建，Spring 容器会负责控制程序之间的关系，而不是由调用者的程序代码直接控制。这样，控制权由应用程序代码转移到了 Spring 容器，控制权发生了反转，这就是 Spring 的控制反转。

从 Spring 容器的角度来看，Spring 容器负责将被依赖对象赋值给调用者的成员变量，这就相当于为调用者注入了它依赖的实例，这就是 Spring 的依赖注入。

Spring 提倡面向接口的编程，依赖注入的基本思想是：明确地定义组件接口，独立开发各个组件，然后根据组件的依赖关系组装运行。

2.3.2 依赖注入的类型

依赖注入的作用就是使用 Spring 框架创建对象时，动态地将其所依赖的对象注入到 Bean 组件中，其实现主要有两种方式，一种是构造方法注入，另一种是属性 setter 方法注入。具体介绍如下。

1. 构造方法注入

构造方法注入是指 Spring 容器使用构造方法注入被依赖的实例，构造方法可以是有参的或者是无参的。在大多数情况下，我们都是通过构造方法来创建类对象，Spring 也可以采用反射的方式，通过使用带参数的构造方法来完成注入，每个参数代表一个依赖，这就是构造方法注入的原理。这种注入方式，如果参数比较少，可读性还是不错的，但若参数很多，那么这种构造方法就比较复杂了，这个时候应该考虑属性 setter 方法注入。

下面通过示例来讲解构造方法注入。在 spring-1 项目中，在 com.ssm.entity 的包中，新建 AdminInfo 类，包括 id、name、pwd 三个属性，其中 id 属性使用 setter 方法注入，name 和 pwd 属性使用构造方法注入，新建带两个参数的构造方法，代码如下：

```java
package com.ssm.entity;
public class AdminInfo {
    private int id;
    private String name;
    private String pwd;
    public void setId(int id) {
        this.id = id;
    }
    // 省略原有 getter/setter 方法
    public AdminInfo() {
    }
    public AdminInfo(String name, String pwd) {
        this.name = name;
        this.pwd = pwd;
    }
    public void print(){
        System.out.println(id+" -- " + name + " -- "+pwd);
    }
}
```

使用 setter 方法注入时，Spring 通过 JavaBean 的无参构造方法实例化对象。当编写带参数构造方法后，Java 虚拟机不会再提供默认的无参构造方法。为了保证使用的灵活性，建议自行添加一个无参构造方法。

修改 Spring 的配置文件 applicationContext.xml，添加代码如下：

```xml
<bean id="adminInfo" class="com.ssm.entity.AdminInfo">
    <property name="id" value="5"></property>
```

```xml
<constructor-arg name="name" value="admin"/>
<constructor-arg name="pwd" value="123456"/>
</bean>
```

一个<constructor-arg>元素表示构造方法的一个参数,且使用时不区分顺序。当构造方法的参数出现混淆,无法区分时,可通过<constructor-arg>元素的 index 属性指定该参数的位置索引,索引从 0 开始。<constructor-arg>元素还提供了 type 属性用来指定参数的类型,避免字符串和基本数据类型的混淆。

新建测试类 TestSpringConstructor,代码如下。

```java
public class TestSpringConstructor {
    public static void main(String[] args) {
        // 加载 applicationContext.xml 配置
        ApplicationContext ctx = new ClassPathXmlApplicationContext("applicationContext.xml");
        // 获取配置中的 adminInfo 实例
        AdminInfo adminInfo = (AdminInfo)ctx.getBean("adminInfo");
        adminInfo.print();
    }
}
```

运行测试类,控制台的运行结果为"5 -- admin -- 123456",通过调用 AdminInfo 类中的 print()方法,打印输出 AdminInfo 类中的属性值,属性值通过在 applicationContext.xml 的配置文件中注入实现。

2. 属性 setter 方法注入

属性 setter 方法注入是指 Spring 容器使用 setter 方法注入被依赖的值或对象,是常见的一种依赖注入方式,这种注入方式具有高度灵活性。属性 setter 方法注入要求 Bean 提供一个默认的构造方法,并为需要注入的属性提供对应的 setter 方法。Spring 先调用 Bean 的默认构造方法实例化 Bean,然后通过反射的方式调用 setter 方法注入属性值。这种方式是 Spring 最主要的方式,在实际工作中使用广泛。在前面 2.2 小节的示例中,userName 属性就是采用属性 setter 方法注入实现的。

Spring 配置文件从 2.0 版本开始采用 schema 形式,使用不同的命名空间管理不同类型的配置,使得配置文件更具扩展性。Spring 基于 schema 的配置方案为许多领域的问题提供了简化的配置方法,大大降低了配置的工作量。下面讲解使用 p 命名空间来简化属性的注入,使用前要先添加 p 命名空间的声明,配置文件中的关键代码如下。

```xml
<?xml version="1.0" encoding="UTF-8"?>
<beans xmlns="http://www.springframework.org/schema/beans"
    xmlns:xsi="http://www.w3.org/2001/XMLSchema-instance"
    xmlns:p="http://www.springframework.org/schema/p"
xsi:schemaLocation="http://www.springframework.org/schema/beans
 http://www.springframework.org/schema/beans/spring-beans.xsd">
    <!-- 使用 p 命名空间法注入值 -->
    <bean id="admin" class="com.ssm.entity.AdminInfo" p:id="8" p:name="yzpc" p:pwd="yzpc" />
</beans>
```

为 AdminInfo 类中的 name 和 pwd 属性添加相应的 setter 方法，并修改 TestSpringConstructor 测试类，测试类的代码修改部分如下：

```
// 获取配置中的 AdminInfo 实例
AdminInfo admin = (AdminInfo)ctx.getBean("admin");
// 调用 print 方法
admin.print();
```

运行测试类，控制台的运行结果为"8 -- yzpc -- yzpc"，使用 p 命名空间简化配置的效果很明显，其使用方式总结如下。

- 对于直接量(基本数据类型、字符串)属性，使用方式如下：p:属性名="属性值"。
- 对于引用 Bean 的属性，使用方式如下：p:属性名-ref="Bean 的 id"。

3．两种注入方式的对比

Spring 同时支持构造方法注入和属性 setter 方法注入两种方式，它们各有优缺点，开发中可以根据实际需要灵活选择，两种方式的特点总结如下。

- 使用 setter 方法时，与传统的 JavaBean 写法更类似，程序开发人员更容易了解和接受，通过 setter 方法设定依赖关系显得更加直观、自然。
- 对于复杂的依赖关系，如果采用构造方法注入，会导致构造器过于臃肿，难以阅读。尤其是在某些属性可选的情况下，多参数的构造器更加笨重。
- 构造方法注入可以在构造器中决定依赖关系的注入顺序，当某些属性的赋值操作有先后顺序时，这点尤为重要。
- 对于依赖关系无须变化的 Bean，构造方法注入更有用处。如果没有 setter 方法，所有的依赖关系全部在构造器内设定，后续代码不会对依赖关系产生破坏。依赖关系只能在构造器中设定，所以只有组件的创建者才能改变组件的依赖关系。而对组件的调用者而言，组件内部的依赖关系完全透明，更符合高内聚的原则。

2.3.3　依赖注入的示例

了解两种注入方式后，下面以属性 setter 方法注入为例，实现一个简单的登录验证，下面讲解 Spring 容器在程序中是如何实现依赖注入的。

(1) 将项目 spring-1 复制并重命名为"spring-2"，再导入到 Eclipse 开发环境中。

(2) 编写 DAO 层。

在项目 spring-2 的 src 目录下，新建包 com.ssm.dao，在包中新建一个接口 UserDAO.java，在接口中添加方法 login()，代码如下：

```
package com.ssm.dao;
public interface UserDAO {
    public boolean login(String loginName,String loginPwd);
}
```

创建接口 UserDAO 的实现类 UserDAOImpl，新建包 com.ssm.dao.impl，创建接口 UserDAO 的实现类 UserDAOImpl，实现 login()方法，代码如下：

```
package com.ssm.dao.impl;
```

```java
import com.ssm.dao.UserDAO;
public class UserDAOImpl implements UserDAO {
    @Override
    public boolean login(String loginName, String loginPwd) {
        if (loginName.equals("admin") && loginPwd.equals("123456")){
            return true;
        }
        return false;
    }
}
```

在登录验证时为了简化 DAO 层代码，暂时没有用到数据库。如果用户名为"admin"，密码为"123456"，则登录成功。

(3) 编写 Service 层。

在 src 目录下新建包 com.ssm.service，在包中新建一个接口 UserService.java，在接口中添加方法 login()，代码如下：

```java
package com.ssm.service;
public interface UserService {
    public boolean login(String loginName,String loginPwd);
}
```

创建接口 UserService 的实现类 UserServiceImpl.java，存放在 com.ssm.service.impl 包中，实现 login()方法，代码如下：

```java
package com.ssm.service.impl;
import com.ssm.dao.UserDAO;
import com.ssm.service.UserService;
public class UserServiceImpl implements UserService{
    // 使用接口 UserDAO 声明对象，添加 setter 方法，用于依赖注入
    UserDAO userDAO;
    public void setUserDAO(UserDAO userDAO) {
        this.userDAO = userDAO;
    }
    // 实现接口中的方法
    @Override
    public boolean login(String loginName, String loginPwd) {
        //调用 userDAO 中的 login()方法
        return userDAO.login(loginName, loginPwd);
    }
}
```

在上述代码中，没有采用传统的 new UserDAOImpl()方式获取数据访问层 UserDAOImpl 类的实例，只是使用 UserDAO 接口声明了对象 userDAO，并为其添加 setter 方法，用于依赖注入。UserDAOImpl 类的实例化和对象 userDAO 的注入将在 applicationContext.xml 配置文件中完成。

(4) 配置 applicationContext.xml 文件。

创建 UserDAOImpl 类和 UserServiceImpl 类的实例，需要添加<bean>标记，并配置其相

关属性，代码如下：

```xml
<!-- 配置创建 UserDAOImpl 的实例 -->
<bean id="userDAO" class="com.ssm.dao.impl.UserDAOImpl"></bean>
<!-- 配置创建 UserServiceImpl 的实例 -->
<bean id="userService" class="com.ssm.service.impl.UserServiceImpl">
    <!-- 属性 setter 方法依赖注入数据访问层组件 -->
    <property name="userDAO" ref="userDAO" />
</bean>
```

<bean>元素用来定义 Bean 的实例化信息，class 属性指定类全名(包名+类名)，id 属性指定生成的 Bean 实例名称。上述配置中，首先通过一个<bean>元素创建 UserDAOImpl 类的实例，在使用另一个<bean>元素创建 UserServiceImpl 类的实例时，使用了<property>元素，该元素是<bean>元素的子元素，用于调用 Bean 实例中的相关 setter 方法完成属性值的赋值，从而实现依赖关系的注入。<property>元素中的 name 属性指定 Bean 实例中的相应属性的名称，这里将 name 属性设置为 userDAO，代表 UserServiceImpl 类中的 userDAO 属性需要注入值。name 属性的值可以通过 ref 属性或者 value 属性指定。当使用 ref 属性时，表示对 Spring IOC 容器中某个 Bean 实例的引用。这里引用了前一个<bean>元素中创建的 UserDAOImpl 类的实例 userDAO，并将该实例赋值给 UserServiceImpl 类中的 userDAO 属性，从而实现了依赖关系的注入。UserServiceImpl 类的 userDAO 属性值是通过调用 setUserDAO()方法完成注入的，这种注入方式称为设值注入，设值注入方式是 Spring 推荐使用的。

(5) 编写测试类。

在 com.ssm 包中创建测试类 TestSpringDI，代码如下：

```java
package com.ssm;
import org.springframework.context.ApplicationContext;
import org.springframework.context.support.ClassPathXmlApplicationContext;
import com.ssm.service.UserService;
public class TestSpringDI {
    public static void main(String[] args) {
        // 加载 applicationContext.xml 配置
        ApplicationContext ctx = new ClassPathXmlApplicationContext("applicationContext.xml");
        // 获取配置中的 UsersServiceImpl 实例
        UserService userService = (UserService)ctx.getBean("userService");
        boolean flag = userService.login("admin", "123456");
        if (flag) {
            System.out.println("登录成功");
        } else {
            System.out.println("登录失败");
        }
    }
}
```

在测试类 TestSpringDI 中，首先通过 ClassPathXmlApplicationContext 类加载 Spring 配置文件 applicationContext.xml，然后从配置文件中获取 UserServiceImpl 类的实例，最后调用 login()方法。运行测试类，当用户名为"admin"、密码为"123456"时，控制台输出"登录成功"，否则输出"登录失败"。

2.4 小　　结

　　本章主要介绍了 Spring 框架入门的一些基础知识。首先讲解了 Spring 框架的概念、优点、体系结构以及下载方法和下载后的目录结构；然后通过一个简单的 HelloSpring 入门程序演示 Spring 框架的简单应用，并演示了依赖注入中的构造方法注入和属性 setter 方法注入；最后以登录验证为例，讲述了 Spring 的核心机制：依赖注入/控制反转。

第 3 章 Spring Bean 的装配模式

作为 Spring 核心机制的依赖注入/控制反转，改变了传统编程习惯，对组件的实例化不再由应用程序完成，转而交由 Spring 容器完成，需要时注入到应用程序中，从而将组件之间的依赖关系进行了解耦。这一切都离不开 Spring 配置文件中使用的<bean>元素，下面我们来深入学习 Spring 中的 Bean。

3.1 Spring IoC 容器

Spring 框架的主要功能是通过其 IoC 容器来实现的，它可以容纳我们所开发的各种 Bean，并且我们可以从中获取各种发布在 Spring IoC 容器里的 Bean，并且通过描述得到它。Spring IoC 容器的设计主要基于 BeanFactory 和 ApplicationContext 两个接口。

3.1.1 Bean 工厂 BeanFactory

Spring IoC 设计的核心是 Bean 容器，BeanFactory 是 Spring IoC 容器的核心接口，采用了 Java 经典的工厂模式，由 org.springframework.beans.factory.BeanFactory 接口定义。作为制造 Bean 的工厂，BeanFactory 接口负责向容器的使用者提供实例，其功能主要是负责初始化各种 Bean，并根据预定的配置完成对象之间依赖关系的组装，最终向使用者提供已完成装配的可用对象。Spring IoC 对容器管理对象没有任何要求，无需继承某个特定类或者实现某些特定接口，这极大地提高了 IoC 容器的可用性。

BeanFactory 接口提供了几个实现类，其中最常用的是 org.springframework.beans.factory.xml.XmlBeanFactory，会根据 XML 配置文件中的定义来装配 Bean，加载配置信息的语法如下：

```
BeanFactory beanFactory=new XmlBeanFactory(new FileSystemResource("D:/bean.xml"));
```

这种加载方式在实际开发中并不多见，读者了解即可。

3.1.2 Bean 工厂 ApplicationContext

ApplicationContext 是 BeanFactory 的子接口之一，换句话说，BeanFactory 是 Spring IoC 容器所定义的最底层接口，而 ApplicationContext 是其高级接口之一，并且对 BeanFactory 功能做了许多有用的扩展，还添加了对国际化、资源访问、事件传播等方面的支持，使之成为 Java EE 应用中首选的 IoC 容器，可应用在 Java APP 和 Java Web 中。所以在大部分的工作场景下，都会使用 ApplicationContext 作为 Spring IoC 容器。

ApplicationContext 的中文含义是"应用上下文"，它继承自 BeanFactory 接口。ApplicationContext 接口有三个常用的实现类，如下所示：

- ClassPathXmlApplicationContext

ClassPathXmlApplicationContext 类从类路径 ClassPath 中寻找指定的 XML 配置文件，找到并装载 ApplicationContext 的实例化工作。例如：

```
ApplicationContext context=new ClassPathXmlApplicationContext(String configLocation);
```

configLocation 参数指定 Spring 配置文件的名称和位置，如"applicationContext.xml"。

- FileSystemXmlApplicationContext

FileSystemXmlApplicationContext 类从指定的文件系统路径中寻找指定的 XML 配置文件，找到并装载 ApplicationContext 的实例化工作。例如：

```
ApplicationContext context=new FileSystemXmlApplicationContext(String configLocation);
```

其与 ClassPathXmlApplicationContext 的区别在于读取 Spring 配置文件的方式，FileSystemXmlApplicationContext 不再从类路径中读取配置文件，而是通过参数指定配置文件的位置，可以获取类路径之外的资源。这种绝对路径的方式，会导致程序的灵活性变差，所以这个方法一般不推荐使用。

- XmlWebApplicationContext

XmlWebApplicationContext 类从 Web 系统中的 XML 文件载入 Bean 定义的信息，Web 应用寻找指定的 XML 配置文件，找到并装载完成 ApplicationContext 的实例化工作，例如：

```
ServletContext servletContext = request.getSession().getServletContext();
ApplicationContext ctx =
WebApplicationContextUtils.getWebApplicationContext(servletContext);
```

在使用 Spring 框架时，可通过实例化其中的任何一个类来创建 Spring 的 ApplicationContext 容器，这些实现类的主要区别在于装载 Spring 配置文件实例化 ApplicationContext 容器的方式不同，在实例化 ApplicationContext 之后，同样通过 getBean 方法从 ApplicationContext 容器中获取装配好的 Bean 实例以供使用。

在 Java 项目中通过 ClassPathXmlApplicationContext 类手工实例化 ApplicationContext 容器通常是不二之选，但对于 Web 项目就不行了。Web 项目的启动是由相应的 Web 服务器负责的，因此在 Web 项目中，ApplicationContext 容器的实例化工作最好交给 Web 服务器来完成，Spring 为此提供了如下两种方式：

1. 基于 ContextLoaderListener 实现

这种方式只适用于 Servlet 2.4 及以上规范的 Servlet，需要在 web.xml 中添加如下代码：

```xml
<!-- 指定 Spring 配置文件的位置，多个配置文件以逗号分隔 -->
<context-param>
    <param-name>contextConfigLocation</param-name>
    <param-value>classpath:applicationContext.xml</param-value>
</context-param>
<!-- 指定以 ContextLoaderListener 方式启动 Spring 容器 -->
<listener>
    <listener-class>
        org.springframework.web.context.ContextLoaderListener
    </listener-class>
</listener>
```

2. 基于 ContextLoaderServlet 实现

该方式需要在 web.xml 中添加如下代码：

```xml
<!-- 指定 Spring 配置文件的位置，多个配置文件以逗号分隔 -->
<context-param>
    <param-name>contextConfigLocation</param-name>
    <param-value>classpath:applicationContext.xml</param-value>
</context-param>
<!-- 指定以 Servlet 方式启动 Spring 容器 -->
<servlet>
    <servlet-name>context</servlet-name>
    <servlet-class>
        org.springframework.web.context.ContextLoaderServlet
    </servlet-class>
    <load-on-startup>1</load-on-startup>
</servlet>
```

在本书后面章节中讲解三大框架 Spring、MyBatis 与 Spring MVC 的整合开发时，将采用基于 ContextLoaderListener 的方式来实现由 Web 服务器实例化 ApplicationContext 容器。

3.2 Bean 的配置

Spring 容器支持 XML 和 Properties 两种格式的配置文件，在实际开发中，最常用的就是 XML 格式的配置方式。这种配置方式通过 XML 文件来注册并管理 Bean 之间的依赖关系。在 Spring 中，XML 配置文件的根元素是<beans>，其下包含<bean>子元素，每个<bean>子元素定义一个 Bean，并描述该 Bean 如何被装配到 Spring 容器中。

<bean>元素中包含多个属性，其常用属性介绍如下。

- id：容器中 Bean 的唯一标识符，Spring 容器对 Bean 的配置、管理通过该属性完成，装配 Bean 时根据 id 值获取对象。
- name：Spring 容器同样可以通过此属性对容器中的 Bean 进行配置和管理，name 属性可以为 Bean 指定多个名称，每个名称之间用逗号或分号隔开。
- class：该属性指定 Bean 的具体实现类，使用对象所在类的全路径。
- scope：设定 Bean 实例的作用范围，其属性值有：singleton(单例)、prototype(原型)、request、session、globalSession、application 和 webSocket。

<bean>元素中同样包含多个子元素，其子元素介绍如下。

- constructor-arg：可以使用此元素传入构造方法的参数进行实例化。该元素的 index 属性设置构造参数的序号(从 0 开始)，type 属性指定构造参数类型，参数值可通过 ref 属性或 value 属性直接指定，也可通过 ref 或 value 子元素指定。
- property：用调用 Bean 实例中的 setter 方法完成属性赋值，从而完成依赖注入。该元素的 name 属性指定 Bean 实例中的相应属性名，ref 属性或 value 属性用于指定参数值。
- ref：<property>、<constructor-arg>等元素的属性或子元素，可用于指定对 Bean 工厂中某个 Bean 实例的引用。
- value：<property>、<constructor-arg>等元素的属性或子元素，可直接用于指定一个常量值。
- list：<property>等元素的子元素，指定 bean 的属性类型为 List 或数组类型的属性值。
- set：<property>等元素的子元素，指定 bean 的属性类型为 Set 类型的属性值。
- map：<property>等元素的子元素，指定 bean 的属性类型为 Map 的属性值。
- entry：<map>元素的子元素，用于设置一个键值对。其 key 属性指定字符串类型的键值，可用 ref 或 value 子元素指定其值，也可通过 value-ref 或 value 属性指定其值。

在 XML 配置文件中，通常一个普通的 Bean 只需定义 id(或者 name)和 class 两个属性。定义 Bean 的示例代码如下：

```xml
<?xml version="1.0" encoding="UTF-8"?>
<beans xmlns="http://www.springframework.org/schema/beans"
    xmlns:xsi="http://www.w3.org/2001/XMLSchema-instance"
    xsi:schemaLocation="http://www.springframework.org/schema/beans
        http://www.springframework.org/schema/beans/spring-beans.xsd">
    <!-- 使用 id 属性定义 bean1，其对应的实现类为 com.ssm.Bean1 -->
    <bean id="bean1" class="com.ssm.Bean1">
    </bean>
    <!-- 使用 name 属性定义 bean2，其对应的实现类为 com.ssm.Bean2 -->
    <bean name="bean2" class="com.ssm.Bean2" />
</beans>
```

上述代码中，分别使用 id 属性和 name 属性定义了两个 Bean，并使用 class 元素指定其对应的实现类。如果在 Bean 中未指定 id 和 name 属性，则 Spring 会将 class 值作为 id 使用。

3.3 Bean 的作用域

容器最重要的任务是创建并管理 JavaBean 的生命周期，创建 Bean 之后，需要了解 Bean 在容器中是如何在不同作用域下工作的。

Bean 的作用域就是指 Bean 实例的生存空间或有效范围，Spring 为 Bean 实例定义了多种作用域来满足不同情况下的应用需求，如下所示。

- singleton：在每个 Spring IoC 容器中，一个 bean 定义对应一个对象实例。
- prototype：一个 bean 定义对应多个对象实例。
- request：在一次 Http 请求中，容器会返回该 Bean 的同一个实例，而对于不同的用户请求，会返回不同的实例。该作用域仅在基于 web 的 Spring ApplicationContext 情形下有效。
- session：在一次 HTTP Session 中，容器会返回该 Bean 的同一个实例。而对于不同的 HTTP Session 请求，会返回不同的实例。该作用域仅在基于 web 的 Spring ApplicationContext 情形下有效。
- global session：在一个全局的 HTTP Session 中，容器会返回该 Bean 的同一个实例。仅在使用 portlet context 时有效。

下面通过示例来具体介绍单实例作用域和原型模式作用域。

1. singleton(单实例)作用域

这是 Spring 容器默认的作用域，当一个 Bean 的作用域为 singleton 时，Spring IoC 容器中只会存在一个共享的 Bean 实例，并且所有对 Bean 的请求，只要 id 与该 Bean 定义相匹配，就只会返回 Bean 的同一实例。换言之，当把一个 Bean 定义设置为 singlton 作用域时，Spring IoC 容器只会创建该 Bean 定义的唯一实例。这个单一实例会被存储到单例缓存 (singleton cache)中，并且所有针对该 Bean 的后续请求和引用都将返回被缓存的对象实例。单实例模式对于无会话状态的 Bean(如 DAO 组件、业务逻辑组件)来说是最理想的选择。

要在 Spring 配置文件 applicationContext.xml 中将 Bean 定义成 singleton，可以这样配置：

```
<bean id="helloSpring" class="com.ssm.HelloSpring" scope="singleton">
    <property name="userName" value="张三"></property>
</bean>
```

将项目 spring-1 复制并重命名为"spring-3"，再导入到 Eclipse 开发环境中。在项目 spring-3 的 com.ssm 包中创建测试类 TestBeanScope，在 main()方法中测试 singleton 作用域，代码如下：

```
package com.ssm;
import org.springframework.context.ApplicationContext;
import org.springframework.context.support.ClassPathXmlApplicationContext;
```

```java
public class TestBeanScope {
    public static void main(String[] args) {
        // 加载 applicationContext.xml 配置
        ApplicationContext context = new 
ClassPathXmlApplicationContext("applicationContext.xml");
        // 获取配置中的实例
        HelloSpring hs1 = (HelloSpring) context.getBean("helloSpring");
        HelloSpring hs2 = (HelloSpring) context.getBean("helloSpring");
        System.out.println(hs1 == hs2);//判断取得的实例值是否相等
    }
}
```

运行测试类 TestBeanScope，控制台输出结果为 true，说明 hs1 和 hs2 的地址是相等的，由此可见在单实例作用域下，只创建了一个 HelloSpring 类的实例。

2. prototype(原型模式)作用域

prototype 作用域的 Bean 在每次对该 Bean 请求时都会创建一个新的 Bean 实例，对需要保持会话状态的 Bean(如 Struts 2 中充当控制器的 Action 类)应该使用 prototype 作用域。Spring 不能对一个原型模式 Bean 的整个生命周期负责，容器在初始化、装配好一个原型模式实例后，将它交给客户端，就不再过问了。因此，客户端要负责原型模式实例的生命周期管理。在 Spring 配置文件中将 Bean 定义成 prototype，配置修改如下：

```xml
<bean id="helloSpring" class="com.ssm.HelloSpring" scope="prototype">
    <property name="userName" value="张三"></property>
</bean>
```

再次运行测试类 TestBeanScope，控制台输出结果为 false，这说明在 prototype 作用域下，创建了两个不同的 HelloSpring 类的实例。

其他作用域，如 request、session 以及 global session 仅在基于 Web 的应用中使用，在以后用到时再作讲解。

3.4 Bean 的装配方式

Spring 容器负责创建应用程序中的 Bean，并通过依赖注入协调这些对象之间的关系。创建应用对象之间协作关系的行为通常称为装配(wiring)，这也是依赖注入(Dependency Injection)的本质，Bean 的装配方式即 Bean 依赖注入。在开发基于 Spring 的应用时，Spring 容器支持多种形式的 Bean 装配方式，如基于 XML 的装配、基于注解(Annotation)的装配和自动装配等，下面介绍这三种装配方式的使用。

3.4.1 基于 XML 的 Bean 装配

Spring 提供了两种基于 XML 的装配方式：属性 setter 方法注入和构造方法注入。在 Spring 实例化 Bean 的过程中，Spring 首先会调用 Bean 的默认构造方法来实例化 Bean 对象，然后通过反射的方式调用 setter 方法来注入属性值。属性 setter 方法注入要求 Bean 必须满足两点：
- Bean 类必须提供一个默认的构造方法。
- Bean 类必须为需要注入的属性提供对应的 setter 方法。

在 Spring 配置文件中，使用属性 setter 方法注入时，在<bean>元素的子元素<property>中为每个属性注入值；而使用构造方法注入时，在<bean>元素的子元素<constructor-arg>中定义构造方法的参数，可使用其 value 属性(或子元素)来设置该参数的值。

在 2.3.2 小节的依赖注入的类型示例中，就是基于 XML 的 Bean 装配，其 Spring 的配置文件代码如下：

```xml
<bean id="adminInfo" class="com.ssm.entity.AdminInfo">
    <property name="id" value="5"></property>
    <constructor-arg name="name" value="admin"/>
    <constructor-arg name="pwd" value="123456"/>
</bean>
```

3.4.2 基于 Annotation 的 Bean 装配

在 Spring 中尽管使用 XML 配置文件可以实现 Bean 的装配工作，但如果应用中 Bean 的数量较多，会导致 XML 配置文件过于臃肿，从而给维护和升级带来一定的困难。从 JDK 5 开始提供了名为 Annotation(注解)的功能，Spring 正是利用这一特性，逐步完善对 Annotation(注解)技术的全面支持，使 XML 配置文件不再臃肿，向"零配置"迈进。

Spring 中定义了一系列的 Annotation(注解)，如下所示。

- @Component 注解

@Component 是一个泛化的概念，使用此注解描述 Spring 中的 Bean，仅仅表示一个组件(Bean)，可以作用在任何层次。使用时只需将该注解标注在相应类上即可。

- @Repository 注解

@Repository 注解用于将数据访问层(DAO 层)的类标识为 Spring 中的 Bean，其功能与@Component 相同。

- @Service 注解

@Service 通常作用在业务层(Service 层)，用于将业务层的类标识为 Spring 中的 Bean，其功能与@Component 相同。

- @Controller 注解

@Controller 通常作用在控制层(如 Spring MVC 的 Controller)，用于将控制层的类标识为 Spring 中的 Bean，其功能与@Component 相同。

- @Autowired 注解

用于对 Bean 的属性变量、属性的 setter 方法及构造方法进行标注，配合对应的注解处理器完成 Bean 的自动配置工作。@Autowired 注解默认按照 Bean 类型进行装配。@Autowired 注解加上@Qualifier 注解，可直接指定一个 Bean 实例名称来进行装配。

- @Resource 注解

作用相当于@Autowired，配置对应的注解处理器完成 Bean 的自动配置工作。区别在于：@Autowired 默认按照 Bean 类型进行装配，@Resource 默认按照 Bean 实例名称进行装配。

@Resource 包括 name 和 type 两个重要属性。Spring 将 name 属性解析为 Bean 实例的名称，将 type 属性解析为 Bean 实例的类型。如果指定 name 属性，则按照实例名称进行装配；如果指定 type，则按照 Bean 类型进行装配。如果都不指定，则先按照 Bean 实例名称装配，如果不能匹配，再按照 Bean 类型进行装配，如果都无法匹配，则抛出 NoSuchBeanDefinitionException 异常。

- @Qualifier 注解

与@Autowired 注解配合，将默认按 Bean 类型装配修改为按 Bean 实例名称进行装配，Bean 的实例名称由@Qualifier 注解的参数指定。

在上面几个注解中，虽然@Repository、@Service 和@Controller 的功能与@Component 注解的功能相同，但为了使类的标注更加清晰，在实际开发中推荐使用@Repository 标注数据访问层(DAO 层)、使用@Service 标注业务逻辑层(Service 层)、使用@Controller 标注控制器层(Controller 层)。

在 2.3.3 小节以登录验证为例讲述依赖注入时，使用了基于 XML 的 Bean 装配。下面将该示例的依赖关系通过注解进行装配，实现过程如下：

(1) 将项目 spring-2 复制并重命名为"spring-4"，再导入到 Eclipse 开发环境中。

(2) 将 spring-aop-5.0.4.RELEASE.jar 文件添加到项目 spring-4 的 lib 目录中，再将该 jar 包添加到项目的构建路径中。

(3) 修改 UserDAO 接口的实现类 UserDAOImpl，如下所示：

```java
package com.ssm.dao.impl;
import org.springframework.stereotype.Repository;
import com.ssm.dao.UserDAO;
@Repository("userDAO")
public class UserDAOImpl implements UserDAO {
    @Override
    public boolean login(String loginName, String loginPwd) {
        if (loginName.equals("admin") && loginPwd.equals("123456")){
            return true;
        }
        return false;
    }
}
```

在 UserDAOImpl 类上使用了@Repository 注解，将数据访问层的类 UserDAOImpl 标识为 Spring Bean，通过 value 属性值标识该 Bean 名称为"userDAO"(value 可以缺省)。

(4) 修改 UserService 接口的实现类 UserServiceImpl，代码如下：

```java
package com.ssm.service.impl;
import org.springframework.beans.factory.annotation.Autowired;
import org.springframework.stereotype.Service;
import com.ssm.dao.UserDAO;
import com.ssm.service.UserService;
@Service("userService")
public class UserServiceImpl implements UserService{
    @Autowired
    UserDAO userDAO;            // 使用接口 UserDAO 声明对象
```

```
    // 实现接口中的方法
    @Override
    public boolean login(String loginName, String loginPwd) {
        //调用 userDAO 中的 login()方法
        return userDAO.login(loginName, loginPwd);
    }
}
```

在 UserServiceImpl 类上使用了@Service 注解，将业务逻辑层的类 UserServiceImpl 标识为 Spring Bean，该 Bean 的名称为"userService"。在 UserDAO 类型的属性 userDAO 上使用了@Autowired 注解，可将步骤(3)中由 Spring 容器实例化的名称为 userDAO 的 Bean 装配到属性 userDAO 中。@Autowired 注解自动装配具有兼容类型的单个 Bean 属性，可以加在构造器、普通字段、一切具有参数的方法上。

（5）修改 Spring 配置文件。

组件扫描默认不是启用的，我们还需要显式修改 Spring 配置文件，让 Spring 能够扫描类路径中的类，并识别出@Component、@Repository、@Service 和@Controller 注解，需要在 Spring 配置文件中启用 Bean 的自动扫描功能，可以通过<context:component-scan />元素，设置属性 base-package 来指定扫描的包名，代码如下。

```xml
<?xml version="1.0" encoding="UTF-8"?>
<beans xmlns="http://www.springframework.org/schema/beans"
    xmlns:xsi="http://www.w3.org/2001/XMLSchema-instance"
    xmlns:context="http://www.springframework.org/schema/context"
    xsi:schemaLocation="http://www.springframework.org/schema/beans
        http://www.springframework.org/schema/beans/spring-beans.xsd
        http://www.springframework.org/schema/context
http://www.springframework.org/schema/context/spring-context.xsd">
    <!-- 配置自动扫描的基包 -->
    <context:component-scan base-package="com.ssm" />
</beans>
```

为了能正常使用<context>元素，需要引入 context 命名空间。base-package 属性指定需要扫描的基包，Spring 容器将会扫描这个基包及其子包中的所有类，当需要扫描多个包时，可以使用逗号分隔。对于扫描到的组件，Spring 有默认的命名策略，使用非限定类名，第一个字母小写，也可以在注解中通过 value 属性值标识组件的名称。<context:component-scan />元素还会自动注册 AutowiredAnnotationBeanPostProcessor 实例可以自动装配具有@Autowired、@Resource 和@Inject 注解的属性。

运行测试类 TestSpringDI，运行效果同项目 spring-2，控制台输出结果为"登录成功"。

3.4.3 自动装配

除了使用 XML 和 Annotation 装配 Bean 外，还有一种常用的装配方式，就是使用自动装配。Spring 的<bean>元素中包含一个 autowire 属性，可通过设置 autowire 属性来自动装配 Bean。所谓自动装配，就是将一个 Bean 注入到其他 Bean 的 Property 中。autowire 属性值及说明如下。

- default：默认值，由<bean>的上级标签<beans>的default-autowire属性值确定。例如<beans default-autowire="byname">，则该<bean>元素中的autowire属性对应的属性值就为byName。
- byName：根据Property的Name自动装配，如果一个Bean的name和另一个Bean中的Property的name相同，则自动装配这个Bean到Property中。
- byType：根据Property的数据类型(TYPE)自动装配，如果一个Bean的数据类型和另一个Bean中的Property的数据类型相同，则自动装配这个Bean到Property中。
- constructor：根据构造函数参数的数据类型，进行byType模式的自动装配。
- autodetect：如果发现默认的构造函数，用constructor模式，否则用byType模式。
- no：默认情况下，不适用自动装配，Bean依赖必须通过ref元素定义。

修改2.4.3小节中的Spring配置文件，将配置文件修改成如下自动装配形式：

```
<!-- 使用bean元素的autowire属性完成自动装配 -->
<bean id="userDAO" class="com.ssm.dao.impl.UserDAOImpl"></bean>
<bean id="userService" class="com.ssm.service.impl.UserServiceImpl"
    autowire="byName" />
```

上述配置文件中，用于配置userService的<bean>元素中除了id和class属性外，还增加了autowire属性，并将其属性值设置为byName。在默认情况下，配置文件中需要通过ref来装配Bean，但设置了autowire="byName"后，Spring会自动寻找userService Bean中的属性，并将其属性名称与配置文件中定义的Bean做匹配。由于UserServiceImpl中定义了userDAO属性及其setter方法，这与配置文件中id为userDAO的Bean相匹配，所以Spring会自动地将id为userDAO的Bean封装到id为userService的Bean中。

对于大型的应用、不鼓励使用自动装配。虽然使用自动装配可减少配置文件的工作量，但大大降低了依赖关系的清晰性和透明性。依赖关系的装配依赖于源文件的属性名，导致Bean与Bean之间的耦合降低到代码层次，不利于高层次解耦。

3.5 小　　结

本章主要介绍了Spring IoC容器、Bean的配置、Bean的作用域和Bean的装配方式，即基于XML的Bean装配、基于Annotation的Bean装配和自动装配。通过本章的学习，读者可以了解Spring IoC容器，Bean的常用属性及其作用，熟悉Bean作用域的种类，掌握Bean的三种装配方式。

第 4 章 Spring AOP(面向方面编程)

AOP(面向方面编程)是一种编程范式,一般适用于具有横切逻辑的场合,如访问控制、事务管理、性能监测等,旨在通过允许横切关注点的分离,提高模块化。目前有许多 AOP 框架,其中最流行的两个框架为 Spring AOP 和 AspectJ。

4.1 AOP 概述

4.1.1 认识 AOP

面向方面编程(Aspect-Oriented Programming,AOP)也称为面向切面编程,是软件编程思想发展到一定阶段的产物,虽然是一种新的编程思想,但却不是面向对象编程(Object-Oriented Programming,OOP)的替代品,它只是 OOP 的有益补充和延伸。

经过几十年的发展,面向对象程序设计方法已成为当前主流,其将程序分解为不同层次的对象,通过封装、继承、多态等特性将对象组织成一个整体来完成功能,但是在一定场合,面向对象编程也暴露出一些问题。在传统的业务处理代码中,通常会进行日志记录、参数合法性验证、异常处理、事务控制等操作。虽然使用 OOP 可以通过组合或者继承的方式重用代码,但要实现某些功能(如日志记录),同样的代码仍然会分散到各个方法中。这样,如果要关闭某个功能,或者对其进行修改,就必须要修改所有的相关方法。这不仅会增加开发者的工作量,相应的代码出错率也会提高。

在业务系统中，总有一些散落、渗透到系统各处且不得不处理的事情，这些穿插在既定业务中的操作就是所谓的"横切逻辑"，也称为切面。怎样才能不受这些附加要求的干扰，专注于真正的业务逻辑呢？将这些重复性的代码抽取出来，放在专门的类和方法中处理，这样就便于管理和维护了。即便如此，依然无法实现既定业务和横切逻辑的彻底解耦，因为业务方法中还要保留这些方法的调用代码，当需要增加或者减少横切逻辑的时候，还是要修改业务方法中的调用代码才能实现。我们所希望的是无需编写显式的调用，而是在需要的时候能够"自动"调用所需的功能，这正是 AOP 所要解决的问题。

AOP 采取横向抽取机制，将分散在各个方法中的重复代码提取出来，然后在程序编译或运行时，再将这些提取出来的代码应用到需要执行的地方。这种采用横向抽取机制的方式，使用传统的 OOP 思想是无法办到的，因为 OOP 只能实现纵向的重用。

面向方面编程，简单地说，就是在不改变原有程序的基础上为代码段增加新的功能，对其进行增强处理，其设计思想来源于代理设计模式。下面以图示的方式进行简单的说明，通常情况下调用对象的方法如图 4-1 所示。

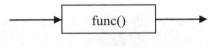

图 4-1　直接调用对象的方法

在代理模式中，可为对象设置一个代理对象，代理对象为 func()提供一个代理方法，当通过代理对象的 func()方法调用源对象的 func()方法时，就可在代理方法中添加新的功能，这就是所谓的增强处理。增强的功能既可以插到源对象的 func()方法前面，也可插到其后面，如图 4-2 所示。

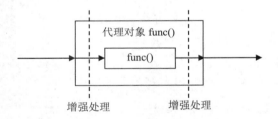

图 4-2　通过代理对象调用方法

在此模式下，就可在原有代码乃至原业务流程都不变的情况下，直接在业务流程汇总切入新代码，增加新功能，这就是所谓的面向切面编程。AOP 的使用，使开发人员在编写业务逻辑时可以专心于核心业务，而不用过多地关注其他业务的实现，这不但提高了开发效率，而且增强了代码的可维护性。

OOP 将应用程序分解成多个层次的对象，而 AOP 将程序分解成多个切面。日志、事务、安全验证等这些"通用的"、散布在系统各处的需要在实现业务逻辑时关注的事情称为"方面"，也可称为"关注点"。如果能将这些"方面"集中处理，然后在具体运行时，再由容器动态织入这些"方面"，至少有以下两个好处：

- 可以减少"方面"代码里的错误，处理策略改变时还能做到统一修改。
- 在编写业务逻辑时可以专心于核心业务。

因此，AOP 要做的事情就是从系统中分离出"方面"，然后集中实现，从而独立地编写业务代码和方面代码，在系统运行时，再将方面"织入"到系统中。

4.1.2 AOP 术语

实际上，AOP 并不是一个新的概念，在一些语言和框架中，早就出现了类似的机制。Java 平台的 EJB 规范、Servlet 规范以及 Struts 2 框架中存在的拦截器机制，实际上与 AOP 要实现的功能非常相似。AOP 是在这些概念基础上的发展，提供了更通用的解决方案。

AOP 中涉及面、通知、切入点、目标对象、代理对象、织入等很多术语，常用的术语简单介绍如下。

- 切面(Aspect)

一个关注点的模块化，该关注点可能会横切多个对象。比如方面(日志、事务、安全验证)的实现，如日志切面、事务切面、权限切面等。在实际应用中，通常存放方面实现的普通 Java 类，该类要被 AOP 容器识别为切面，需要在配置中通过<bean>标记指定。

- 连接点(JoinPoint)

程序执行中的某个具体的执行点，比如某方法调用的时候或者处理异常的时候。在 Spring AOP 中，一个连接点总是表示一个方法的执行，如图 4-2 所示原对象的 func()方法就是一个连接点。

- 切入点(Pointcut)

切入点是指切面与程序流程的交叉点，即那些需要处理的连接点，如图 4-3 所示。当某个连接点满足预先指定的条件时，AOP 框架能够定位到这个连接点，该连接点将被添加增强处理，该连接点也就变成了切入点。通常在程序中，切入点指的是类或者方法名，如某个通知要应用到所有以 add 开头的方法中，那么所有满足这一规则的方法都是切入点。

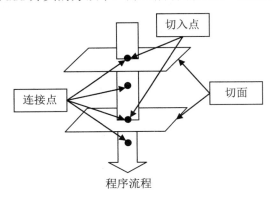

图 4-3 切面、连接点和切入点

- 通知/增强处理(Advice)

在切面的某个特定的连接点上执行的动作(一段程序代码)，是切面的具体实现。以目标方法为参照点，根据放置的位置不同，可以分为前置通知、后置通知、异常通知、环绕通知和最终通知等 5 种。例如图 4-2 中，在原对象的 func()方法之前插入的增强处理为前置通知，在该方法正常执行完以后插入的增强处理为后置通知。切面类中的某个方法具体属于哪类通知，需要在配置中指定。许多 AOP 框架(包括 Spring)都是以拦截器作通知模型，并维护一个以连接点为中心的拦截器链。Advice 直译为"通知",但这种说法并不确切,在此翻译为"增强处理"便于理解。

- 目标对象(Target Object)

目标对象是指被一个或者多个切面所通知的对象,也被称为被通知对象。如果 AOP 框架采用的是动态的 AOP 实现,那么该对象就是一个被代理的对象。这些对象中只包含核心业务逻辑代码,所有日志、事务、安全验证等方面的功能等待 AOP 容器的织入。

- 代理对象(Proxy Object)

代理对象是指将通知应用到目标对象之后,被动态创建的对象。代理对象的功能相当于目标对象中实现的核心业务逻辑功能加上方面(日志、事务、安全验证)代码实现的功能。

- 织入(Weaving)

织入是指生成代理对象并将切面内容放入到流程中的过程,即将切面代码插入到目标对象上,从而创建一个新的代理对象的过程。

AOP 的概念比较生涩难懂,切面可以理解为由增强处理和切入点组成,既包含了横切逻辑的定义,也包含了连接点的定义。面向切面编程主要关心两个问题,即在什么位置执行什么功能。Spring AOP 是负责实施切面的框架,即由 Spring AOP 完成织入工作。

4.2 基于 XML 配置文件的 AOP 实现

Spring AOP 通知包括前置通知、返回通知、正常返回通知、异常通知和环绕通知。使用 AOP 框架时,开发者需要做的主要工作是定义切入点和通知(增强处理),通常采用 XML 配置文件或注解的方式,配置好切入点和增强的信息后,AOP 框架会自动生成 AOP 代理。本节将基于 XML 配置文件的方式实现前置通知、返回通知、异常通知和环绕通知。

4.2.1 前置通知

前置通知在连接点(所织入的业务方法)前面执行,不会影响连接点的执行,除非此处抛出异常。下面通过示例演示如何实现前置通知,其过程如下:

(1) 将 spring-1 项目复制并重命名为"spring-5",再导入到 Eclipse 开发环境中。

(2) 在前面核心包的基础上,向项目中导入所需的 jar 包:spring-aop-5.0.4.RELEASE.jar、spring-aspects-5.0.4.RELEASE.jar、aopalliance-1.0.jar 和 aspectjweaver-1.9.1.jar,把文件添加到项目 spring-5 的 lib 目录中,再将前述 4 个 jar 包添加到项目的构建路径中。

- spring-aop-5.0.4.RELEASE.jar:Spring AOP 提供的实现包,Spring 包中已经提供。
- spring-aspects-5.0.4.RELEASE.jar:提供对 AspectJ 的支持,以便可以方便地将面向方面的功能集成进 IDE 中,Spring 包中已经提供。
- aopalliance-1.0.jar:AOP 联盟提供的规范包,该 jar 包可以通过地址 http://mvnrepository.com/artifact/aopalliance/aopalliance/ 下载。
- aspectjweaver-1.9.1.jar:如果使用 @Aspect 注解方式,可以在类上直接加一个 @Aspect 注解,不用费事在 xml 里配了,但是这需要额外的 jar 包(aspectjweaver-1.9.1.jar)。因为 spring 直接使用 AspectJ 的注解功能,注意只是使用了它的注解功能而已。该 jar 包可通过地址 http://mvnrepository.com/artifact/org.aspectj/aspectjweaver/ 下载。

(3) 创建包 com.ssm.service，在包中创建接口 ProductService，添加方法 browse，模拟用户浏览商品的业务。

```java
package com.ssm.service;
public interface ProductService {
    // 定义抽象方法 browse，模拟某用户浏览某商品
    public void browse(String loginName,String ProductName);
}
```

(4) 创建包 com.ssm.service.impl，在包中创建接口 ProductService 的实现类 ProductServiceImpl，实现模拟用户浏览商品的 browse 方法。

```java
package com.ssm.service.impl;
import com.ssm.service.ProductService;
public class ProductServiceImpl implements ProductService {
    // 实现方法 browse，模拟某用户浏览某商品
    @Override
    public void browse(String loginName, String ProductName) {
        System.out.println("执行业务方法 browse");
    }
}
```

(5) 创建包 com.ssm.aop，在包中创建日志通知类 AllLogAdvice，在类中编写用于生成日志记录的方法 myBeforeAdvice，如下所示：

```java
package com.ssm.aop;
import java.text.SimpleDateFormat;
import java.util.Arrays;
import java.util.Date;
import java.util.List;
import org.aspectj.lang.JoinPoint;
public class AllLogAdvice {
    // 此方法将作为前置通知
    public void myBeforeAdvice(JoinPoint joinPoint) {
        // 获取业务方法参数
        List<Object> args = Arrays.asList(joinPoint.getArgs());
        // 日志格式字符串
        String logInfoText = "前置通知: "
                + new SimpleDateFormat("yyyy-MM-dd HH:mm:ss")
                .format(new Date()) + " " + args.get(0).toString()
            + " 浏览商品 " + args.get(1).toString();
        // 将日志信息输出到控制台
        System.out.println(logInfoText);
    }
}
```

这里我们把 myBeforeAdvice 作为前置通知使用，即将该方法添加到目标方法之前执行，为了能够在通知方法中获得当前连接点的信息，以便实施相关的判断和处理，可在通知方

法中声明一个 JoinPoint 接口类型的参数 jionPoint，Spring 会自动注入实例。通过 jionPoint 的 getArgs()方法，myBeforeAdvice 就能获得业务方法 browse 的参数 loginName 和 productName。

(6) 编辑 Spring 配置文件。

在 Spring 配置文件 applicationContext.xml 中，采用 AOP 配置方式将日志类 AllLogAdvice 与业务组件 ProductService 原本两个互不相关的类和接口通过 AOP 元素进行装配，从而将日志通知类 AllLogAdvice 中的日志通知织入到 ProductService 中，以实现预期的日志记录。applicationContext.xml 配置文件的内容如下所示：

```xml
<?xml version="1.0" encoding="UTF-8"?>
<beans xmlns="http://www.springframework.org/schema/beans"
    xmlns:xsi="http://www.w3.org/2001/XMLSchema-instance"
    xmlns:aop="http://www.springframework.org/schema/aop"
    xsi:schemaLocation="http://www.springframework.org/schema/beans
        http://www.springframework.org/schema/beans/spring-beans.xsd
        http://www.springframework.org/schema/aop
        http://www.springframework.org/schema/aop/spring-aop.xsd">
    <!-- 实例化业务类的 Bean -->
    <bean id="productService" class="com.ssm.service.impl.ProductServiceImpl"></bean>
    <!-- 实例化日志通知/增强处理类(切面)的 Bean -->
    <bean id="allLogAdvice" class="com.ssm.aop.AllLogAdvice"/>
    <!-- 配置 aop -->
    <aop:config>
        <!-- 配置日志切面 -->
        <aop:aspect id="logaop" ref="allLogAdvice">
        <!-- 定义切入点，切入点采用正则表达式，含义是对 browse 的方法进行拦截 -->
            <aop:pointcut expression="execution(public void browse(String,String))" id="logpointcut"/>
            <!-- 将日志通知类中的 myBeforeAdvice 方法指定为前置通知 -->
            <aop:before method="myBeforeAdvice" pointcut-ref="logpointcut"/>
        </aop:aspect>
    </aop:config>
</beans>
```

由于 Spring 的 AOP 配置标签是放置在 aop 命名空间之下的，需要在 Spring 配置文件的<beans>元素中，导入 AOP 命名空间及其配套的 schemaLocation。在配置文件中，首先实例化业务类 ProductServiceImpl 的 Bean，然后实例化日志通知/增强处理类(切面) AllLogAdvice 的 Bean，最后通过<aop:config>元素进行 AOP 的配置。在配置 AOP 时，通过<aop:aspect>子元素配置日志切面；在配置日志切面时，先通过<aop:pointcut>子元素定义切入点，切入点采用正则表达式 execution(public void browse(String,String))，含义是对 browse(String,String)的方法进行拦截。再通过<aop:before>子元素将日志通知类中的 myBeforeAdvice 方法指定为前置通知。

上面的配置代码中的 execution 是切入点指示符，括号中是一个切入点表达式，用于配置需要切入增强处理的方法的特征。切入点表达式支持模糊匹配，下面介绍几种常用的模糊匹配。
- public * browse(String,String)："*"表示匹配所有类型的返回值。
- public void *(String,String)："*"表示匹配所有方法名。
- public void browse(..)：".."表示匹配所有参数个数和类型。
- * com.ssm.service.*.*(..)：表示匹配 com.ssm.service 包下所有类的所有方法。
- * com.ssm.service..*.*(..)：表示匹配 com.ssm.service 包及其子包下所有类的所有方法。

具体使用时可以根据自己的需求来设置切入点的匹配规则。当然，匹配的规则和关键字还有很多，可参考 Spring 的开发手册学习。

(7) 在 com.ssm 包中创建测试类 TestAOP.java，代码如下所示：

```java
package com.ssm;
import org.springframework.context.ApplicationContext;
import org.springframework.context.support.ClassPathXmlApplicationContext;
import com.ssm.service.ProductService;
public class TestAOP {
    public static void main(String[] args) {
        // 初始化 spring 容器，加载 applicationContext.xml 配置
        ApplicationContext ctx = new
ClassPathXmlApplicationContext("applicationContext.xml");
        // 通过容器获取配置中 productService 的实例
        ProductService productService =
(ProductService)ctx.getBean("productService");
        // 调用 productService 中的 browse 方法
        productService.browse("张三", "Lenovo 天逸 310");
    }
}
```

执行测试类 TestAOP，控制台输出如下所示：

```
前置通知：2018-06-24 22:10:24 张三 浏览商品 Lenovo 天逸 310
执行业务方法 browse
```

从控制台输出可以看出，在业务方法 browse 执行前，先输出了日志通知类 AllLogAdvice 中 myBeforeAdvice 方法产生的日志记录。

4.2.2 返回通知

返回通知是指在连接点正常执行后实施增强，不管是正常执行完成，还是抛出异常，都会执行返回通知中的内容。下面通过示例演示如何实现返回通知，其过程如下所示：

(1) 在日志通知类 AllLogAdvice 中添加方法 myAfterReturnAdvice，作为返回通知。

```java
public void myAfterReturnAdvice(JoinPoint joinPoint) {
    // 获取方法参数
    List<Object> args = Arrays.asList(joinPoint.getArgs());
    // 日志格式字符串
    String logInfoText = "返回通知："
```

```
            + new SimpleDateFormat("yyyy-MM-dd HH:mm:ss")
            .format(new Date()) + " " + args.get(0).toString()
            + " 浏览商品 " + args.get(1).toString();
        // 将日志信息输出到控制台
        System.out.println(logInfoText);
    }
```

(2) 在 Spring 配置文件 applicationContext.xml 中的 <aop:aspect> 元素内添加 <aop:after-returning>元素，将 AllLogAdvice 日志通知类中的 myAfterReturnAdvice 方法指定为返回通知。

```
<aop:after-returning method="myAfterReturnAdvice"
    pointcut-ref="logpointcut" />
```

将 applicationContext.xml 中前置通知的<aop:before>配置加以注释，再执行测试类 TestAOP，控制台输出如下所示：

执行业务方法browse
返回通知：2018-06-25 13:34:18 张三 浏览商品 Lenovo 天逸 310

从控制台输出可以看出，在业务方法 browse 执行后才输出日志通知类 AllLogAdvice 中 myAfterReturnAdvice 方法产生的日志记录。

4.2.3 异常通知

异常通知在连接点抛出异常后执行，下面通过示例演示如何实现异常通知，其过程如下所示：

(1) 修改 ProductServiceImpl 类中的 browse 方法，人为抛出一个异常。

```
package com.ssm.service.impl;
import com.ssm.service.ProductService;
public class ProductServiceImpl implements ProductService {
    @Override
    public void browse(String loginName, String productName) {
        System.out.println("执行业务方法browse");
        // 演示异常通知时，人为抛出该异常
        throw new RuntimeException("这是特意抛出的异常信息！");
    }
}
```

(2) 在日志通知类 AllLogAdvice 中添加方法 myThrowingAdvice，作为异常通知。

```
public void myThrowingAdvice(JoinPoint joinPoint, Exception e) {
    // 获取被调用的类名
    String targetClassName = joinPoint.getTarget().getClass().getName();
    // 获取被调用的方法名
    String targetMethodName = joinPoint.getSignature().getName();
    // 日志格式字符串
    String logInfoText = "异常通知：执行" + targetClassName + "类的 "
        + targetMethodName + "方法时发生异常";
    // 将日志信息输出到控制台
```

(3) 在 Spring 配置文件 applicationContext.xml 中的 <aop:aspect> 元素内添加 <aop:after-throwing> 元素，将 AllLogAdvice 日志通知类中的 myThrowingAdvice 方法指定为异常通知。

```
<aop:after-throwing method="myThrowingAdvice"
    pointcut-ref="logpointcut" throwing="e" />
```

将 applicationContext.xml 中前置通知的 <aop:before> 和返回通知的 <aop:after-returning> 配置加以注释，再执行测试类 TestAOP，控制台输出如下所示：

```
执行业务方法 browse
异常通知：执行 com.ssm.service.impl.ProductServiceImpl 类的 browse 方法时发生异常
Exception in thread "main" java.lang.RuntimeException: 这是特意抛出的异常信息！
```

从控制台输出可以看出，在执行业务方法 browse 时，输出日志通知类 AllLogAdvice 中 myThrowingAdvice 方法产生的日志记录。

4.2.4 环绕通知

环绕通知围绕在连接点前后，比如一个方法调用的前后，这是最强大的通知类型，能在方法调用前后自定义一些操作。环绕通知还需要负责决定是继续处理 joinPoint，还是中断执行。下面通过示例演示如何实现环绕通知，其过程如下所示：

(1) 修改 ProductServiceImpl 类中的 browse 方法，通过 while 循环延长方法的执行时间。

```java
public void browse(String loginName, String mealName) {
    System.out.println("执行业务方法 browse");
    int i = 100000000;
    while (i > 0) {
        i--;
    }
}
```

(2) 在日志通知类 AllLogAdvice 中添加方法 myAroundAdvice，作为环绕通知。

```java
public void myAroundAdvice(ProceedingJoinPoint joinPoint) throws Throwable
{
    long beginTime = System.currentTimeMillis();
    joinPoint.proceed();
    long endTime = System.currentTimeMillis();
    // 获取被调用的方法名
    String targetMethodName = joinPoint.getSignature().getName();
    // 日志格式字符串
    String logInfoText = "环绕通知：" + targetMethodName + "方法调用前时间" + beginTime  + "毫秒," + "调用后时间" + endTime + "毫秒。";
    // 将日志信息输出到控制台
    System.out.println(logInfoText);
}
```

ProceedingJoinPoint 对象是 JoinPoint 的子接口，该对象只用在@Around 的切面方法中。

(3) 在 Spring 配置文件 applicationContext.xml 中的<aop:aspect>元素内添加<aop:around>元素，将 AllLogAdvice 日志通知类中的 myAroundAdvice 方法指定为环绕通知。

```
<aop:around method="myAroundAdvice" pointcut-ref="logpointcut" />
```

将 applicationContext.xml 中前置通知的<aop:before>、返回通知的<aop:after-returning>和异常通知的<aop:after-throwing>元素配置加以注释，再执行测试类 TestAOP，控制台输出如下所示：

执行业务方法 browse
环绕通知：browse 方法调用前时间 1529906757688 毫秒，调用后时间 1529906757702 毫秒。

从控制台输出可以看出，通过环绕通知可以记录业务方法 browse 执行前后的时间。

4.3 基于@AspectJ 注解的 AOP 实现

基于 XML 配置文件的 AOP 实现免不了在 Spring 配置文件中配置大量的信息，不仅配置麻烦，而且会造成配置文件臃肿。为了解决这个问题，AspectJ 框架为 AOP 的实现提供了一套注解，用以取代 Spring 配置文件中为实现 AOP 功能所配置的臃肿代码。

AspectJ 是一个面向切面的框架，它扩展了 Java 语言、定义了 AOP 语法，能够在编译期提供代码的织入，并提供了一个专门的编译器用来生成遵守字节编码规范的 Class 文件。@AspectJ 是 AspectJ 5 新增的功能，使用 JDK 5.0 注解技术和正规的 AspectJ 切点表达式语言描述切面。因此在使用@AspectJ 之前，需要保证 JDK 是 5.0 或更高版本，否则将无法使用注解技术。Spring 通过集成 AspectJ 实现了以注解的方式定义切面，大大减轻了配置文件的工作量，此外，因为 Java 的反射机制无法获取方法参数名，Spring 还需要利用轻量级的字节码处理 asm(已集成在 Spring Core 模块中)来处理@AspectJ 中所描述的方法参数名。关于 AspectJ 注解的说明如下所示。

- @Aspect：用于定义一个切面。
- @Pointcut：用于定义一个切入点，切入点的名称由一个方面名称定义。在使用时还需要定义一个包含名字和任意参数的方法签名来表示切入点名称。实际上，这个方法签名就是一个返回值为 void 且方法体为空的普通方法。
- @Before：用于定义一个前置通知，相当于 BeforeAdvice。在使用时，通常需要指定一个 value 属性值，该属性值用于指定一个切入点表达式(可以是已有的切入点，也可以直接定义切入点表达式)。
- @AfterReturning：用于定义一个后置通知，相当于 AfterReturningAdvice。在使用时可以指定 pointcut、value 和 returning 属性，其中 pointcut 和 value 这两个属性的作用一样，都用于指定切入点表达式。returning 属性用于表示 Advice 方法中可定义与此同名的形参，该形参可用于访问目标方法的返回值。
- @AfterThrowing：用于定义一个异常通知，相当于 ThrowAdvice。在使用时可指定 pointcut、value 和 throwing 属性，其中 pointcut 和 value 属性用于指定切入点表达式，而 throwing 属性用于访问目标方法抛出的异常，该属性值与异常通知方法中

第 4 章　Spring AOP(面向方面编程)

同名的形参一致。
- @Around：用于定义一个环绕通知，相当于 MethodInterceptor。在使用时需要指定一个 value 属性，该属性用于指定该通知被植入的切入点。
- @After：用于定义最终 final 通知，不管是否异常，该通知都会执行。使用时需要指定一个 value 属性，该属性用于指定该通知被植入的切入点。

为了使读者快速掌握这些注解，接下来使用@AspectJ 注解重新实现 4.2 小节中的 AllLogAdvice 日志类功能，步骤如下所示：

(1) 将项目 spring-5 复制并重命名为 "spring-6"，再导入到 Eclipse 开发环境中。

(2) 在项目 spring-6 中，修改 ProductService 接口的实现类 ProductServiceImpl，在类上添加@Component("productService")注解，在 Spring 容器中自动创建 ProductServiceImpl 类的 Bean 实例。

```
@Component("productService")
public class ProductServiceImpl implements ProductService {
    ……
}
```

(3) 修改日志通知类 AllLogAdvice，使用注解定义 Bean、切面、切点和 4.2 小节中的四种类型通知。

```
package com.ssm.aop;
import java.text.SimpleDateFormat;
import java.util.Arrays;
import java.util.Date;
import java.util.List;
import org.aspectj.lang.JoinPoint;
import org.aspectj.lang.ProceedingJoinPoint;
import org.aspectj.lang.annotation.AfterReturning;
import org.aspectj.lang.annotation.AfterThrowing;
import org.aspectj.lang.annotation.Around;
import org.aspectj.lang.annotation.Aspect;
import org.aspectj.lang.annotation.Before;
import org.aspectj.lang.annotation.Pointcut;
import org.springframework.stereotype.Component;
/**
 * 定义切面类，在此类中编写通知
 */
@Aspect
@Component
public class AllLogAdvice {
    // 定义切入点表达式
    @Pointcut("execution(* com.ssm.service.ProductService.*(..))")
    // 使用一个返回值为 void、方法体为空的方法来命名切入点
    private void allMethod() {
    }
    // 此方法将作为前置通知
    @Before("allMethod()")
```

```java
    public void myBeforeAdvice(JoinPoint joinPoint) {
        // 获取业务方法参数
        List<Object> args = Arrays.asList(joinPoint.getArgs());
        // 日志格式字符串
        String logInfoText = "前置通知："
                + new SimpleDateFormat("yyyy-MM-dd HH:mm:ss").format(new Date()) + " " + args.get(0).toString()
                + " 浏览商品 " + args.get(1).toString();
        // 将日志信息输出到控制台
        System.out.println(logInfoText);
    }
    // 此方法将作为返回通知
    @AfterReturning("allMethod()")
    public void myAfterReturnAdvice(JoinPoint joinPoint) {
        // 获取方法参数
        List<Object> args = Arrays.asList(joinPoint.getArgs());
        // 日志格式字符串
        String logInfoText = "返回通知："
                + new SimpleDateFormat("yyyy-MM-dd HH:mm:ss").format(new Date()) + " " + args.get(0).toString()
                + " 浏览餐品 " + args.get(1).toString();
        // 将日志信息输出到控制台
        System.out.println(logInfoText);
    }
    // 此方法将作为异常通知
    @AfterThrowing(pointcut="allMethod()",throwing="e")
    public void myThrowingAdvice(JoinPoint joinPoint, Exception e) {
        // 获取被调用的类名
        String targetClassName = joinPoint.getTarget().getClass().getName();
        // 获取被调用的方法名
        String targetMethodName = joinPoint.getSignature().getName();
        // 日志格式字符串
        String logInfoText = "异常通知：执行" +targetClassName + "类的 "+ targetMethodName + "方法时发生异常";
        // 将日志信息输出到控制台
        System.out.println(logInfoText);
    }
    // 此方法将作为环绕通知
    @Around("allMethod()")
    public void myAroundAdvice(ProceedingJoinPoint joinPoint) throws Throwable {
        long beginTime = System.currentTimeMillis();
        joinPoint.proceed();
        long endTime = System.currentTimeMillis();
        // 获取被调用的方法名
        String targetMethodName = joinPoint.getSignature().getName();
        // 日志格式字符串
```

```
            String logInfoText = "环绕通知：" + targetMethodName + "方法调用前时
间" + beginTime + "毫秒," + "调用后时间" + endTime + "毫秒";
            // 将日志信息输出到控制台
            System.out.println(logInfoText);
    }
}
```

在 AllLogAdvice 类上，首先使用@Aspect 注解定义了切面类，由于该类在 Spring 中是作为组件使用的，所以还需要添加@Component 注解才能生效，使用@Component 注解在 Spring 容器中自动创建 AllLogAdvice 类的 Bean 实例；然后使用@Pointcut 注解定义一个切入点，切入点的名字为 allMethod()，切入点的正则表达式 execution(* com.ssm.service.ProductService.*(..))的含义是对 com.ssm.service.ProductService 接口中的所有方法进行拦截；再分别使用@Before、@AfterReturning、@AfterThrowing 和@Around 注解定义前置通知、返回通知、异常通知和环绕通知，这些通知中的代码含义与前一小节相同，此处不再赘述。

(4) 修改 Spring 配置文件，配置自动扫描的包，并在配置文件中开启基于@AspectJ 切面的注解处理器。

```
<?xml version="1.0" encoding="UTF-8"?>
<beans xmlns="http://www.springframework.org/schema/beans"
    xmlns:xsi="http://www.w3.org/2001/XMLSchema-instance"
    xmlns:aop="http://www.springframework.org/schema/aop"
    xmlns:context="http://www.springframework.org/schema/context"
    xsi:schemaLocation="http://www.springframework.org/schema/beans
        http://www.springframework.org/schema/beans/spring-beans.xsd
        http://www.springframework.org/schema/aop
        http://www.springframework.org/schema/aop/spring-aop.xsd
        http://www.springframework.org/schema/context
 http://www.springframework.org/schema/context/spring-context.xsd">
    <!-- 配置自动扫描的包 -->
    <context:component-scan base-package="com.ssm" />
    <!-- 开启基于@AspectJ 切面的注解处理器 -->
    <aop:aspectj-autoproxy />
</beans>
```

由于 Spring 配置文件中使用了<context>和<aop>元素，因此需要引入 context 和 aop 命名空间及其配套的 schemaLocation。

在日志通知类 AllLogAdvice 中，依次启用一个通知方法进行测试，将其他通知方法加以注释，并根据所测试通知类型的需要，修改 ProductServiceImpl 类中 browse 方法的代码(参考 4.2 小节)，再运行测试类 TestAOP，效果与 4.2 小节相同。

基于@AspectJ 注解的 AOP 实现效果与基于 XML 配置文件的 AOP 实现效果相同，相对来说，使用注解的方式更加简单、方便，Spring 配置文件变得更为简洁，所以在实际开发中推荐使用注解的方式进行 AOP 开发。

4.4 小　　结

本章主要介绍了 Spring AOP 的相关概念，并以日志通知为例先后讲解了基于 XML 配置文件的 AOP 实现和基于@AspectJ 注解的 AOP 实现。通过对比分析可知，使用 Spring 为 AOP 实现提供的一组注解，极大地简化了 Spring 的配置。

第 5 章 Spring 的数据库编程

通过前面几章的学习,读者应该对 Spring 框架核心技术中的几个重要模块有了一定的了解。Spring 框架降低了 Java EE 的使用难度,Spring 为开发者提供了 JDBC 模板模式,那就是 JdbcTemplate,它降低了 JDBC 的使用难度。本章将对 Spring 中的 JDBC 知识进行详细讲解。

5.1 Spring JDBC

传统的 JDBC 即使执行一条简单的 SQL 语句,其过程也不简单,要先打开数据库连接执行 SQL 语句,然后组装结果,最后关闭数据库资源,但太多的 try…catch…finally…语句,造成了代码泛滥。在 Spring 出现之后,为了解决这些问题,Spring 提供了自己的方案,那就是 JdbcTemplate 模板。Spring JDBC 是 Spring 所提供的持久层技术,它的主要目的是降低 JDBC API 的使用难度,以一种更直接、更简洁的方式使用 JDBC API。Spring 中的 JDBC 模块负责数据库资源管理,可以省去连接和关闭数据库的代码,简化了对数据库的操作,使得开发人员无需在数据库操作上花更多精力,可以从烦琐的数据库操作中解脱出来,从而将更多的精力投入到编写业务逻辑中。

5.1.1 Spring JdbcTemplate 类

Spring 框架在数据库开发中的应用主要使用的是 JdbcTemplate 类,它是 Spring 针对

JDBC 代码失控提供的解决方案。该类作为 Spring JDBC 的核心类，提供了对所有数据库操作功能的支持。该类是在原始 JDBC 的基础上，构建一个抽象层，提供许多使用 JDBC 的模板和驱动模块，为 Spring 应用操作关系数据库提供了更大的便利。

JdbcTemplate 类的继承关系十分简单，它继承了抽象类 JdbcAccessor，同时实现了接口 JdbcOperations。

在抽象类 JdbcAccessor 的设计中，该类为其子类提供了一些访问数据库时的公共属性，具体如下。

- DataSource：其主要功能是获取数据库连接，具体实现时还可以引入对数据库连接的缓冲池和分布式事务的支持，它可以作为访问数据库资源的标准接口。
- SQLExceptionTranslator：org.springframework.jdbc.support.SQLExceptionTranslator 接口负责对 SQLException 进行转译。通过必要的设置或者获取 SQLExceptionTranslator 中的方法，可以使 JdbcTemplate 在需要处理 SQLException 时，委托 SQLExceptionTranslator 的实现类来完成相关的转译工作。

在 JdbcOperation 接口中，定义了通过 Jdbc 操作数据库的基本操作方法，而 JdbcTemplate 类提供了这些接口方法的实现，包括添加、修改、查询和删除等操作。

5.1.2 Spring JDBC 的配置

Spring JDBC 模块主要由 4 个包组成，分别是 core(核心包)、object(对象包)、dataSource(数据源包)和 support(支持包)。JdbcTemplate 类就在核心包中，该类包含所有数据库操作的基本方法。关于这 4 个包的具体说明如下。

- core：核心包，包含 JDBC 的核心功能，包括 JdbcTemplate 类、SimpleJdbcInsert 类、SimpleJdbcCall 类以及 NamedParameterJdbcTemplate 类。
- dataSource：数据源包，访问数据源的实用工具类，它有多种数据源的实现，可以在 JavaEE 容器外部测试 JDBC 代码。
- object：对象包，以面向对象的方式访问数据库，它允许执行查询并返回结果作为业务对象，可以在数据表的列和业务对象的属性之间映射查询结果。
- support：支持包，包含 core 和 object 包的支持类，例如，提供异常转换功能的 SQLException 类。

Spring 对数据库的操作都封装在这几个包中，要想使用 Spring JDBC，就需要对其进行配置。在 Spring 中，JDBC 的配置是在配置文件 applicationContext.xml 中完成的，其配置模板如下。

```
<?xml version="1.0" encoding="UTF-8"?>
<beans xmlns="http://www.springframework.org/schema/beans"
    xmlns:xsi="http://www.w3.org/2001/XMLSchema-instance"
    xsi:schemaLocation="http://www.springframework.org/schema/beans
 http://www.springframework.org/schema/beans/spring-beans.xsd">
    <!-- 1.配置数据源   -->
    <bean id="dataSource"
class="org.springframework.jdbc.datasource.DriverManagerDataSource">
        <!-- 数据库驱动名称，不同类型数据库的名称-->
```

```xml
        <property name="driverClassName" value="com.mysql.jdbc.Driver"/>
        <!-- 连接数据库的数据源所在的 url 地址 -->
        <property name="url" value="jdbc:mysql://localhost:3306/eshop"/>
        <!-- 连接数据库的用户名 -->
        <property name="username" value="root"/>
        <!-- 连接数据库的密码 -->
        <property name="password" value="123456"/>
    </bean>
    <!-- 2.配置 JDBC 模板 -->
    <bean id="jdbcTemplate"
  class="org.springframework.jdbc.core.JdbcTemplate">
        <!-- 默认必须使用数据源 -->
        <property name="dataSource" ref="dataSource"></property>
    </bean>
    <!-- 3.配置注入的实体类,以下是模拟代码 -->
    <bean id="xxx" class="Xxx">
        <property name="jdbcTemplate" ref="jdbcTemplate"></property>
    </bean>
</beans>
```

上述代码中,dataSource 的配置就是 JDBC 连接数据库时所需要的四个属性,这四个属性需要根据数据库类型或者机器配置的不同设置相应的属性值。如果数据库类型不同,需要更改驱动名称;如果数据库不是本机的数据库,则需要将地址中的 localhost 替换成相应主机的 IP 地址;如果修改过 MySQL 数据库的端口号(默认为 3306),则需要改为修改后的端口号;同时连接数据库的用户名和密码需要与数据库创建时设置的用户名和密码保持一致。

定义 JdbcTemplate 时,需要将 dataSource 注入到 JdbcTemplate 中,而其他需要使用 JdbcTemplate 的 Bean,也需要将 JdbcTemplate 注入到该 Bean 中(通常注入到数据访问层 Dao 类中,在 Dao 类中进行与数据库的相关操作)。

5.2 JdbcTemplate 的常用方法

在 JdbcTemplate 类中,提供了大量的查询和更新数据库的方法,Spring JDBC 就是使用这些方法来操作数据库的。下面分别介绍 execute()方法、update()方法和 query()方法。

5.2.1 execute()方法

execute(String sql)方法能够完成执行 SQL 语句的功能,下面以创建和删除数据库表的 SQL 语句为例,来讲解此方法的使用,具体步骤如下。

(1) 启动前端工具 SQLyog,用来管理 MySQL 数据库,选中左侧的 root@localhost 并右击,在弹出的快捷菜单中选择"执行 SQL 脚本"命令,在弹出的对话框中选择执行源代码中提供的 eshop.sql 脚本,就可导入 eshop 数据库,如图 5-1 所示。

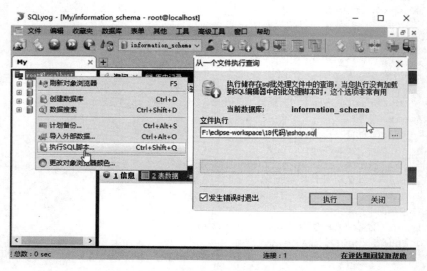

图 5-1 执行 sql 脚本导入数据库

（2）将项目 spring-1 复制并重命名为"spring-7"，再导入到 Eclipse 开发环境中。

（3）在前面已经添加核心包的基础上，向项目中添加 spring-jdbc-5.0.4.RELEASE.jar、spring-tx-5.0.4.RELEASE.jar 和 MySQL 数据库驱动 mysql-connector-java-5.1.38-bin.jar。把 3 个 jar 文件添加到项目 spring-7 的 lib 目录中，再将这些 jar 包添加到项目的构建路径中。

（4）修改 src 路径下的 applicationContext.xml 文件，在该文件中配置 id 为 dataSource 的数据源 Bean 和 id 为 jdbcTemplate 的 JDBC 模板 Bean，并将数据源注入到 JDBC 模板中。applicationContext.xml 配置文件的内容如下：

```xml
<?xml version="1.0" encoding="UTF-8"?>
<beans xmlns="http://www.springframework.org/schema/beans"
    xmlns:xsi="http://www.w3.org/2001/XMLSchema-instance"
    xsi:schemaLocation="http://www.springframework.org/schema/beans
    http://www.springframework.org/schema/beans/spring-beans.xsd">
    <!-- 1.配置数据源 -->
    <bean id="dataSource"
class="org.springframework.jdbc.datasource.DriverManagerDataSource">
        <!-- 数据库驱动名称，不同类型数据库的名称-->
        <property name="driverClassName" value="com.mysql.jdbc.Driver"/>
        <!-- 连接数据库的数据源所在的 url 地址 -->
        <property name="url" value="jdbc:mysql://localhost:3306/eshop"/>
        <!-- 连接数据库的用户名 -->
        <property name="username" value="root"/>
        <!-- 连接数据库的密码 -->
        <property name="password" value="123456"/>
    </bean>
    <!-- 2.配置 JDBC 模板 -->
    <bean id="jdbcTemplate"
class="org.springframework.jdbc.core.JdbcTemplate">
        <!-- 默认必须使用数据源 -->
        <property name="dataSource" ref="dataSource"></property>
```

```
        </bean>
</beans>
```

(5) 在 com.ssm 包中创建测试类 TestJdbcTemplate，在 main()主方法中通过 Spring 容器获取在配置文件中定义的 JdbcTemplate 实例，使用该实例的 execute(String sql)方法执行创建数据表的 SQL 语句，如下所示：

```java
package com.ssm;
import org.springframework.context.ApplicationContext;
import org.springframework.context.support.ClassPathXmlApplicationContext;
import org.springframework.jdbc.core.JdbcTemplate;
public class TestJdbcTemplate {
    public static void main(String[] args) {
        // 初始化 spring 容器，加载 applicationContext.xml 配置
        ApplicationContext ctx = new ClassPathXmlApplicationContext("applicationContext.xml");
        // 通过容器，获取 JdbcTemplate 的实例
        JdbcTemplate jdbcTemplate = (JdbcTemplate)ctx.getBean("jdbcTemplate");
        String sql = "create table user(id int primary key auto_increment, userName varchar(20),password varchar(32))";
        // 使用 execute()方法执行 SQL 语句，创建用户表 user
        jdbcTemplate.execute(sql);
        System.out.println("用户表 user 创建成功！");
    }
}
```

执行测试类 TestJdbcTemplate，在控制台输出"用户表 user 创建成功！"的提示，查看 SQLyog，在 eshop 数据库下的表中能够看到 user 表，如图 5-2 所示。

图 5-2　user 数据表创建成功

5.2.2　update()方法

update()方法可以完成插入、更新和删除操作。在 JdbcTemplate 类中，update 方法中存在多个重载的方法，其常用方法具体介绍如下：

- int update(String sql)：该方法是最简单的 update 方法重载形式，可以直接传入 SQL 语句并返回受影响的行数。
- int update(PreparedStatementCreator psc)：该方法执行从 PreparedStatementCreator 返回的语句，然后返回受影响的行数。
- int update(String sql, PreparedStatementSetter pss)：该方法通过 PreparedStatementSetter 设置 SQL 语句中的参数，并返回受影响的行数。
- int update(String sql, Object...args)：该方法使用 Object...args 设置 SQL 语句中的参数，要求参数不能为空，并返回受影响的行数。

接下来，我们通过一个用户账户管理的实例来实现对用户信息的插入、修改和删除操作，具体步骤如下。

(1) 在 spring-7 项目中，新建 com.ssm.entity 包，在包中新建 User 类。在 User 类中定义 id、userName 和 password 属性，并为属性添加 setter/getter 方法，重写 toString()方法，代码如下：

```java
package com.ssm.entity;
public class User {
    private int id;              //用户 id
    private String userName;     //用户名
    private String password;     //用户密码
    // 此处省略相应属性的 setter/getter 方法
    ......
    // 重写 toString 方法
    @Override
    public String toString() {
        return "User 对象: "+ id +" -- "+userName+" -- "+password;
    }
}
```

(2) 新建 com.ssm.dao 包，在该包中新建 UserDAO 接口，并在接口中定义添加、修改、删除用户的方法，代码如下：

```java
package com.ssm.dao;
import com.ssm.entity.User;
public interface UserDAO {
    // 添加用户
    public int addUser(User user);
    // 修改用户
    public int updateUser(User user);
    // 删除用户
    public int deleteUser(int id);
}
```

(3) 新建 com.ssm.dao.impl 包，在该包中创建 UserDAO 接口的实现类 UserDAOImpl，并在类中实现添加、修改和删除的方法，代码如下：

```java
package com.ssm.dao.impl;
import org.springframework.jdbc.core.JdbcTemplate;
import com.ssm.dao.UserDAO;
import com.ssm.entity.User;
```

```java
public class UserDAOImpl implements UserDAO{
    // 声明 JdbcTemplate 属性及其 setter 方法
    private JdbcTemplate jdbcTemplate;
    public void setJdbcTemplate(JdbcTemplate jdbcTemplate) {
        this.jdbcTemplate = jdbcTemplate;
    }
    // 添加用户
    @Override
    public int addUser(User user) {
        String sql="insert into user(userName,password) values(?,?)";
        // 使用数组来存储 SQL 语句中的参数
        Object[] object=new Object[] {user.getUserName(),user.getPassword()};
        // 执行添加操作，返回的是受 SQL 语句影响的记录条数
        int result=jdbcTemplate.update(sql,object);
        return result;
    }
    // 修改用户
    @Override
    public int updateUser(User user) {
        String sql="update user set userName=?,password=? where id=?";
        // 使用数组来存储 SQL 语句中的参数
        Object[] params=new Object[] {user.getUserName(),user.getPassword(),user.getId()};
        // 执行修改操作，返回的是受 SQL 语句影响的记录条数
        int result=jdbcTemplate.update(sql,params);
        return result;
    }
    // 删除用户
    @Override
    public int deleteUser(int id) {
        String sql="delete from user where id=?";
        // 执行删除操作，返回的是受 SQL 语句影响的记录条数
        int result=jdbcTemplate.update(sql,id);
        return result;
    }
}
```

在上述的添加、修改和删除的代码中可以看出它们实现的步骤类似，只是定义的 SQL 语句有所不同。

(4) UserDAOImpl 类中有对 JdbcTemplate 类的引用，因此要在 applicationContext.xml 文件中实现 UserDAOImpl 对 JdbcTemplate 类的依赖注入，需修改 applicationContext.xml 配置文件，定义一个 id 为 userDAO 的 Bean，该 Bean 用于将 jdbcTemplate 注入到 userDAO 实例中，其代码如下：

```xml
<!-- 配置一个 id 为 userDAO 的 Bean -->
<bean id="userDAO" class="com.ssm.dao.impl.UserDAOImpl">
    <!-- 将 jdbcTemplate 注入到 userDAO 实例中 -->
    <property name="jdbcTemplate" ref="jdbcTemplate"/>
</bean>
```

(5) 在测试类 TestJdbcTemplate 中,添加一个 JUnit4 类型的单元测试方法 addUserTest(),该单元测试方法主要用于添加用户信息,具体步骤如下。

① 选择要添加单元测试的项目并右击,选择 Build Path → Configure Build Path 命令,如图 5-3 所示。

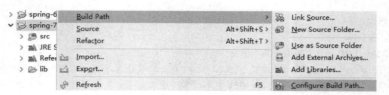

图 5-3　Build Path 构建路径

② 在弹出的 Java Build Path 对话框中,选择 Libraries 选项卡,然后选中 Classpath,单击右侧的 Add library 按钮,在 Add Library 界面中,选择 JUnit,单击 Next 按钮,如图 5-4 所示。

③ 在 JUnit Library 界面中,在 JUnit library version 下拉列表框中选择单元测试的版本,这里我们选择 JUnit 4,单击 Finish 按钮完成,如图 5-5 所示。

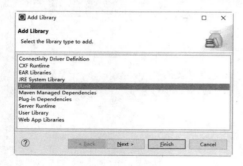
图 5-4　选择 JUnit

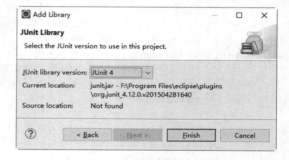
图 5-5　选择 JUnit 4 版本

④ 回到 Java Build Path 对话框,可以看到添加了单元测试包,单击 Apply and Close 按钮,如图 5-6 所示。

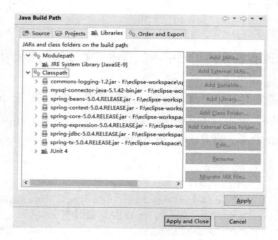

图 5-6　添加 JUnit 成功

⑤ 添加一个 JUnit4 类型的单元测试方法 addUserTest()，代码如下：

```java
@Test
public void addUserTest() {
    // 初始化 spring 容器, 加载 applicationContext.xml 配置
    ApplicationContext ctx = new
    ClassPathXmlApplicationContext("applicationContext.xml");
    // 通过 Spring 容器, 获取 UserDAO 的实例
    UserDAO userDAO = (UserDAO)ctx.getBean("userDAO");
    // 创建 User 对象, 并向 User 对象中添加数据
    User user=new User();
    user.setUserName("yzpc");
    user.setPassword("yzpc");
    // 执行 addUser()方法, 并获取返回结果
    int result=userDAO.addUser(user);
    if (result>0) {
        System.out.println("成功往数据表中插入了"+result+"条数据！");
    }else {
        System.out.println("往数据表中插入数据失败！");
    }
}
```

在上述代码中，获取了 UserDAO 的实例后，又创建了 User 对象，并为 User 对象的属性赋值。而后调用 UserDAO 对象的 addUser()方法向数据表中添加一条数据。最后，通过返回的受影响的行数来判断数据是否插入成功。

⑥ 用鼠标右键单击 addUserTest()方法，在弹出的快捷菜单中选择 Run As → JUnit Test 命令来运行测试方法，如图 5-7 所示。

图 5-7　运行 JUnit

⑦ 选择 JUnit Test 命令后，Eclipse 中会出现一个名为 JUnit 的视窗窗口，如图 5-8 所示。

图 5-8　JUnit 控制台

在图 5-8 中，JUnit 视窗窗口的进度条为绿色表明运行结构正确，如果进度条为红色则表示有错误，并且会在窗口中显示所报的错误信息。

测试执行通过后，从 Console 控制台的输出结果可以看出，addUserTest()方法已经执行成功，如图 5-9 所示。

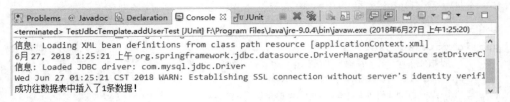

图5-9 添加记录运行结果

此时，可以通过 SQLyog 查看数据中的 user 表，user 表中新插入了一条数据。

（6）执行完添加用户操作后，接下来使用 JdbcTemplate 类的 update()方法执行更新操作，在测试类 TestJdbcTemplate 中，添加一个测试方法 updateUserTest()，其代码如下：

```java
@Test
public void updateUserTest() {
    // 初始化spring容器，加载applicationContext.xml配置
    ApplicationContext ctx = new
    ClassPathXmlApplicationContext("applicationContext.xml");
    // 通过容器，获取UserDAO的实例
    UserDAO userDAO = (UserDAO)ctx.getBean("userDAO");
    // 创建User对象，并向User对象中添加数据
    User user=new User();
    user.setId(1);
    user.setUserName("yzpc");
    user.setPassword("123456");
    // 执行updateUser()方法，并获取返回结果
    int result=userDAO.updateUser(user);
    if (result>0) {
        System.out.println("成功修改了"+result+"条数据！");
    }else {
        System.out.println("修改操作执行失败！");
    }
}
```

与 addUserTest()方法相比，修改操作的代码增加了 id 属性值的设置，并将密码修改为 123456 后，调用了 UserDAO 对象中的 updateUser()方法执行对数据表的修改操作。

使用 JUnit4 运行 updateUserTest()方法后，从 Console 控制台的输出结果看出，updateUserTest()方法已经执行成功，如图 5-10 所示。

图 5-10 修改记录运行结果

再次查询数据库中的 User 表，其结果如图 5-11 所示。

id	userName	password
1	yzpc	123456

图 5-11 修改后的 User 表

(7) 修改记录后，在测试类 TestJdbcTemplate 中，添加一个测试方法 deleteUserTest()来执行删除操作，其代码如下：

```
@Test
public void deleteUserTest() {
    // 初始化 spring 容器，加载 applicationContext.xml 配置
    ApplicationContext ctx = new
    ClassPathXmlApplicationContext("applicationContext.xml");
    // 通过容器，获取 UserDAO 的实例
    UserDAO userDAO = (UserDAO)ctx.getBean("userDAO");
    // 执行 deleteUser()方法，并获取返回结果
    int result=userDAO.deleteUser(1);
    if (result>0) {
        System.out.println("成功删除了"+result+"条数据！");
    }else {
        System.out.println("删除操作执行失败！");
    }
}
```

上述代码中，获取了 UserDAO 的实例后，执行实例中的 deleteUser()方法来删除 id 为 1 的数据(id=1 的数据要在数据表中存在)。使用 JUnit4 运行 deleteUserTest()方法后，从 Console 控制台的输出结果看出，deleteUserTest()方法已经执行成功，如图 5-12 所示。

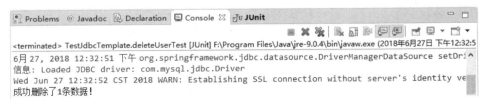

图 5-12 删除记录运行结果

再次查询数据库中的 User 表，发现 id 为 1 的记录已经被删除。

5.2.3 query()方法

JdbcTemplate 对 JDBC 的流程做了封装，提供了大量的 query()方法来处理各种对数据库表的查询操作，常用的 query()方法如下。

- List query(String sql, PreparedStatementSetterpss, RowMapper rowMapper)：该方法根据 String 类型参数提供的 SQL 语句创建 PreparedStatement 对象，通过 RowMapper 将结果返回到 List 中。
- List query(String sql,Object[] args, RowMapper rowMapper)：该方法使用 Object[]的值来设置 SQL 中的参数值，采用 RowMapper 回调方法可以直接返回 List 类型的数据。

- queryForObject(String sql,Object[]args, RowMapper rowMapper)：该方法将 args 参数绑定到 SQL 语句中，通过 RowMapper 返回单行记录，并转换为一个 Object 类型返回。
- queryForList(String sql, Object[] args, class<T>elementType)：该方法可以返回多行数据的结果，但必须是返回列表，elementType 参数返回的是 List 元素类型。

了解了几个常用的 query()方法后，接下来，我们尝试从 user 表中查询数据，在 UserDAO 接口中增加按照 id 查询的方法和查询所有用户的方法，在 UserDAOImpl 中具体实现两个方法，实现步骤如下。

(1) 首先通过 SQLyog 工具向 user 表中插入几条数据，插入后 user 表中的数据如图 5-13 所示。

id	userName	password
1	yzpc	yzpc
2	zhangsan	123456
3	lisi	123
4	wangwu	1234

图 5-13　User 表

(2) 在 UserDAO 中，分别创建一个通过 id 查询单个用户信息和查询所有用户信息的方法，代码如下：

```
// 通过 id 查询
public User findUserById(int id);
// 查询所有用户
public List<User> findAllUser();
```

(3) 在 UserDAO 接口的实现类 UserDAOImpl 中，实现接口中的方法，并使用 query()方法分别进行查询，代码如下：

```
// 通过 id 查询用户信息
@Override
public User findUserById(int id) {
    // 定义单个查询的 SQL 语句
    String sql="select * from user where id=?";
    // 创建一个新的 BeanPropertyRowMapper 对象,将结果集通过 Java 的反射机制映射到
Java 对象中
    RowMapper<User> rowMapper=new BeanPropertyRowMapper<User>(User.class);
    // 将 id 绑定到 SQL 语句中，并通过 RowMapper 返回一个 Object 类型的对象
    return this.jdbcTemplate.queryForObject(sql, rowMapper, id);
}
// 查询所有用户信息
@Override
public List<User> findAllUser() {
    // 定义查询所有用户的 SQL 语句
    String sql="select * from User";
    // 创建一个新的 BeanPropertyRowMapper 对象
    RowMapper<User> rowMapper=new BeanPropertyRowMapper<User>(User.class);
    //执行静态的 SQL 查询，并通过 RowMapper 返回结果
    return this.jdbcTemplate.query(sql, rowMapper);
}
```

在 UserDAOImpl 实现类的方法中，BeanPropertyRowMapper 是 BeanMapper 接口的实现类，它可以自动地将数据表中的数据映射到用户定义的类中(需要用户自定义类中的字段要与数据表中的字段相对应)。BeanPropertyRowMapper 对象创建后，在 findUserById()方法中通过 queryForObject()方法返回一个 Object 类型的单行记录，而在 findAllUser()方法中通过

query()方法返回一个结果集合。

(4) 在 TestJdbcTemplate 测试类中，添加一个测试方法 findUserByIdTest()来测试相应的条件查询，代码如下：

```java
public void findUserByIdTest() {
    // 初始化 spring 容器，加载 applicationContext.xml 配置
    ApplicationContext ctx = new ClassPathXmlApplicationContext("applicationContext.xml");
    // 通过容器，获取 UserDAO 的实例
    UserDAO userDAO = (UserDAO)ctx.getBean("userDAO");
    // 执行 findUserById()方法，获取 User 对象
    User user=userDAO.findUserById(1);
    System.out.println(user);
}
```

在以上代码中，通过执行 findUserById()方法获取了 id 为 1 的对象信息，并通过输出语句输出。使用 JUnit4 测试运行后，控制台输出结果如图 5-14 所示。

图 5-14　查询单个用户的运行结果

(5) 接下来测试查询所有用户信息的方法。在 TestJdbcTemplate 测试类中，添加一个测试方法 findAllUserTest()来查询所有用户，代码如下：

```java
@Test
public void findAllUserTest() {
    // 初始化 spring 容器，加载 applicationContext.xml 配置
    ApplicationContext ctx = new ClassPathXmlApplicationContext("applicationContext.xml");
    // 通过容器，获取 UserDAO 的实例
    UserDAO userDAO = (UserDAO)ctx.getBean("userDAO");
    // 执行 findAllUser()方法，获取 User 对象的集合
    List<User> users = userDAO.findAllUser();
    // 循环输出集合中对象
    for (User user : users) {
        System.out.println(user);
    }
}
```

在上述代码中，调用 UserDAO 对象的 findAllUser()方法查询所有用户信息集合，并通过 for 循环查询结果。使用 JUnit4 成功运行 findAllUserTest()方法后，控制台的显示信息如图 5-15 所示。

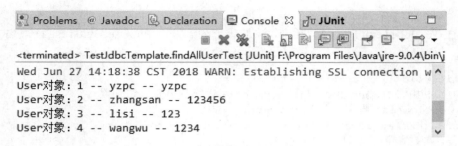

图 5-15　查询所有用户运行结果

5.3　小　　结

　　本章对 Spring 框架中使用 Spring JDBC 数据操作进行了详细讲解。首先讲解了 Spring JDBC 中的核心类以及如何在 Spring 中配置 Spring JDBC，然后通过案例讲解了 Spring JDBC 核心类 JdbcTemplate 中常用方法的使用。通过本章的学习，读者能够学会如何使用 Spring 框架进行数据库开发，并能深切地体会到 Spring 框架的强大功能。

第 6 章 Spring MVC 简介

对 Web 应用来说，表示层是不可或缺的重要环节。传统的 Struts 2 框架就是一个优秀的 Web 框架。除了 Struts 2 框架外，Spring 框架也为表示层提供了一个优秀的 Web 框架，即 Spring MVC。由于 Spring MVC 采用了松耦合可插拔组件结构，因此比其他 MVC 框架具有更大的扩展性和灵活性。通过注解，Spring MVC 使得 POJO 成为处理用户请求的控制器，无需实现任何接口。

6.1 MVC 模式概述

Java Web 应用的结构经历了 Model I 和 Model II 两个时代，从 Model I 发展到 Model II 是技术发展的必然。

6.1.1 Model I 和 Model II

在早期的 Java Web 应用开发中，JSP 文件既要负责处理业务逻辑和控制程序的运行流程，还要负责数据的显示，即用 JSP 文件来独立自主地完成系统功能的所有任务。传统的 Model I 模式如图 6-1 所示。

改进的 Model I 利用 JSP 页面与 JavaBean 组件共同协作来完成系统功能的所有任务，JSP 文件负责程序的流程控制和数据显示逻辑任务，JavaBean 负责处理业务逻辑任务。改进的 Model I 模式如图 6-2 所示。

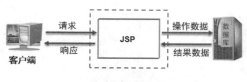

图 6-1 传统的 Model I 模式

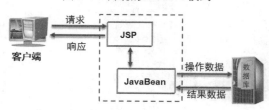

图 6-2 改进的 Model I 模式

Model II 模式基于 MVC 架构的设计模式。在 Model II 模式下，利用 JSP 页面、Servlet 和 JavaBean 组件分工协作共同完成系统功能的所有任务。其中，JSP 负责数据显示逻辑任务，Servlet 负责程序流程控制逻辑任务，JavaBean 负责处理业务逻辑任务。Model II 模式如图 6-3 所示。

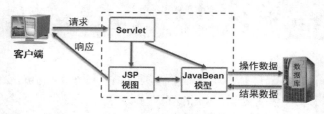

图 6-3 Model II 模式

引入 MVC 模式，使得 Model II 模式具有组件化的特点，从而更有利于大规模应用的开发，但也增加了应用开发的复杂程度。MVC 设计模式简单地说，就是将数据显示、流程控制和业务逻辑处理分离，使之相互独立。

6.1.2 MVC 模式及其优势

MVC 思想不是哪个语言所特有的设计思想，也并不是 Web 应用所特有的思想，而是一种规范。MVC 思想将一个应用分成三个基本部分：Model(模型)、View(视图)和 Controller(控制器)，这三个部分以最少的耦合协同工作，从而提高了应用的可扩展性和可维护性。MVC 设计模式中模型、视图和控制器三者之间的关系如图 6-4 所示。

概括起来，MVC 模式具有如下特点。

(1) 各司其职，互不干涉。在 MVC 模式中，3 层各司其职，所以如果哪一层的需求发生了变化，就只需要更改相应层中的代码，而不会影响其他层。

(2) 有利于开发中的分工。在 MVC 模式中，由于按层把系统分开，因此能更好地实现开发中的分工。网页设计人员可以开发 JSP 页面，对业务熟悉的开发人员可以开发模型中相关业务处理的方法，而其他开发人员可开发控制器，以进行程序控制。

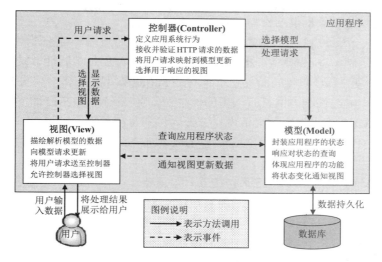

图 6-4　MVC 模式各层的关系

(3) 有利于组件的重用。分层后更有利于组件的重用，如控制层可独立成一个通用的组件，视图层也可做成通用的操作界面。MVC 最重要的特点就是把显示和数据分离，这样就增加了各个模块的可重用性。

6.2　Spring MVC 概述

Spring MVC 是 Spring 框架中用于 Web 应用开发的一个模块，是 Spring 提供的一个基于 MVC 设计模式的轻量级 Web 框架。Spring 框架提供了构建 Web 应用程序的全功能 MVC 模块。Spring MVC 框架本质上相当于 Servlet，提供了一个 DispatcherServlet 作为前端控制器来分派请求，同时提供灵活的配置处理程序映射、视图解析、语言环境和主题解析，并支持文件上传。

在 MVC 设计模式中，Spring MVC 作为控制器(Controller)来建立模型与视图的数据交互，是一个典型的 MVC 框架，是结构最清晰的 MVC Model II 实现，如图 6-5 所示。

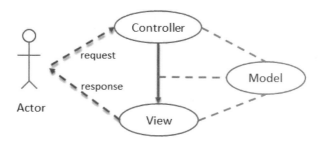

图 6-5　Spring MVC-Model II 实现

在 Spring MVC 框架中，Controller 替代 Servlet 担负控制器的职能，Controller 接收请求，调用相应的 Model 进行处理，处理器完成业务处理后返回处理结果。Controller 调用相应的 View 并对处理结果进行视图渲染，最终传送响应消息到客户端。由于 Spring MVC 的结构较为复杂,上述只是对其框架结构的一个简单描述。

Spring MVC 分离了控制器、模型对象、分派器以及处理程序对象的角色,这种分离让它们更容易进行定制。Spring MVC 框架无论是在框架设计还是扩展性、灵活性等方面都全面超越了 Struts 2 等 MVC 框架,而且它本身就是 Spring 框架的一部分,与 Spring 框架的整合可以说是无缝集成,性能方面具有天生的优越性。Spring MVC 具有如下特点。

- Spring MVC 拥有强大的灵活性、非侵入性和可配置性。
- Spring MVC 提供了一个前端控制器 DispatcherServlet,开发者无须额外开发控制器对象。
- Spring MVC 分工明确,包括控制器、验证器、命令对象、模型对象、处理器映射器、视图解析器等,每一个功能实现由一个专门的对象负责。
- Spring MVC 可以自动绑定用户输入,并正确地转换数据类型。
- Spring MVC 使用一个名称/值的 Map 对象实现更加灵活的模型数据类型。
- Spring MVC 内置了常见的校验器,可以检验用户输入,如果校验不同,则重定向回输入表单。输入校验是可选的,并且支持编程方式及声明方式。
- Spring MVC 支持国际化,支持根据用户区域显示多国语言,并且国际化的配置非常简单。
- Spring MVC 支持多种视图技术,最常见的有 JSP 技术以及其他技术,包括 Velocity 和 FreeMarker。
- Spring MVC 提供了一个简单而强大的 JSP 标签库,支持数据绑定功能,使得编写 JSP 页面更加容易。

6.3 Spring MVC 环境搭建

Spring MVC 框架所需的 jar 文件包含在 Spring 框架的资源包中,如下所示:
- spring-web-5.0.4.RELEASE.jar:在 Web 应用开发时使用 Spring 框架所需的核心类。
- spring-webmvc-5.0.4.RELEASE.jar:Spring MVC 框架相关的所有类,包含框架的 Servlet、Web MVC 框架,以及对控制器和视图的支持。

下面搭建 Spring MVC 的开发环境,建立一个简单的 Spring MVC 程序帮助读者理解 Spring MVC 程序的开发步骤。

(1) 创建 Web 项目,添加所需要的 jar 包。

在 Eclipse 中,创建一个名为"springmvc-1"的 Web 项目,在前面章节已经下载过 spring 的资源文件(spring-framework-5.0.4.RELEASE-dist.zip),解压后的 libs 文件夹中包含如图 6-6 所示的 12 个 jar 包,将这几个 jar 包以及 aopalliance-1.0.jar、aspectjweaver-1.9.1.jar、commons-logging-1.2.jar 和 cglib-3.2.0.jar 这 4 个 jar 包添加到项目 springmvc-1 的 WebContent\WEB-INF\lib 路径中。

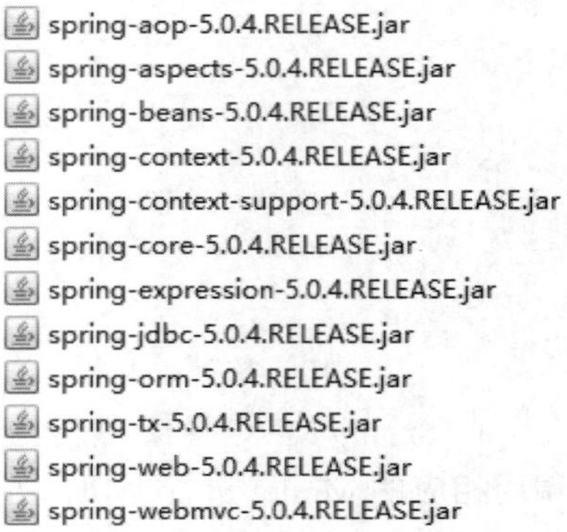

- spring-aop-5.0.4.RELEASE.jar
- spring-aspects-5.0.4.RELEASE.jar
- spring-beans-5.0.4.RELEASE.jar
- spring-context-5.0.4.RELEASE.jar
- spring-context-support-5.0.4.RELEASE.jar
- spring-core-5.0.4.RELEASE.jar
- spring-expression-5.0.4.RELEASE.jar
- spring-jdbc-5.0.4.RELEASE.jar
- spring-orm-5.0.4.RELEASE.jar
- spring-tx-5.0.4.RELEASE.jar
- spring-web-5.0.4.RELEASE.jar
- spring-webmvc-5.0.4.RELEASE.jar

图 6-6 Spring MVC 所依赖的 jar 包

(2) 在 web.xml 文件中，配置 Spring MVC 的前端控制器 DispatcherServlet。

Spring MVC 是基于 Servlet 的框架，DispatcherServlet 是整个 Spring MVC 框架的核心，它负责接受请求并将其分派给相应的处理器处理，关键配置代码如下：

```xml
<!-- 配置 Spring MVC 的前端控制器 DispatcherServlet -->
<servlet>
    <servlet-name>dispatcherServlet</servlet-name>
    <servlet-class>org.springframework.web.servlet.DispatcherServlet
    </servlet-class>
    <!-- 初始化参数，配置 Spring MVC 配置文件的位置及名称 -->
    <init-param>
        <param-name>contextConfigLocation</param-name>
        <param-value>classpath:springmvc.xml</param-value>
    </init-param>
    <!-- 表示容器在启动时，立即加载 dispatcherServlet -->
    <load-on-startup>1</load-on-startup>
</servlet>
<!-- 让 Spring MVC 的前端控制器拦截所有的请求 -->
<servlet-mapping>
    <servlet-name>dispatcherServlet</servlet-name>
    <url-pattern>/</url-pattern>
</servlet-mapping>
```

上述配置的目的在于，让 Web 容器使用 Spring MVC 的 DispatcherServlet，并通过设置 url-pattern 为 "/"，将所有的 URL 请求都映射到这个前端控制器 DispatcherServlet。在配置 DispatcherServlet 的时候，通过设置 contextConfigLocation 参数来指定 Spring MVC 配置文件的位置，此处使用 Spring 资源路径的方式进行指定。

(3) 创建 Spring MVC 的配置文件。

在项目 springmvc-1 的 src 目录下创建 Spring MVC 配置文件 springmvc.xml，在该配置文件中，我们使用 Spring MVC 最简单的配置方式进行配置，主要配置如下：

```xml
<?xml version="1.0" encoding="UTF-8"?>
<beans xmlns="http://www.springframework.org/schema/beans"
    xmlns:xsi="http://www.w3.org/2001/XMLSchema-instance"
    xmlns:aop="http://www.springframework.org/schema/aop"
    xmlns:context="http://www.springframework.org/schema/context"
    xmlns:mvc="http://www.springframework.org/schema/mvc"
    xsi:schemaLocation="http://www.springframework.org/schema/beans
    http://www.springframework.org/schema/beans/spring-beans.xsd
        http://www.springframework.org/schema/context
    http://www.springframework.org/schema/context/spring-context.xsd
        http://www.springframework.org/schema/mvc
        http://www.springframework.org/schema/mvc/spring-mvc.xsd">
    <!-- 配置处理器 Handle,映射为"/hello"请求 -->
    <bean name="/hello" class="com.springmvc.controller.HelloController"/>
    <!-- 配置视图解析器,将控制器方法返回的逻辑视图解析为物理视图 -->
    <bean class="org.springframework.web.servlet.view
.InternalResourceViewResolver">
    </bean>
</beans>
```

在springmvc.xml文件中,首先要引入beans、aop、context和mvc命名空间,然后主要完成配置处理器映射和配置视图解析器。

① 配置处理器映射。

在前面的web.xml里配置了DispatcherServlet,并配置了哪些请求需要通过此Servlet进行处理,接下来DispatcherServlet要将一个请求交给哪个特定的Controller处理?它需要咨询一个名为HandlerMapping的Bean,之后把URL请求指定给一个Controller处理(就像web.xml文件使用<servlet-mapping>将URL映射到相应的Servlet上)。Spring提供了多种处理器映射(HandlerMapping)的支持,例如:

- org.springframework.web.servlet.handler.BeanNameUrlHandlerMapping
- org.springframework.web.servlet.SimpleUrlHandlerMapping
- org.springframework.web.servlet.mvc.annotation.DefaultAnnotationHandlerMapping
- org.springframework.web.servlet.mvc.method.annotation.RequestMappingHandlerMapping

可以根据需求选择处理器映射,这里我们选择BeanNameUrlHandlerMapping,若没有明确声明任何处理器映射,Spring会默认使用BeanNameUrlHandlerMapping,即在Spring容器中查找与请求URL同名的Bean,通过声明HelloController业务控制器类,将其映射到/hello请求。

② 配置视图解析器。

处理请求的最后一件事就是解析输出,该任务由视图(这里使用JSP)实现,那么需要确定:指定的请求需要使用哪个视图进行请求结果的解析输出?DispatcherServlet会查找到一个视图解析器,将控制器返回的逻辑视图名称转换成渲染结果的实际视图。Spring提供了多种视图解析器,例如:

- org.springframework.web.servlet.view.InternalResourceViewResolver
- org.springframework.web.servlet.view.ContentNegotiatingViewResolver

在springmvc.xml配置文件中,并没有配置处理器映射和处理器适配器,当用户没有配置这两项时,Spring会使用默认的处理器映射和处理器适配器处理请求。

(4) 创建处理请求的控制器类。

在项目的src目录下创建包com.springmvc.controller,在包中创建类HelloController.java,并实现Controller接口中的handleRequest方法,用来处理hello请求,代码如下:

```java
package com.springmvc.controller;
import javax.servlet.http.HttpServletRequest;
import javax.servlet.http.HttpServletResponse;
import org.springframework.web.servlet.ModelAndView;
import org.springframework.web.servlet.mvc.Controller;
public class HelloController  implements Controller{
    @Override
    public ModelAndView handleRequest(HttpServletRequest req,
HttpServletResponse res) throws Exception {
        System.out.println("Hello,Spring MVC!");    //控制台输出
        ModelAndView mv=new ModelAndView();
        mv.addObject("msg","这是第一个Spring MVC程序!");
        mv.setViewName("/ch06/first.jsp");
```

```
        return mv;
    }
}
```

上述代码中，HelloController 是一个实现 Controller 接口的控制器，它可以处理一个单一的请求动作。handleRequest 是 Controller 接口必须实现的方法，该方法必须返回一个包含视图名或视图名和模型的 ModelAndView 对象，该对象既包含视图信息，也包含模型数据信息。这样 Spring MVC 就可以使用视图对模型数据进行解析。本例返回的模型中包含一个名为"msg"的字符串对象，返回的视图路径为/ch06/first.jsp，因此，请求将被转发到 ch06 路径下的 first.jsp 页面。

ModelAndView 对象代表 Spring MVC 中呈现视图界面时所使用的 Model(模型数据)和 View(逻辑视图名称)。由于 Java 一次只能返回一个对象，所以 ModelAndView 的作用就是封装这两个对象，一次返回我们所需要的 Model 和 View。当然，返回的模型和视图也都是可选的，在一些情况下，模型中没有任何数据，那么只返回视图即可，或者只返回模型，让 Spring MVC 根据请求 URL 来决定。后面章节还会对 ModelAndView 对象进行讲解。

(5) 创建视图页面。

在项目的 WebContext 路径下创建 ch06 文件夹，在 ch06 文件夹中创建 JSP 视图页面 first.jsp，并在该视图页面上通过 EL 表达式输出"msg"中的信息，代码如下：

```
<%@ page language="java" contentType="text/html; charset=UTF-8"
pageEncoding="UTF-8"%>
<!DOCTYPE html PUBLIC "-//W3C//DTD HTML 4.01 Transitional//EN"
"http://www.w3.org/TR/html4/loose.dtd">
<html>
<head>
<meta http-equiv="Content-Type" content="text/html; charset=UTF-8">
<title>Spring MVC 的入门程序</title>
</head>
<body>
    ${msg}
</body>
</html>
```

(6) 部署项目，启动 Tomcat 测试。

将项目 springmvc-1 发布到 Tomcat 中，并启动 Tomcat 服务器，在浏览器地址栏中访问 http://localhost:8080/springmvc-1/hello，其运行效果如图 6-7 所示。

图 6-7　第一个 Spring MVC 程序

从图 6-7 可以看到，浏览器中已经显示出了模型对象的字符串信息，控制台窗口中输出了"这是第一个 Spring MVC 程序!"提示，这也就说明第一个 Spring MVC 程序执行成功。

使用 MVC 框架就应该遵守 MVC 思想，MVC 框架不赞成浏览器直接访问 Web 应用的视图页面，用户的所有请求都只应向控制器发送，由控制器调用模型组件、视图组件向用户呈现数据。

6.4　Spring MVC 请求流程

通过前面示例，简单总结 Spring MVC 的处理流程。当用户发送 URL 请求 http://localhost:8080/springmvc-1/hello 时，根据 web.xml 中对 DispatcherServlet 的配置，该请求被 DispatcherServlet 截获，并根据 HandlerMapping 找到处理相应请求的 Controller 控制器(HelloController)；Controller 处理完成后，返回 ModelAndView 对象；该对象告诉 DispatcherServlet 需要通过哪个视图来进行数据模型的展示，DispatcherServlet 根据视图解析器把 Controller 返回的逻辑视图名渲染成真正的视图并输出，呈现给用户。

接下来深入了解 Spring MVC 框架的请求处理流程，如图 6-8 所示。

按照图 6-8 可以知道，Spring MVC 的请求处理流程如下。

(1) 用户通过客户端向服务器发起一个 request 请求，此请求会被前端控制器(DispatcherServlet)所拦截。

(2) 前端控制器请求处理器映射器(HandlerMapping)去查找 Handler，可以依据 XML 配置或注解去查找。

(3) 处理器映射器根据请求 URL 找到具体的处理器，生成处理器对象及处理器拦截器(如果有则生成)，并返回给前端控制器。

(4) 前端控制器请求处理器适配器(HandlerAdapter)去执行相应的 Handler(常称为 Controller)。

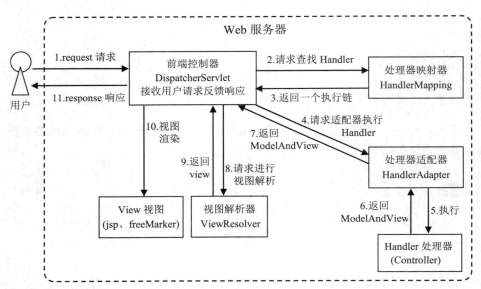

图 6-8　Spring MVC 请求处理流程

(5) 处理器适配器会调用并执行 Handler 处理器，这里的处理器指的是程序中编写的 Controller 类，也被称为后端控制器。在请求信息到达真正调用 Handler 的处理方法之前的这段时间内，Spring MVC 还完成了很多工作。

- 消息转换：将请求消息(如 Json、xml 等数据)转换成一个对象，将对象转换为指定的响应信息。
- 数据转换：对请求消息进行数据转换，如 String 转换成 Integer、Double 等。
- 数据格式化：对请求消息进行数据格式化，如将字符串转换成格式化数字或格式化日期等。
- 数据验证：验证数据的有效性(长度、格式等)，验证结果存储到 BindingResult 或 Error 中。

(6) Controller 执行完毕后会返回给处理器适配器一个 ModelAndView 对象(Spring MVC 底层对象)，该对象中会包含 View 视图信息或包含 Model 数据模型和 View 视图信息。

(7) 处理器适配器接收到 Controller 返回的 ModelAndView 后，将其返回给前端控制器。

(8) 前端控制器接收到 ModelAndView 后，选择一个合适的视图解析器(ViewReslover)对视图进行解析。

(9) 视图解析器解析后，会根据 View 视图信息匹配到相应的视图结果，反馈给前端控制器。

(10) 前端控制器收到 View 视图后，进行视图渲染，将模型数据(在 ModelAndView 对象中)填充到 request 域。

(11) 前端控制器向用户响应结果。

以上就是 Spring MVC 的整个请求处理流程，其中用到的组件有前端控制器(DispatcherServlet)、处理器映射器(HandlerMapping)、处理器适配器(HandlerAdapter)、Handler 处理器(Controller)、视图解析器(ViewResolver)、视图(View)。其中，DispatcherServlet、HandlerMapping、HandlerAdapter 和 ViewResolver 对象的工作是在框架内部执行的，开发人员并不需要关心这些对象内部的实现过程，只需要配置 DispatcherServlet，完成 Handler 处理器(Controller)中的业务处理，并在视图中展示相应信息即可。

6.5 小　　结

本章首先对 MVC 模式进行了简单介绍，然后介绍了 Spring MVC 框架。通过入门案例搭建 Spring MVC 环境，对 Spring MVC 的工作流程进行了详细讲解。

第 7 章　Spring MVC 常用注解

上一章学习了 Spring MVC 的基本开发环境搭建以及 Spring MVC 的请求流程。在 Spring 2.5 之前，只能使用实现 Controller 接口的方式开发控制器，在 Spring 2.5 之后，新增加了基于注解的控制器以及其他一些常用注解。到目前为止，Spring 的版本虽然变化较大，但注解的特性一直被延续下来，并不断扩展，极大地减少了程序员的开发工作，让广大开发者的工作变得更为轻松。本章将对 Spring MVC 中的常用注解进行讲解。

7.1　基于注解的控制器

7.1.1　@Controller 注解

对于上一章所讲解的 springmvc-1 示例，如果有多个请求，需要在 springmvc.xml 文件中配置多个映射关系，并且还需要建立多个 JavaBean 作为控制器来进行请求的处理。若业务复杂，这样并不合适。最常用的解决方式是使用 Spring MVC 提供的一键式配置方法 <mvc:annotation-driven/>，通过注解的方式来进行 Spring MVC 开发，配置此标签后，Spring MVC 会帮助我们自动做一些注册组件之类的事情。这种配置方法相对简单，适用于初学者快速搭建 Spring MVC 环境。

接下来改造 springmvc-1 的示例，讲解基于注解的控制器，只需修改 springmvc.xml 文件和 Controller 类的实现，具体修改过程如下。

(1) 将项目 springmvc-1 复制并重命名为 springmvc-2，再导入到 Eclipse 开发环境中。选中 springmvc-2 项目并右击，在快捷菜单中选择 Properties 命令，进入属性设置界面，找到 Web Project Settings 选项，将右侧 Context root 文本框中的 springmvc-1 改为 springmvc-2，单击 Apply and Close 按钮，如图 7-1 所示。

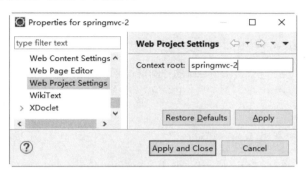

图 7-1　修改项目发布路径

(2) 修改 springmvc.xml 的配置文件，关键代码如下：

```xml
<?xml version="1.0" encoding="UTF-8"?>
<beans xmlns="http://www.springframework.org/schema/beans"
    ……
    xsi:schemaLocation="http://www.springframework.org/schema/beans
    …… http://www.springframework.org/schema/mvc/spring-mvc.xsd">
    <!-- 配置自动扫描的包 -->
    <context:component-scan base-package="com.springmvc"/>
    <mvc:annotation-driven/>
    <!-- 配置视图解析器，将控制器方法返回的逻辑视图解析为物理视图 -->
    <bean class="org.springframework.web.servlet.view
.InternalResourceViewResolver">
        <property name="prefix" value="/ch06/"></property>
        <property name="suffix" value=".jsp"></property>
    </bean>
</beans>
```

删除了<bean name="/hello" class="com.springmvc.controller.HelloController"/>，增加了两个标签。

- <mvc:annotation-driven/>：配置该标签会自动注册 DefaultAnnotationHandlerMapping(处理器映射器)与 AnnotationMethodHandlerAdapter(处理器适配器)两个 Bean。Spring MVC 需要通过这两个 Bean 实例来完成对@Controller 和@RequestMapping 等注解的支持，从而找出 URL 与 handler method 的关系并予以关联。换言之，完成在 Spring 容器中这两个 Bean 的注册是 Spring MVC 为@Controller 分发请求的必要支持。
- <context:component-scan …/>：该标签是对包进行扫描，实现注解驱动 Bean 的定义，同时将 Bean 自动注入容器中使用，即标注了 Spring MVC 注解(如@Controller)的 Bean 生效。换句话说，若没有配置此标签，那么标注@Controller 的 Bean 仅仅是一个普通的 JavaBean，而不是一个可以处理请求的控制器。

这里还是使用 InternalResourceViewResolver 定义该视图解析器，通过配置 prefix(前缀)和 suffix(后缀)，将控制器方法返回的逻辑视图名渲染为物理视图。

(3) 修改 HelloController.java，代码如下：

```java
package com.springmvc.controller;
import org.springframework.stereotype.Controller;
import org.springframework.web.bind.annotation.RequestMapping;
import org.springframework.web.servlet.ModelAndView;
/**
 * HelloController 是一个基于注解的控制器，可同时处理多个请求，并无须实现任何接口，
 * org.springframework.stereotype.Controller 注解用于指示该类是一个控制器
 */
@Controller
public class HelloController{
    /**
     * org.springframework.web.bind.annotation.RequestMapping 注解
     * 用来映射请求的 URL 和请求的方法等。本例用来映射"/hello"
     * hello()是普通方法，返回一包含视图名或视图名和模型的 ModelAndView
     */
    @RequestMapping(value="/hello")
    public ModelAndView hello() {
        System.out.println("Hello,Annotation Spring MVC!");
        //创建 ModelAndView 对象，该对象包含返回视图名、模型名称以及模型对象
        ModelAndView mv=new ModelAndView();
        //添加模型数据，可以是任何 POJO 对象
        mv.addObject("msg","这是基于注解的 Spring MVC 程序!");
        //设置逻辑视图名，视图解析器会根据该名字解析到具体的视图页面
        mv.setViewName("first");
        //返回 ModelAndView 对象
        return mv;
    }
}
```

上述代码中，使用@Controller 对 HelloController 类进行标注，使其成为一个可处理 http 请求的控制器；使用@RequestMapping 映射一个请求和请求的方法，对 HelloController 类中的 hello()方法进行标注，确定 hello()对应的请求 URL。如果还有其他的业务 URL 请求，只需在该类下添加方法即可，当然方法要用@RequestMapping 标注，确定方法对应的请求 URL。

(4) 部署运行。

在地址栏输入请求 http://localhost:8080/springmvc-2/hello 后，运行结果如图 7-2 所示。

图 7-2　基于注解的控制器实现

与实现 Controller 接口的方式相比，使用注解的方式显得更加简单。同时 Controller 实现类只能处理单一的请求动作，而基于注解的控制器可以同时处理多个请求动作，这样就

解决了之前在配置文件中创建多个 Bean 的问题，而无需再多建 JavaBean 作为 Controller 去满足业务需求。因此在实际开发中通常都会使用基于注解的形式。

7.1.2 @RequestMapping 注解

Spring 通过@Controller 注解找到相应的控制器类以后，还需要知道控制器内部对每一个请求是如何处理的，这就需要 org.springframework.web.bind.annotation.RequestMapping 注解类型(在上面小节已有讲解)。@RequestMapping 注解的作用是为控制器指定可以处理哪些 URL 请求，可以使用该注解标注在一个方法或一个类上。标注在类上时，该类的所有方法都将映射为相对于类级别的请求；标注在方法上时，该方法将成为一个请求处理方法，它会在程序接收到对应的 URL 请求时被调用。示例代码如下：

```java
package com.springmvc.controller;
import org.springframework.stereotype.Controller;
import org.springframework.web.bind.annotation.RequestMapping;
@Controller
@RequestMapping(value="/user")
public class UserController {
    @RequestMapping(value="/register")
    public String register() {
        return "register";
    }
    @RequestMapping(value="/login")
    public String login() {
        return "login";
    }
}
```

由于 UserController 类中添加了 value="/user"的@RequestMapping 注解，因此所有相关路径都要加上"/user"，此方法被映射到如下请求 URL：

http://localhost:8080/springmvc-2/user/register

http://localhost:8080/springmvc-2/user/login

@RequestMapping 注解除了可以指定 value 属性外，还可指定其他一些属性，所有属性都是可选的，具体如下。

- value 属性

该属性是@RequestMapping 注解的默认属性，因此如果有唯一的属性，则可以省略属性名，用来映射一个请求和一种方法。可以使用@RequestMapping 注释一个类或方法。

@RequestMapping("/hello")和@RequestMapping(value="/hello")标注含义相同。

但如果超过一个属性，就必须写上 value 属性。

- method 属性

该属性用来指示方法仅仅处理哪些 http 请求方式，可支持一个或多个请求方式。

@RequestMapping("/hello",method=RequestMethod.POST)表示只支持 POST 请求方式。

@RequestMapping("/hello",method={RequestMethod.POST,RequestMethod.GET})表示支持 POST 和 GET 请求方式。

如果没有指定 method 属性，则请求处理方法可以处理任意请求处理方式。
- params 属性

该属性指定 request 中必须包含某些参数值时，才让方法处理。例如，@RequestMapping(value="/hello",method=RequestMethod.POST,params="myParam=myValue")表示方法仅处理名为 myParam、值为 myValue 的请求。
- headers 属性

该属性指定 request 中必须包含某些指定的 header 值，才能让方法请求处理。例如，@RequestMapping(value="/hello",method=RequestMethod.POST,headers={"content-type=text/*","Referer=http://www.yzu.edu.cn/"})表示方法将处理 request 的 header 中具有 text/html、text/plain 等内容，并且包含指定 Referer 请求头和对应值为 http://www.yzu.edu.cn 的请求。
- consumes 属性

该属性指定处理请求的提交内容类型(Context-Type)。@RequestMapping(value="/hello",method=RequestMethod.POST,consumes="application/json")表示方法仅处理 request Context-Type 为 application/json 类型的请求。
- produces 属性

该属性指定返回的内容类型，返回的内容类型必须是 request 请求头中所包含的类型。@RequestMapping(value="/hello",method=RequestMethod.POST,produces="application/json")表示方法仅处理 request 请求头中包含 application/json 类型的请求，同时指明了返回的内容类型为 application/json。

7.2 请求映射方式

在 spring MVC 的控制器中，我们经常使用@RequestMapping 来完成请求映射，我们可以在类定义上和方法定义上使用注解，其配置的路径将为类中定义的所有方法的父路径。Spring MVC 可以根据请求方式、Ant 风格的 URL 路径和 REST 风格的 URL 路径进行映射。

7.2.1 根据请求方式进行映射

@RequestMapping 注解除了可以根据请求的 URL 进行映射外，还可以根据请求方式进行映射。如果想根据请求方式进行映射，可通过设置 method 属性来实现。

(1) 在 springmvc-2 的项目下，新建 UserController 类，使用@Controller 注解，并在类中添加方法 requestMethod()，使用@RequestMapping 注解的 method 属性指定该方法的请求方式为 POST，代码如下：

```
package com.springmvc.controller;
import org.springframework.stereotype.Controller;
import org.springframework.web.bind.annotation.RequestMapping;
import org.springframework.web.bind.annotation.RequestMethod;
@Controller
@RequestMapping(value="/user")
public class UserController {
```

```
    @RequestMapping(value = "/requestMethod", method = RequestMethod.POST)
    public String requestMethod() {
        return "success";
    }
}
```

（2）在 WebContent 的路径下新建一个"ch07"文件夹，并在该文件夹中新建 index.jsp 页面，在页面中添加一个 Request Method 超链接，超链接的代码如下：

```
<a href="../user/requestMethod">GET Request Method</a>
```

在 ch07 文件夹中，新建一个 success.jsp 页面，在页面中给出提示信息"欢迎来学习 Spring MVC！"。

（3）修改 springmvc.xml 配置文件，将视图解析器中的 prefix(前缀)"ch06"改为"ch07"，代码如下：

```
<bean class="org.springframework.web.servlet.view.
InternalResourceViewResolver">
    <property name="prefix" value="/ch07/"></property>
    <property name="suffix" value=".jsp"></property>
</bean>
```

（4）重启 Tomcat，浏览页面 index.jsp，单击 GET Request Method 超链接，浏览器会显示错误信息"HTTP Status 405 - Request method 'GET' not supported"，如图 7-3 所示。

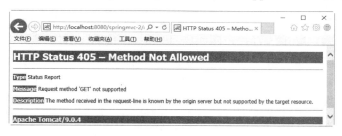

图 7-3 请求处理提示错误

错误原因在于：requestMethod 方法被设置为处理 POST 方式的请求，而通过超链接发出请求的方式为 GET 方式。

（5）在 index.jsp 页面中添加一个表单，通过表单提交请求，代码如下：

```
<form action="../user/requestMethod" method="post">
    <input type="submit" value="POST Request Method">
</form>
```

浏览页面 index.jsp，单击 POST Request Method 按钮提交请求，页面成功转发到 success.jsp。

7.2.2 Ant 风格的 URL 路径映射

Ant 风格的 URL 支持"？""*"和"**"三种匹配符，"？"符号匹配文件名中的一个字符，"*"符号匹配文件名中的任意字符，"**"符号匹配多层路径。

下面以"*"符号为例演示 Ant 风格的 URL 路径映射。在 UserController 类中添加方法 pathAnt()，代码如下：

```
@RequestMapping("/*/pathAnt")
public String pathAnt(){
    System.out.println("Path Ant");
    return "success";
}
```

在 index.jsp 页面中添加一个 Path Ant 超链接，代码如下：

```
<a href="../user/my/pathAnt">Path Ant</a><br><br>
```

重启 Tomcat，浏览页面 index.jsp，单击 Path Ant 链接，页面成功转发到 success.jsp，并且在控制台输出"Path Ant"。

7.2.3 REST 风格的 URL 路径映射

REST 是 Representational State Transfer 的缩写，意思是表现层状态转化，是当前流行的一种互联网软件架构，采用 REST 风格可以有效降低开发的复杂性，提高系统的可伸缩性。

在 Web 开发中，REST 使用 HTTP 协议连接器来标识对资源的操作(获取/查询、创建、删除、修改)，用 HTTP Method(请求方法)标识操作类型，HTTP GET 标识获取和查询资源，HTTP POST 标识创建资源，HTTP PUT 标识修改资源，HTTP DELETE 标识删除资源。

这样，URI 加上 HTTP Method 构成了 REST 风格数据处理的核心，URI 确定操作的对象，HTTP Method 确定操作的方式。例如，"/User/1 HTTP GET"表示获取 id 为 1 的 User 对象，"/User/1 HTTP DELETE"表示删除 id 为 1 的 User 对象，"/User/1 HTTP PUT"表示更新 id 为 1 的 User 对象，"/User HTTP POST"表示新增 User 对象。

由于 form 表单只支持 GET 和 POST 请求，而不支持 DELETE 和 PUT 等请求方式，Spring 提供了一个过滤器 HiddenHttpMethodFilter，可以将 DELETE 和 PUT 请求转换为标准的 HTTP 方式，即能将 POST 请求转为 DELETE 或 PUT 请求。

下面通过一个简单的示例演示 REST 风格的 URL 路径映射，实现过程如下：

(1) 在 web.xml 文件中配置过滤器 HiddenHttpMethodFilter，如下所示：

```
<!-- 配置HiddenHttpMethodFilter，可将POST请求转为DELETE或PUT请求 -->
<filter>
    <filter-name>HiddenHttpMethodFilter</filter-name>
    <filter-class>org.springframework.web.filter.HiddenHttpMethodFilter
</filter-class>
</filter>
<filter-mapping>
    <filter-name>HiddenHttpMethodFilter</filter-name>
    <url-pattern>/*</url-pattern>
</filter-mapping>
```

(2) 处理 GET 请求。

① 在 index.jsp 页面中添加 Rest GET 超链接，如下所示：

```html
<a href="../user/rest/1">Rest GET</a>
```

② 在 UserController 类中添加方法 restGET，处理 GET 方式请求，如下所示：

```java
@RequestMapping(value = "/rest/{id}", method = RequestMethod.GET)
public String restGET(@PathVariable("id") Integer id) {
    System.out.println("Rest GET:" + id);
    return "success";
}
```

重启 Tomcat，浏览 index.jsp 页面，单击 Rest GET 链接，页面成功转发到 success.jsp，控制台输出如下所示：

```
Rest GET:1
```

(3) 处理 POST 请求。

① 在 index.jsp 页面中添加一个表单，代码如下：

```html
<form action="../user/rest" method="post">
    <input type="submit" value="Rest POST">
</form>
```

② 在 UserController 类中添加方法 restPOST，处理 POST 方式请求，如下所示：

```java
@RequestMapping(value = "/rest", method = RequestMethod.POST)
public String restPOST() {
    System.out.println("Rest POST");
    return "success";
}
```

重启 Tomcat，浏览 index.jsp 页面，单击 Rest POST 按钮，页面成功转发到 success.jsp，控制台输出如下所示：

```
Rest POST
```

(4) 处理 DELETE 请求。

① 在 index.jsp 页面中添加一个表单，如下所示：

```html
<form action="../user/rest/1" method="post">
    <input type="hidden" name="_method" value="DELETE">
    <input type="submit" value="Rest DELETE">
</form>
```

表单中使用了一个名称为"_method"的隐藏域，并给其赋值 DELETE。过滤器 HiddenHttpMethodFilter 正是通过"_method"的值，将 POST 请求转为 DELETE。

② 在 UserController 类中添加方法 restDELETE，处理 DELETE 方式请求，如下所示：

```java
@RequestMapping(value="/rest/{id}",method=RequestMethod.DELETE)
public String restDELETE(@PathVariable("id") Integer id) {
    System.out.println("Rest DELETE:" + id);
    return "redirect:/user/doTransfer";
}
```

restDELETE 方法返回值使用了重定向，将请求重定向到"user/doTransfer"，该映射对应的处理方法如下所示：

```
@RequestMapping("/doTransfer")
public String doTransfer() {
    return "success";
}
```

重启 Tomcat，浏览 index.jsp 页面，单击 Rest DELETE 按钮，页面成功转发到 success.jsp，控制台输出如下所示：

```
Rest DELETE:1
```

从控制台输出结果可以看出，单击 Rest DELETE 按钮发出的请求被成功提交给 UserController 类中的 restDELETE 方法来处理。

(5) 处理 PUT 请求。

① 在 index.jsp 页面中添加一个表单，如下所示：

```
<form action="../user/rest/1" method="post">
    <input type="hidden" name="_method" value="PUT">
    <input type="submit" value="Rest PUT">
</form>
```

表单中使用了一个名称为"_method"的隐藏域，并给其赋值 PUT。过滤器 HiddenHttpMethodFilter 正是通过"_method"的值，将 POST 请求转为 PUT。

② 在 HelloController 类中添加方法 restPUT，处理 PUT 方式请求，如下所示：

```
@RequestMapping(value = "/rest/{id}", method = RequestMethod.PUT)
public String restPUT(@PathVariable("id") Integer id) {
    System.out.println("Rest PUT:" + id);
    return "redirect:/user/doTransfer";
}
```

重启 Tomcat，浏览 index.jsp 页面，单击 Rest PUT 按钮，页面成功转发到 success.jsp，控制台输出如下所示：

```
Rest PUT:1
```

从控制台输出结果可以看出，单击 Rest PUT 按钮发出的请求被成功提交给 UserController 类中的 restPUT 方法来处理。

7.3 绑定控制器类处理方法入参

我们知道，当用户在页面触发某种请求时，一般会将一些参数(key/value)带到后台。在 Spring MVC 中可以通过参数绑定，将客户端请求的 key/value 数据绑定到 Controller 处理方法的形参上。Spring MVC 支持将多种途径传递的参数绑定到控制器类的处理方法的输入参数中。

1. 映射 URL 绑定的占位符到方法入参

使用 @PathVariable 注解可以将 URL 中的占位符绑定到控制器方法的入参中。在

UserController 类中添加 pathVariable()方法，使用@PathVariable 注解映射 URL 中的占位符到目标方法的参数中，代码如下：

```
@RequestMapping("/pathVariable/{id}")
public String pathVariable(@PathVariable("id") Integer id) {
    System.out.println("Path Variable:" + id);
    return "success";
}
```

在 index.jsp 页面中添加一个 Path Variable 超链接，代码如下：

```
<a href="../user/pathVariable/1">Path Variable</a><br><br>
```

URL 中的占位符{id}通过注解@PathVariable("id")绑定到 pathVariable 方法的入参 id 中，重启 Tomcat，浏览页面 index.jsp，单击 Path Variable 链接，页面成功转发到 success.jsp，控制台输出"Path Variable:1"。

2. 绑定请求参数到控制器方法参数

在控制器方法入参处使用@RequestParam 注解可以将请求参数传递给方法，通过@RequestParam 注解的 value 属性指定参数名，required 属性指定参数是否必需，默认为 true，表示请求参数中必须包含对应的参数，如果不存在，则抛出异常。

在 UserController 类中添加 requestParam()方法，使用@RequestParam 注解绑定请求参数到控制器方法参数，代码如下：

```
@RequestMapping("/requestParam")
public String requestParam(
    @RequestParam(value = "loginName") String loginName,
    @RequestParam(value = "loginPwd") String loginPwd) {
    System.out.println("Request Param:" + loginName + " " + loginPwd);
    return "success";
}
```

在 index.jsp 页面中添加一个 Request Param 超链接，代码如下：

```
<a href="../user/requestParam?loginName=admin&loginPwd=123456">
Request Param</a> <br><br>
```

重启 Tomcat，浏览页面 index.jsp，单击 Request Param 链接，页面成功转发到 success.jsp，控制台输出"Request Param:admin 123456"。

3. 将请求参数绑定到控制器方法的表单对象

Spring MVC 会按照参数名和属性名进行自动匹配，自动为该对象填充属性值，并且支持级联。在项目的 src 目录下创建包 com.springmvc.entity，在包中创建用户实体类 User.java 和地址实体类 Address.java。实体类 Address 的代码如下：

```
package com.springmvc.entity;
public class Address {
    private String province;
    private String city;
    // 此处省略属性的 getter 和 setter 方法
```

```
    // 此处省略构造方法
    // 重写toString()方法
    @Override
    public String toString() {
        return "Address[province="+province+",city="+city+"]";
    }
}
```

实体类 User 的代码如下：

```
package com.springmvc.entity;
public class User {
    private String loginName;
    private String loginPwd;
    private Address address;
    // 此处省略属性的getter和setter方法
    // 此处省略构造方法
    // 重写toString()方法
    @Override
    public String toString() {
        return "User[loginName="+loginName+",loginPwd="
                +loginPwd+",address="+address+"]";
    }
}
```

然后在 UserController 类中添加 saveUser() 方法，将请求参数绑定到控制器方法的表单对象 User 中。

```
@RequestMapping("/saveUser")
public String saveUser(User user) {
    System.out.println(user);
    return "success";
}
```

最后在 index.jsp 页面中创建一个表单，代码如下：

```
<form action="../user/saveUser" method="post">
    loginName:<input type="text" name="loginName"><br>
    loginPwd:<input type="password" name="loginPwd"><br>
    province:<input type="text" name="address.province"><br>
    city:<input type="text" name="address.city"><br>
    <input type="submit" value="提交">
</form>
```

重启 Tomcat，浏览 index.jsp 页面，在表单中输入用户名"admin"、密码"123456"、省份"JiangSu"和城市"YangZhou"，单击"提交"按钮，控制台输出如下所示：

```
User[loginName=admin,loginPwd=123456,address=Address
[province=JiangSu,city=YangZhou]]
```

4. 将请求参数绑定到控制器方法的 Map 对象

Spring MVC 注解可以将表单数据传递到控制器方法中的 Map 类型的入参中。在

com.springmvc.entity 包中创建 UserMap 类,定义 Map<String, User>类型的属性 uMap,并为该属性提供 getter 和 setter 方法,代码如下:

```java
package com.springmvc.entity;
import java.util.Map;
public class UserMap {
    private Map<String, User> uMap;
    public Map<String, User> getuMap() {
        return uMap;
    }
    public void setuMap(Map<String, User> uMap) {
        this.uMap = uMap;
    }
}
```

在 UserController 类中添加 getUser()方法,实现将请求参数绑定到控制器方法的 Map 对象,并遍历 Map,将 Map 中的内容输出到控制台。

```java
@RequestMapping("/getUser")
public String getUser(UserMap uMap) {
    Set set = uMap.getuMap().keySet();
    Iterator iterator = set.iterator();
    while (iterator.hasNext()) {
        Object keyName = iterator.next();
        User u = uMap.getuMap().get(keyName);
        System.out.println(u);
    }
    return "success";
}
```

在 index.jsp 页面中创建一个表单,代码如下:

```html
<form action="../user/getUser" method="post">
    loginName1:<input type="text" name="uMap['u1'].loginName"><br>
    loginPwd1:<input type="password" name="uMap['u1'].loginPwd"><br>
    province1:<input type="text" name="uMap['u1'].address.province">
    <br>
    city1:<input type="text" name="uMap['u1'].address.city"><br>
    loginName2:<input type="text" name="uMap['u2'].loginName"><br>
    loginPwd2:<input type="password" name="uMap['u2'].loginPwd"><br>
    province2:<input type="text" name="uMap['u2'].address.province">
    <br>
    city2:<input type="text" name="uMap['u2'].address.city"><br>
    <input type="submit" value="提交">
</form>
```

重启 Tomcat,浏览页面 index.jsp,在表单中分别输入两个用户的信息。第一个用户的用户名"my"、密码"123456"、省份"JiangSu"和城市"NanJing";第二个用户的用户名"sj"、密码"123456"、省份"JiangSu"和城市"YangZhou"。单击"提交"按钮,控制台输出如下所示:

```
User[loginName=my,loginPwd=123456,address=Address[province=JiangSu,
city=NanJing]]
User[loginName=sj,loginPwd=123456,address=Address[province=JiangSu,
city=YangZhou]]
```

从控制台输出可以看出，表单中输入的两个用户的信息成功传递到控制器方法 getUser 的 UserMap 类型的参数 uMap 中。

7.4 控制器类处理方法的返回值类型

在前面的示例中，当控制器处理完请求时，会以字符串的形式返回逻辑视图名。除了 String 类型外，Spring MVC 返回的类型还有 ModelAndView、Model、ModelMap、Map 等。

如果返回类型是 ModelAndView，则其中可包含视图和模型信息，且 Spring MVC 会将模型信息存放到 request 域中。在 UserController 类中添加 returnModelAndView()方法，返回 ModelAndView 类型。代码如下：

```
@RequestMapping("/returnModelAndView")
public ModelAndView returnModelAndView() {
    String viewName="success";
    ModelAndView mv=new ModelAndView(viewName);
    User user = new User("zhangsan", "123456", new Address("jiangsu", "nanjing"));
    mv.addObject("user", user);
    return mv;
}
```

在 index.jsp 页面中添加一个 ModelAndView 超链接，如下所示：

```
<a href="../user/returnModelAndView">ModelAndView</a><br><br>
```

在 success.jsp 页面中添加用于访问 ModelAndView 对象中保存的 User 对象的代码，如下所示：

```
ModelAndView:${requestScope.user }
```

重启 Tomcat，浏览页面 index.jsp，单击 ModelAndView 链接，success.jsp 页面显示如下：

欢迎来学习 Spring MVC! ModelAndView:User [loginName=zhangsan, loginPwd=123456, address=Address [province=jiangsu, city=nanjing]]

存入 ModelAndView、Model、ModelMap、Map 中的数据对象，可以通过 request 作用域来访问。

7.5 保存模型属性到 HttpSession

通过在控制器类的相应方法上标注@SessionAttributes 注解，可将模型数据保存到 HttpSession 中，以便多个请求之间共用该模型属性。在 UserController 类中添加方法

sessionAttributes，先将模型属性 user 保存到 ModelMap 中，并在控制器类 UserController 上标注@SessionAttributes 注解，将 User 类型的模型属性 user 存入 HttpSession 中，代码如下：

```java
package com.springmvc.controller;
……
import org.springframework.web.bind.annotation.SessionAttributes;
@Controller
@RequestMapping(value="/user")
@SessionAttributes(value={"user"})
public class UserController {
    @RequestMapping("/sessionAttributes")
    public String sessionAttributes(ModelMap model) {
        User user = new User("lisi", "123456", new Address("jiangsu", "yangzhou"));
        model.put("user", user);
        return "success";
    }
}
```

在 index.jsp 页面中添加一个 Session Attributes 超链接，如下所示：

```
<a href="../user/sessionAttributes">Session Attributes</a><br><br>
```

在 success.jsp 页面中添加用于访问保存到 HttpSession 中的 User 对象的代码，如下所示：

```
HttpSeesion 中保存的 user: ${sessionScope.user }
```

重启 Tomcat，浏览页面 index.jsp，单击 Session Attributes 链接，success.jsp 页面显示如下：

```
HttpSeesion 中保存的 user: Users [loginName=lisi, loginPwd=123456, address=Address [province=jiangsu, city=yangzhou]]
```

7.6 在控制器类的处理方法执行前执行指定的方法

如果想让一个方法在控制器类的所有处理方法之前执行，可以通过在该方法上标注@ModelAttribute 注解来实现。

在 UserController 类中添加 getUser()方法，在方法上标注@ModelAttribute 注解。在 getUser()方法中实例化 User 对象 user，保存到 Model 中，getUser()方法的返回值为对象 user。

```java
@ModelAttribute
public User getUser(Model model) {
    User user = new User("wangwu", "123456", new Address("JiangSu", "SuZhou"));
    model.addAttribute("user", user);
    return user;
}
```

然后在 UserController 类中添加 modelAttribute()方法，并删除 UserController 类前面添加的@SessionAttributes(value={"user"})注解，代码如下：

```
@RequestMapping("/modelAttribute")
public String modelAttribute(User user) {
    System.out.println(user);
    return "success";
}
```

在 index.jsp 页面中添加一个 Model Attribute 超链接,如下所示:

```
<a href="../user/modelAttribute">Model Attribute</a><br><br>
```

重启 Tomcat,浏览页面 index.jsp,单击 Model Attribute 超链接,控制台输出如下:

```
User[loginName=wangwu, loginPwd=123456, address=Address [province=JiangSu, city=SuZhou]]
```

单击 Model Attribute 链接时,请求被 UserController 类中的 modelAttribute 方法处理,方法中没有对对象 user 进行初始化,而控制台输出显示对象 user 已经被初始化过,这一初始化过程显然是在使用@ModelAttribute 注解标注的 getUser()方法中完成的。也就是说在执行方法 modelAttribute()前,先调用了 getUser()方法,实例化对象 user 后将其存入 Model,又因为 getUser()方法返回了该对象 user,被传递给 modelAttribute 方法的参数 user。

success.jsp 页面显示如下所示:

```
欢迎来学习 Spring MVC! ModelAndView:User[loginName=wangwu,loginPwd=123456,
address=Address[province=JiangSu,city=SuZhou]]
```

在 getUser()方法中,对象 user 被存入 Model,访问 request 作用域可以获得对象 user 的值。由于对象 user 没有保存在 HttpSeesion 中,访问 session 作用域无法获取对象 user 的值。

7.7 直接页面转发、自定义视图与页面重定向

1. 直接页面转发

如果想不经过控制器类的处理方法直接转发到页面,可以通过使用<mvc:view-controller>元素来实现。在 Spring MVC 配置文件 springmvc.xml 中,添加<mvc:view-controller>元素,其配置如下所示:

```
<mvc:view-controller path="/success" view-name="success"/>
<mvc:view-controller path="/index" view-name="index"/>
```

重启 Tomcat,在浏览器中直接输入地址"http://localhost:8080/springmvc-2/success",页面成功转发到 success.jsp,而地址栏并没有变化。

需要注意,如果在 springmvc.xml 文件中没有添加<mvc:annotation-driven />元素,浏览 index.jsp 页面中的超链接时会出现问题。使用<mvc:annotation-driven>元素后,会自动注册 RequestMappingHandlerMapping、RequestMappingHandlerAdapter 和 ExceptionHandlerExceptionResolver 这三个 Bean,就可以解决问题。

2. 自定义视图

通过使用 BeanNameViewResolver 类可以实现用户自定义的视图。创建一个类 MyView,

实现 View 接口，存放在新建的 com.springmvc.view 包中，实现一个简单的自定义视图。

```
package com.springmvc.view;
import java.util.Map;
import javax.servlet.http.HttpServletRequest;
import javax.servlet.http.HttpServletResponse;
import org.springframework.stereotype.Component;
import org.springframework.web.servlet.View;
@Component
public class MyView implements View {
    @Override
    public String getContentType() {
        return "text/html";
    }
    @Override
    public void render(Map<String, ?> arg0, HttpServletRequest request,
            HttpServletResponse response) throws Exception {
        response.getWriter().println("hello,this is my view");
    }
}
```

在 MyView 类上标注@Component 注解，Spring 会为该类创建 Bean 实例。在 render 方法中，向浏览器输出一个简单的字符串"hello,this is my view"作为自定义页面内容。

在 springmvc.xml 文件中配置视图解析器，使用视图名称实现视图解析。

```
<bean class="org.springframework.web.servlet.view.
BeanNameViewResolver">
    <property name="order" value="50" />
</bean>
```

在 UserController 类中添加方法 beanNameViewResolver，来使用自定义视图。

```
@RequestMapping("/beanNameViewResolver")
public String beanNameViewResolver(){
    return "myView";
}
```

由于 BeanNameViewResolver 类是根据 Bean 的名称来解析视图的，自定义的视图类 MyView 在 Spring IOC 中的 Bean 实例名为"myView"，因此 beanNameViewResolver 方法的返回值应该使用 Bean 实例名"myView"。

在 index.jsp 页面中添加一个 BeanNameViewResolver 超链接，如下所示：

```
<a href="../user/beanNameViewResolver">BeanNameViewResolver</a><br>
```

重启 Tomcat，浏览页面 index.jsp，单击 BeanNameViewResolver 链接，显示自定义的页面内容，如下所示：

```
Hello, This is my view?
```

3. 页面重定向

在前面的示例中，控制器类的方法返回的字符串默认通过转发的方式跳转到目标页面，如果返回的字符串中带 redirect 前缀，则会采用重定向的方式跳转到目标页面。

在 UserController 类中添加 redirect()方法，在返回字符串中使用重定向。

```
@RequestMapping("/redirect")
public String redirect(){
    return "redirect:/ch07/index.jsp";
}
```

在 index.jsp 页面中添加一个 Redirect 超链接，如下所示：

```
<a href="../user/redirect">Redirect</a><br><br>
```

重启 Tomcat，浏览页面 index.jsp，单击 Redirect 链接，页面重定向到 index.jsp。

7.8　Spring MVC 返回 JSON 数据

现如今由于移动互联网的兴起，简洁的 JSON 格式成为很多系统之间进行交互的主要格式。移动端的前台系统(Android 或 IOS)与后台交互时，普遍使用 HTTP 协议进行 JSON 格式信息的传输，以实现移动系统前后端之间的信息交互。还有一些网页异步加载功能，也是利用 JavaScript 语言或相关插件(jQuery 等前端脚本框架)实现 Ajax 异步数据请求，与后台进行 JSON 格式的 HTTP 信息交互。

在 Spring MVC 中，为开发者提供了一种简洁的实现不同数据格式交互的机制(JSON、XML 以及其他数据格式)，其会将前台传来的 JSON/XML 等格式信息自动转换为相应的包装类，或者将输出的信息转换为 JSON/XML 等格式的数据。

一般情况下，利用注解@ResponseBody 都会在异步获取数据时使用，被其标注的处理方法返回的数据将输出到响应流中，由客户端获取并显示数据。如果想让控制器类的处理方法返回 JSON 数据，可以使用 HttpMessageConverter 类来实现。如果不想显式地创建 HttpMessageConverter 的 Bean 实例，可以在 Spring MVC 配置文件 springmvc.xml 中添加 <mvc:annotation-driven>元素。

```
<mvc:annotation-driven />
```

在 JSON 的使用中，四种常见的 JSON 技术如下。

- json-lib：json-lib 是最早也是应用最广泛的 json 解析工具，它的缺点就是依赖很多第三方包，如 commons-beanutils.jar、commons-collections-3.2.jar、commons-lang-2.6.jar、commons-logging-1.1.1.jar、ezmorph-1.0.6.jar。对于复杂类型的转换，json-lib 对于 JSON 转换成 Bean 还有缺陷，比如一个类里面若出现另一个类的 List 或者 Map 集合，json-lib 从 JSON 到 Bean 的转换就会出现问题。所以 json-lib 在功能和性能上都不能满足现在互联网化的需求。
- Jackson：开源的 Jackson 是 Spring MVC 内置的 JSON 转换工具。相比 json-lib 框架，Jackson 所依赖的 jar 包较少，简单易用并且性能也要相对高些。而且 Jackson 社区比较活跃，更新速度也比较快。但是 Jackson 对于复杂类型的从 JSON 转换到 Bean 会出现问题，一些集合 Map、List 的转换也不能正常实现。而 Jackson 对于复杂类型的从 Bean 转换到 JSON，转换的 JSON 格式不是标准的 JSON 格式。
- Gson：Gson 是目前功能最全的 JSON 解析器，Gson 最初是应 Google 公司内部需求而由 Google 自行研发的，自从 2008 年 5 月公开发布第 1 版后已被许多公司和用户应用。Gson 的应用主要是 toJson 与 fromJson 两个转换函数，它们无依赖且不需要额外的 jar 文件，能够直接在 JDK 上运行。Gson 可以完成复杂类型的从 JSON 到 Bean 或从 Bean 到 JSON 的转换，是 JSON 解析的利器。其在功能上无可挑剔，但是性能比 FastJson 稍差。
- FastJson：FastJson 是一个用 Java 语言编写的高性能的 JSON 处理器，由阿里巴巴公司开发。它的特点是无依赖，不需要额外的 jar 文件，能够直接在 JDK 上运行。但是 FastJson 在复杂类型的从 Bean 转换到 JSON 上会出现一些问题，可能会因为引用的类型不当，导致 JSON 转换出错，因而需要指定引用。FastJson 采用独创的算法，将 parse 的速度提升到极致，超过所有 JSON 库。

下面通过示例介绍如何使用 Spring MVC 内置的 JackSon 技术返回 JSON 数据，实现步骤如下。

(1) 下载并添加 JSON 的相关 jar 包。

通过浏览器访问 http://mvnrepository.com/open-source/json-libraries 网址，下载三个 jar 包，如图 7-4 所示。

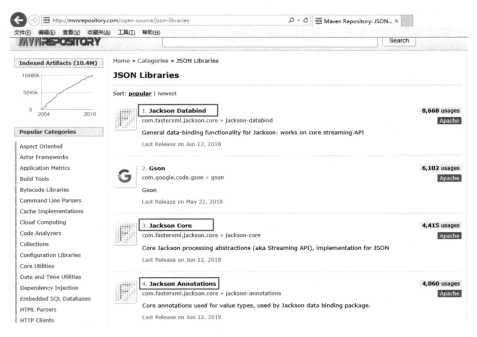

图 7-4　下载 JSON 的相关 jar 包

下载完成后将 jackson-annotations-2.9.6.jar、jackson-core-2.9.6.jar 和 jackson-databind-2.9.6.jar 这 3 个 jar 包复制到 springmvc-2 项目的 WebContent\WEB-INF\lib 目录下。

(2) 下载并引入 jQuery 资源文件。

访问 https://jquery.com/download/ 网址，下载 jquery-3.3.1.min.js 文件，如图 7-5 所示。

图 7-5　下载 jQuery 资源文件

在 springmvc-2 项目的 WebContent 目录下创建一个文件夹 scripts，将 jQuery 资源文件 jquery-3.3.1.min.js 复制到其中。

(3) 处理静态资源文件 jquery-3.3.1.min.js。

在 Spring MVC 配置文件 springmvc.xml 中添加<mvc:default-servlet-handler />元素，如下所示：

```
<mvc:default-servlet-handler />
```

该元素将在 Spring MVC 上下文中定义一个 DefaultServletHttpRequestHandler，它会对进入 DispatcherServlet 的请求进行筛查，如果发现是没有经过映射的请求，就将该请求交由 Web 应用服务器默认的 Servlet 处理。如果不是静态资源的请求，才由 DispatcherServlet 继续处理。如果没有添加<mvc:default-servlet-handler />元素，Web 容器启动时会抛出如下异常：

```
警告: No mapping found for HTTP request with URI [/springmvc-2/scripts/jquery-3.3.1.min.js] in DispatcherServlet with name 'dispatcherServlet'
```

(4) 在 UserController 类中添加 returnJson()方法，在该方法前添加@ResponseBody 注解，如下所示：

```
@ResponseBody
@RequestMapping("/returnJson")
public Collection<User> returnJson() {
    Map<Integer, User> us = new HashMap<Integer, User>();
    us.put(1, new User("zhangsan", "123456",
            new Address("Jiangsu","NanJing")));
    us.put(2, new User("lisi", "123456",
            new Address("Jiangsu", "YangZhou")));
    us.put(3, new User("wangwu", "123456",
            new Address("Jiangsu", "SuZhou")));
    return us.values();
}
```

returnJson 方法的返回类型为 Collection<User>，但在标注@ResponseBody 注解后，返回类型就转变为 JSON 格式了。

(5) 在 index.jsp 页面的<head></head>标签中通过<script>标签，引入 jQuery 资源文件 jquery-3.3.1.min.js，如下所示：

```
<script type="text/javascript" src="../scripts/jquery-3.3.1.min.js">
</script>
```

在 index.jsp 页面中添加一个 Test Json 超链接，如下所示：

```
<a href="javascript:void(0)" id="returnJson" onclick="getUserJson()">
Test Json</a><br><br>
```

单击 Test Json 链接，将执行一个 JavaScript 脚本函数 getUserJson()。在 index.jsp 页面中新建<script>脚本标签，并在其中创建函数 getUsersJson()，如下所示：

```
<script type="text/javascript">
    function getUserJson() {
        var url = "../user/returnJson";
        var args = {};
        $.post(url, args, function(data) {
        });
    }
</script>
```

在 getUserJson 函数中，使用$.post 将请求提交到控制器类 UserController 中的 returnJson() 方法，参数 data 就是 returnJson 方法执行后返回的 JSON 格式的数据。

(6) 重启 Tomcat，使用 IE 浏览器浏览页面 index.jsp，按 F12 键，进入开发人员模式，单击"网络"，单击 Test Json 链接，在名称路径上就会出现 returnJson 的请求，单击右侧出现的"正文"选项卡标签，单击"响应正文"，在下方就会出现相应的 JSON 格式的数据，如图 7-6 所示。

图 7-6　查看 JSON 格式的数据

注：不同浏览器的调试方式也不同。

7.9 小　　结

本章首先介绍了 Spring MVC 的常用注解，包括@Controller、@RequestMapping 两个最重要的注解，以及 3 种请求映射方式。接着介绍常用参数绑定注解，通过绑定控制器类处理方法入参、控制器类处理方法的返回值类型、保存模型属性到 HttpSession、在控制器类的处理方法执行前执行指定方法等来讲解绑定参数注解。最后介绍了在 Spring MVC 中将数据转换成 JSON 格式的方法。

第 8 章 Spring MVC 标签库

我们在进行 Spring MVC 项目开发时，一般会使用 EL 表达式和 JSTL 标签来完成页面视图，其实 Spring MVC 也有一套自己的表单标签库。Spring 从 2.0 开始，通过 Spring MVC 表单标签库，可以很容易地将模型数据中的表单对象绑定到 HTML 表单元素中。相比其他的标签库，Spring 的标签库集成在 Spring Web MVC 框架中，因此它们可以直接使用命令对象和其他控制器处理的数据对象，这样 JSP 更容易开发、阅读和维护。

8.1 Spring MVC 表单标签库概述

表单标签库的实现类在 spring-webmvc-5.0.4.RELEASE.jar 文件中，要使用 Spring MVC 的表单标签库，首先要和使用 JSTL 标签一样，必须在 JSP 页面中添加一行引用 Spring 标签库的 taglib 指令声明，如下所示：

```
<%@ taglib prefix="fm" uri="http://www.springframework.org/tags/form" %>
```

引入标签声明之后就可以使用 Spring 表单标签了。表 8-1 显示了表单标签库中的相关标签。

表 8-1 表单标签库中的相关标签

标 签	说 明
\<fm:form /\>	渲染表单元素
\<fm:input /\>	渲染输入框\<input type="text"\>元素

续表

标　签	说　明
<fm:password />	渲染密码框<input type="password">元素
<fm:hidden />	渲染隐藏框<input type="hidden">元素
<fm:textarea />	渲染多行输入框 textarea 元素
<fm:checkbox />	渲染复选框<input type="checkbox">元素
<fm:radiobutton />	渲染单选按钮<input type="radio">元素
<fm:select />	渲染下拉列表元素
<fm:option />	渲染一个可选元素
<fm:errors />	在 span 元素中渲染字段错误

Spring 提供了十多个表单标签，基本上这些标签都拥有以下共有属性。

- path：属性路径，表示表单对象属性。
- cssClass：表单组件对应的 CSS 样式类名。
- cssErrorClass：当提交表单后报错(服务端错误)时采用的 CSS 样式类。
- cssStyle：表单组件对应的 CSS 样式。
- htmlEscape：绑定的表单属性值是否要对 HTML 特殊字符进行转换，默认为 true。

此外，表单组件标签也拥有 HTML 标签的各种属性，如 id、onclick 等属性，可以根据需要灵活使用。

8.2　Spring MVC 表单标签库

8.2.1　form 标签

Spring MVC 中的 form 标签主要有两个作用：

- 自动绑定 Model 中的一个属性值到当前 form 对应的实体对象，默认为 command 属性，这样我们就可以在 form 表单体中方便地使用该对象的属性了。
- 支持我们在提交表单时使用除了 GET 和 POST 之外的其他方法进行提交，包括 DELETE 和 PUT 等。

form 标签除了前面介绍的共有属性外，还有以下常用的属性，不包含 HTML 中如 method 和 action 等属性，如下所示。

- modelAttribute：form 绑定的模型属性名称，默认为 command。
- commandName：form 绑定的也是模型属性名称，默认为 command。其作用与 modelAttribute 属性相同。
- acceptCharset：定义服务器接受的字符编码。

commandName 属性是其中最重要的属性，它定义了模型属性的名称，其中包含一个绑定的 JavaBean 对象，该对象的属性将用于填充所生成的表单。如 commandName 属性存在，则必须在返回包含该表单的视图的请求处理方法中添加响应的模型属性。

通常我们都会指定 commandName 或 modelAttribute 属性，指定绑定到的 JavaBean 的名

称,这两个属性功能基本一致。

8.2.2 input 标签

Spring MVC 的 input 标签会被渲染为一个类型为 text 的普通 HTML input 标签。使用 Spring MVC 的 input 标签的唯一作用就是绑定表单数据,通过 path 属性来指定要绑定的 Model 中的值。

下面通过示例来讲解 form 和 input 标签的使用,具体步骤如下。

(1) 将项目 springmvc-2 复制并重命名为 springmvc-3,再导入到 Eclipse 开发环境中。选中 springmvc-3 项目并右击,在快捷菜单中选择 Properties 命令,进入属性设置界面,找到 Web Project Settings 选项,将右侧 Context Root:文本框中的 springmvc-2 改为 springmvc-3,单击 Apply and Close 按钮。删除 src 下的相关包,删除 WebContent 下的 ch06 和 ch07 文件夹,保留相关 jar 包以及 springmvc.xml 配置文件。

(2) 在项目的 WebContent 路径下新建一个 ch08 文件夹,在该文件夹中新建 register.jsp 页面,并在页面中添加一行引用 Spring 标签库的声明,如下所示:

```
<%@ taglib prefix="fm" uri="http://www.springframework.org/tags/form" %>
……
<body>
    <h3>注册页面</h3>
    <fm:form action="register" method="post">
        姓名:<fm:input path="name"/><br><br>
        性别:<fm:input path="sex"/><br><br>
        年龄:<fm:input path="age"/><br><br>
    </fm:form>
</body>
```

如果 Model 中存在一个属性名称为 command 的 JavaBean,而且该 JavaBean 拥有属性 name、sex 和 age,则在渲染上面代码时就会取 command 的对应属性值赋给对应标签的属性。

(3) web.xml 配置不变,修改 springmvc.xml 配置文件,将视图解析器中的 prefix(前缀)ch07 改为 ch08,代码如下:

```
<property name="prefix" value="/ch08/"></property>
```

(4) 在 src 下新建 com.springmvc.entity 包,并在包中新建类 User,如下所示:

```
package com.springmvc.entity;
public class User {
    private String name;
    private String sex;
    private int age;
    // 省略属性的 setter、getter 方法
    // 省略构造方法
}
```

(5) 在 src 下新建 com.springmvc.controller 包,在包中新建 UserController 类,并通过 @Controller 和 @RequestMapping 注解实现一个 register()方法,如下所示:

```
@Controller
public class UserController {
    @RequestMapping(value="/register", method = RequestMethod.GET)
    public String register(Model model) {
        User user=new User("zhangsan","男",23);
        //model 中添加属性 command，值为 user 对象
        model.addAttribute("command",user);
        return "register";
    }
}
```

在上述代码中，将 user 设置到 model 中，属性名为 command。

(6) 部署 springmvc-3 这个 Web 项目，启动 Tomcat，在浏览器的地址栏中访问 http://localhost:8080/springmvc-3/register 进行测试，运行效果如图 8-1 所示。

在上述代码中，假设 Model 中存在一个属性名为 command 的 JavaBean，且它的 name、sex 和 age 属性分别为 "zhangsan" "男" 和 "23"，则在浏览器页面单击右键，选择 "查看源" 命令，可以看到 Spring MVC 的标签渲染时生成的 html 代码如下所示：

图 8-1　测试 form 和 input 标签

```
<form id="command" action="register" method="post">
    姓名:<input id="name" name="name" type="text" value="zhangsan"/><br><br>
    性别: <input id="sex" name="sex" type="text" value="男"/><br><br>
    年龄: <input id="age" name="age" type="text" value="23"/><br><br>
</form>
```

从生成的代码可以看出，当没有指定 form 标签的 id 属性时，它会自动获取 form 标签绑定的 Model 中的对应属性名称 command 作为 id；而对于 input 标签，在没有指定 id 的情况下它会自动获取 path 指定的属性值作为它的 id 和 name。

Spring MVC 指定 form 标签默认自动绑定的是 Model 的 command 属性值，那么当 form 对象对应的属性名称不是 command 时，怎么办？对于这种情况，Spring 提供了 commandName 属性，可通过该属性来指定将使用 Model 中哪个属性作为 form 标签需要绑定的 command，除了 commandName 属性外，指定 modelAttribute 属性也可以达到相同的效果。

修改 register.jsp 页面中的 form 标签，添加 modelAttribute 属性值为 "user"，如下所示：

```
<fm:form modelAttribute="user" action="register" method="post">
......
</fm:form>
```

修改 UserController 类中的 register 方法，将 model 属性值 "command" 修改为 "user"，如下所示：

```
model.addAttribute("user",user);
```

重启 Tomcat，在浏览器的地址栏中访问 http://localhost:8080/springmvc-3/register 进行测

试，其运行效果与修改前一致，查看页面源文件，form 表单的 id 属性变为 user，不是 model 默认的 command，如下所示：

```
<form id="user" action="register" method="post">
    姓名:<input id="name" name="name" type="text" value="zhangsan"/><br><br>
    性别: <input id="sex" name="sex" type="text" value="男"/><br><br>
    年龄: <input id="age" name="age" type="text" value="23"/><br><br>
</form>
```

8.2.3 password 标签

Spring MVC 的 password 标签会被渲染为一个类型为 password 的普通 HTML input 标签。password 标签的用法和 input 标签相似，也能绑定表单数据，只是它生成的是一个密码框，并且多了一个 showPassword 属性，表示显示或遮盖密码，默认值为 false。

下面是 password 标签的示例：

```
<fm:password path="password"/>
```

上面代码运行时，password 标签会被渲染成下面的 HTML 元素：

```
<input id="password" name="password" type="password" value=""/>
```

8.2.4 hidden 标签

Spring MVC 的 hidden 标签会被渲染为一个类型为 hidden 的普通 HTML input 标签。其用法和 input 标签类似，也能绑定表单数据，只是生成的是一个隐藏域，界面上看不到任何内容。

下面是 hidden 标签的示例：

```
<fm:hidden path="hid"/>
```

上面代码运行时，hidden 标签会被渲染成下面的 HTML 元素：

```
<input id="hid" name="hid" type="hidden" value=""/>
```

8.2.5 textarea 标签

Spring MVC 的 textarea 标签会被渲染为一个类型为 textarea 的普通 HTML 标签。textarea 是一个支持多行输入的 HTML 元素。

下面是 textarea 标签的示例：

```
<fm:textarea path="remark" rows="5" cols="20"/>
```

上面代码运行时，textarea 标签会被渲染成下面的 HTML 元素：

```
<textarea id="remark" name="remark" rows="5" cols="20"></textarea>
```

8.2.6 checkbox 标签

Spring MVC 的 checkbox 标签会被渲染为一个类型为 checkbox 的普通 HTML 标签。checkbox 标签除了 8.1 小节介绍的共有属性外，还有一个 label 属性，将其作为 label 被渲染的复选框的值。

checkbox 标签绑定的数据类型如下。

- 绑定 boolean 数据：当 checkbox 绑定的是一个 boolean 数据时，其状态与被绑定的 boolean 数据的状态是一样的，即为 false 时复选框不选中，为 true 时复选框选中。
- 绑定列表数据：列表数据主要包括数组、List 和 Set。假设有一个 User 类，User 类中有一个类型为 List 的属性 courses。当要显示该 User 的 courses 时，可以使用 checkbox 标签来绑定 courses 数据进行显示。当 checkbox 标签的 value 属性在我们绑定的列表数据中存在时，该 checkbox 将为选中状态。

下面通过示例来讲解 checkbox 标签的使用，具体步骤如下。

(1) 在 springmvc-3 项目中，为 User 类添加一个 boolean 类型的变量和 List<String> 类型的变量，代码如下：

```java
package com.springmvc.entity;
import java.util.List;
public class User {
    // 添加讲解 checkbox 标签所用属性
    private boolean reader;
    private List<String> courses;
    public boolean isReader() {
        return reader;
    }
    public void setReader(boolean reader) {
        this.reader = reader;
    }
    public List<String> getCourses() {
        return courses;
    }
    public void setCourses(List<String> courses) {
        this.courses = courses;
    }
    // 省略 User 类的构造方法
}
```

(2) 在 UserController 类中，添加一个 checkbox 方法，并通过 @RequestMapping 注解实现映射，代码如下：

```java
@RequestMapping(value="/checkbox", method = RequestMethod.GET)
public String checkbox(Model model) {
    User user=new User();
    user.setReader(true);
    List<String> list=new ArrayList<String>();
    list.add("Java 程序设计");
```

```
         list.add("JavaEE 框架技术");
         user.setCourses(list);
         model.addAttribute("user",user);
         return "checkbox";
}
```

在 checkbox 的方法中创建了 User 对象，并分别设置了变量 reader 和 courses 的值，设置 boolean 变量 reader 的值为 true，页面的 checkbox 复选框"已经阅读相关协议"会被选中；为集合变量 courses 添加"Java 程序设计"和"JavaEE 框架技术"，页面的两个 checkbox 复选框会被选中，并将它们添加到 model 中和页面进行绑定。

(3) 在 WebContent/ch08 的路径下，新建 checkbox.jsp 页面，并在页面中添加一行引用 Spring 标签库的声明，如下所示：

```
<%@ taglib prefix="fm" uri="http://www.springframework.org/tags/form" %>
……
<body>
    <h3>fm:checkbox 标签测试</h3>
    <fm:form modelAttribute="user" method="post" action="checkbox">
        选择课程：
        <fm:checkbox path="courses" value="Java 程序设计" label="Java 程序设计"/> 
        <fm:checkbox path="courses" value="Java Web 程序设计" label="Java Web 程序设计"/> 
        <fm:checkbox path="courses" value="JavaEE 框架技术" label="JavaEE 框架技术"/> 
        <br><br>
        <fm:checkbox path="reader" value="true"/>已经阅读相关协议
    </fm:form>
</body>
```

(4) 重启 Tomcat，在浏览器的地址栏中访问 http://localhost:8080/springmvc-3/checkbox 进行测试，运行效果如图 8-2 所示。

图 8-2　测试 checkbox 标签

8.2.7　radiobutton 标签

Spring MVC 的 radiobutton 标签会被渲染为一个类型为 radio 的普通 HTML input 标签。radiobutton 标签除了 8.1 小节介绍的共有属性外，还有一个 label 属性，将其作为 label 被渲染的单选按钮的值。下面通过示例来讲解 radiobutton 标签的使用，具体步骤如下。

(1) 在 UserController 的类中，添加一个 radiobutton 方法，并通过@RequestMapping 注解实现映射，代码如下：

```
@RequestMapping(value="/radiobutton", method = RequestMethod.GET)
public String radiobutton(Model model) {
    User user=new User();
    user.setSex("女");
    model.addAttribute("user",user);
    return "radiobutton";
}
```

前面我们在 User 类中定义过 sex 属性，用来绑定页面的 radiobutton 标签数据。在 UserController 类的 radiobutton 方法中，设置 sex 变量的值为"女"，则页面的 radio 单选按钮的 value="女"会被选中。

(2) 在 WebContent/ch08 的路径下，新建 radiobutton.jsp 页面，并在页面中添加一行引用 Spring 标签库的声明，如下所示：

```
<%@ taglib prefix="fm" uri="http://www.springframework.org/tags/form" %>
……
<body>
    <h3>fm:radiobutton 标签测试</h3>
    <fm:form modelAttribute="user" method="post" action="radiobutton">
        性别：
        <fm:radiobutton path="sex" value="男"/>男  
        <fm:radiobutton path="sex" value="女"/>女  
    </fm:form>
</body>
```

(3) 重启 Tomcat，在浏览器的地址栏中访问 http://localhost:8080/springmvc-3/radiobutton 网址进行测试，运行效果如图 8-3 所示。

图 8-3　测试 radiobutton 标签

8.2.8　select 标签

Spring MVC 的 select 标签会渲染一个 HTML select 元素，被渲染元素的选项可能来自 items 属性的一个 Collection、Map 及 Array，或者来自一个嵌套的 option 或者 options 标签。select 标签除了 8.1 小节介绍的共有属性外，还包括以下属性。

- items：用于指定 select 元素的数据源类型，如 Collection 接口、Map 接口或者 Array 数组。

- itemLabel：item 属性中定义的 Collection、Map 或者 Array 中的对象属性，为每个 select 元素提供 label。
- itemValue：item 属性定义的 Collection、Map 或者 Array 中的对象属性，为每个 select 元素提供值。

其中，items 属性特别有用，因为它可以绑定到对象的 Collection、Map、Array，为 select 元素生成选项。

8.2.9 option 标签

Spring MVC 的 option 标签会渲染为一个 HTML option 元素。option 标签的主要属性是 8.1 小节介绍的共有属性。

8.2.10 options 标签

Spring MVC 的 options 标签会渲染 select 元素中使用的一个 HTML option 元素列表。options 标签除了 8.1 小节介绍的共有属性外，还包括以下属性。

- items：用于生成 option 列表元素的对象的 Collection、Map 或者 Array。
- itemLabel：item 属性中定义的 Collection、Map 或者 Array 中的对象属性，为每个 option 元素提供 label。
- itemValue：item 属性定义的 Collection、Map 或者 Array 中的对象属性，为每个 option 元素提供值。

下面我们通过示例来演示 select、option 和 options 标签的使用，步骤如下：

(1) 在 User 类中，添加 deptId 属性，并生成 setter 和 getter 方法，代码如下：

```java
public class User {
    // 省略前面已定义过的属性
    ……
    private int deptId;
    // deptId 属性的 setter、getter 方法
    public int getDeptId() {
        return deptId;
    }
    public void setDeptId(int deptId) {
        this.deptId = deptId;
    }
}
```

(2) 在 UserController 类中，添加一个 select 方法，并通过@RequestMapping 注解实现映射，代码如下：

```java
@RequestMapping(value="/select", method = RequestMethod.GET)
public String select(Model model) {
    User user=new User();
    user.setDeptId(3);
    model.addAttribute("user",user);
```

```
        return "select";
}
```

设置 deptId 的值，页面上 select 下拉列表框对应的 option 项会被选中。

（3）在 WebContent/ch08 的路径下，新建 select.jsp 页面，并在页面中添加一行引用 Spring 标签库的声明，如下所示：

```
<%@ taglib prefix="fm" uri="http://www.springframework.org/tags/form"%>
……
<body>
    <h3>fm:select 标签添加 fm:option 标签</h3>
    <fm:form modelAttribute="user" method="post" action="select">
        部门：
        <fm:select path="deptId">
            <fm:option value="1">机械工程学院</fm:option>
            <fm:option value="2">电气工程学院</fm:option>
            <fm:option value="3">信息工程学院</fm:option>
            <fm:option value="4">土木工程学院</fm:option>
        </fm:select>
    </fm:form>
</body>
```

select.jsp 中使用 Spring MVC 的 selec 标签，path 属性绑定 model 的 deptId 属性，select 标签中使用 Spring MVC 的 option 标签直接添加部门数据。由于被绑定的 deptId 属性的值是 3，所以下拉列表框中的 value="3"的信息工程学院会被默认选中。

（4）重启 Tomcat，在浏览器的地址栏中访问 http://localhost:8080/springmvc-3/select 网址进行测试，运行效果如图 8-4 所示。

图 8-4　测试 select 标签

select.jsp 中使用了 Spring MVC 的 option 标签直接添加部门数据。除此之外，还可以使用 select 标签的 items 属性自动加载后台传递过来的数据并将其显示在下拉列表框中。修改前面 UserController 类中的 select 方法，如下所示：

```
@RequestMapping(value="/select", method = RequestMethod.GET)
public String select(Model model) {
    User user=new User();
    user.setDeptId(3);
    //页面展现的可选择的 select 下拉列表框内容
    Map<Integer, String> deptMap=new HashMap<Integer,String>();
    deptMap.put(1, "机械工程学院");
    deptMap.put(2, "电气工程学院");
    deptMap.put(3, "信息工程学院");
```

```
    deptMap.put(4, "土木工程学院");
    model.addAttribute("deptMap", deptMap);
    model.addAttribute("user",user);
    return "select";
}
```

修改 select.jsp 页面，添加 select 标签的 items 属性绑定 Map，如下所示：

```
<fm:form modelAttribute="user" method="post" action="select">
    ……  <!-- 省略已运行过的代码 -->
    <h3>fm:select 标签 items 属性绑定 Map</h3>
    部门：<fm:select path="deptId" items="${deptMap}"/>
</fm:form>
```

select.jsp 页面中 select 标签的 items 属性会加载 model 中的 deptMap 数据并将其显示在页面上。

重启 Tomcat，在浏览器的地址栏中访问 http://localhost:8080/springmvc-3/select 网址进行测试，运行效果如图 8-5 所示。

图 8-5　测试 select 标签的 items 属性

还可使用 options 标签的 items 属性自动加载后台传递过来的数据并将其显示在下拉列表框中，修改 select.jsp 的页面，添加代码如下：

```
<h3>使用 fm:options 标签 items 属性绑定 Map</h3>
部门：
<fm:select path="deptId">
    <fm:options items="${deptMap}"/>
</fm:select>
```

重启 Tomcat，在浏览器的地址栏中访问 http://localhost:8080/springmvc-3/select 网址进行测试，运行效果如图 8-6 所示。

图 8-6　测试 options 标签

在实际开发中，经常会出现 select 下拉列表框中的数据来自于数据库的表数据，并且获取的数据被封装在 JavaBean 中。这时，就可使用 select 标签或者 options 标签的 items、itemLabel 和 itemValue 属性来加载数据。

在 com.springmvc.entity 包中，新建 Dept 类，包含 id 和 name 属性，如下所示：

```java
package com.springmvc.entity;
public class Dept {
    private int id;
    private String name;
    // 省略属性的 setter、getter 方法
    // 省略构造方法
}
```

修改前面 UserController 类中的 select 方法，如下所示：

```java
@RequestMapping(value="/select", method = RequestMethod.GET)
public String select(Model model) {
    User user=new User();
    user.setDeptId(3);
    model.addAttribute("user",user);
    ……
    List<Dept> deptList=new ArrayList<Dept>();
    deptList.add(new Dept(1,"机械工程学院"));
    deptList.add(new Dept(2,"电气工程学院"));
    deptList.add(new Dept(3,"信息工程学院"));
    deptList.add(new Dept(4,"土木工程学院"));
    model.addAttribute("deptList",deptList);
    return "select";
}
```

在 select 方法中模拟从数据库中获取部门信息，并将其封装到 Dept 对象中，且将多个部门信息装载到 List 集合对象中，最后添加到 model 中。

修改 select.jsp 页面，添加 options 标签的 items 属性绑定 Object，并使用 itemLabel 和 itemValue 属性来加载数据，如下所示：

```jsp
<h3>fm:select 标签使用 fm:options 绑定 Object</h3>
部门：
<fm:select path="deptId">
    <fm:options items="${deptList}" itemLabel="name" itemValue="id"/>
</fm:select>
```

在 select.jsp 页面的 options 标签的 items 属性中加载 model 中的 deptList，并将集合中的元素及 Dept 对象的 name 属性设置为 option 的 label，id 属性设置为 option 的 value。

重启 Tomcat，在浏览器的地址栏中访问 http://localhost:8080/springmvc-3/select 网址进行测试，看到如图 8-7 所示的界面。

图 8-7　测试 options 标签绑定对象

8.2.11　errors 标签

Spring MVC 的 errors 标签对应于 Spring MVC 的 Errors 对象，它的作用就是显示 Errors 对象中包含的错误信息。如果 Errors 不为 null，则会渲染一个 HTML span 元素，用来显示错误信息。errors 标签除了 8.1 小节介绍的共有属性外，还有 delimiter 属性，用来定义两个 input 元素之间的分隔符，默认没有分隔符。

errors 标签的 path 属性绑定一个错误信息，可通过 path 属性显示两种类型的错误信息：
- 所有的错误信息，这个时候 path 的值应该是 "*"。
- 当前对象的某一个属性的错误信息，这个时候 path 的值应为所需显示的属性名称。

下面我们通过示例来演示 errors 标签的用法，步骤如下。

(1) 在 springmvc-3 项目的 src 下新建 com.springmvc.validator 包，并在该包中新建 UserValidator 类实现 org.springframework.validation.Validator 接口，完成验证功能，如下所示：

```
package com.springmvc.validator;
import org.springframework.validation.Errors;
import org.springframework.validation.ValidationUtils;
import org.springframework.validation.Validator;
import com.springmvc.entity.User;
public class UserValidator implements Validator {
    @Override
    public boolean supports(Class<?> clazz) {
        return User.class.equals(clazz);
    }
    @Override
    public void validate(Object object, Errors errors) {
        // 验证 User 类中的 name、sex 和 age 属性是否为空
        ValidationUtils.rejectIfEmpty(errors, "name", null, "用户名不能为空");
        ValidationUtils.rejectIfEmpty(errors, "sex", null, "性别不能为空");
        ValidationUtils.rejectIfEmpty(errors, "age", null, "年龄不能为空");
    }
}
```

(2) 在 UserController 类中，使用@InitBinder 注解绑定验证对象，如下所示：

```
@Controller
public class UserController {
    @RequestMapping(value="/registerForm", method = RequestMethod.GET)
    public String registerForm(Model model) {
        User user=new User();
        //在 model 中添加属性 user，值为 user 对象
        model.addAttribute("user",user);
        return "registerForm";
    }
    @InitBinder
    public void initBinder(DataBinder binder) {
        // 设置验证的类为 UserValidator
```

```
        binder.setValidator(new UserValidator());
    }
    @RequestMapping(value="/registerValidator", method = RequestMethod.POST)
    public String registerValidator(@Validated User user,Errors errors)
    {
        //如果Errors对象有Field错误，重新跳回到注册页面，否则正常提交
        if (errors.hasFieldErrors()) {
            return "registerForm";
        }
        return "submit";
    }
}
```

（3）复制 register.jsp 页面，重命名为 registerForm.jsp，增加 errors 标签，如下所示：

```
<fm:form modelAttribute="user" action="registerValidator" method="post">
    姓名：<fm:input path="name"/>
    <font color="red"><fm:errors path="name"/></font><br><br>
    性别：<fm:input path="sex"/>
    <font color="red"><fm:errors path="sex"/></font><br><br>
    年龄：<fm:input path="age"/>
    <font color="red"><fm:errors path="age"/></font><br><br>
    <input type="submit" value="注册">
</fm:form>
```

在 registerForm.jsp 页面中，在每个需要输入的空间后面增加了一个 errors 标签，用来显示错误信息。

（4）重启 Tomcat，在浏览器的地址栏中访问 http://localhost:8080/springmvc-3/registerForm 网址进行测试。跳转到注册页面，如果文本框不输入任何信息，单击"注册"按钮，提交请求。因为没有提交注册信息，故验证出错。registerForm 请求处理方法会将请求重新转发到注册页面，errors 标签会显示错误信息，如图 8-8 所示。

图 8-8　测试 errors 标签

8.3　小　　结

本章介绍了 Spring MVC 的表单标签以及如何使用表单标签绑定数据，表单标签的功能强大，需要读者好好掌握。

第 9 章 Spring MVC 类型转换、数据格式化和数据校验

Spring MVC 会根据请求方法的签名不同,将请求消息中的信息以一定的方式转换并绑定到请求方法的参数中。其实在请求信息真正到达处理方法之前,Spring MVC 还完成了许多工作,包括数据类型转换、数据格式化以及数据校验等。

9.1 数据绑定简介

在执行程序时,Spring MVC 会根据客户端请求参数的不同,将请求消息中的信息以一定的方式转换并绑定到控制器类的方法参数中。这种将请求消息数据与后台方法参数建立连接的过程就是 Spring MVC 的数据绑定。

在数据绑定过程中,Spring MVC 框架会通过数据绑定的核心部件 DataBinder 将请求参数串的内容进行类型转换,然后将转换后的值赋给控制器类中方法的形参,这样后台方法就可以正确绑定并获取客户端请求携带的参数了。数据绑定涉及以下几个主要部分:

- DataBinder:数据绑定的核心部件,它在整个流程中起到核心调度的作用。
- ConversionService:可以利用 org.springframework.context.support.ConversionServiceFactoryBean 在 Spring 的上下文中定义一个 ConversionService,它是 Spring 类型转换体系的核心接口。
- BindingResult:包含已完成数据绑定的入参对象和相应的校验错误对象,Spring MVC 会抽取 BindingResult 中的入参对象及校验错误对象,将它们赋给处理方法的相应入参。

Spring MVC 进行数据绑定的运行机制如图 9-1 所示。

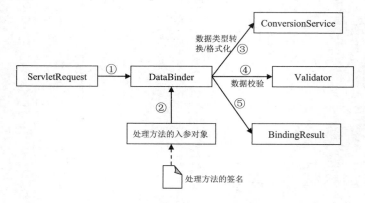

图 9-1　Spring MVC 数据绑定机制

Spring MVC 数据绑定信息处理过程的步骤描述如下。

(1) SpringMVC 框架将 ServletRequest 对象传递给 DataBinder。

(2) 将处理方法入参对象实例传递给 DataBinder。

(3) DataBinder 调用装配在 SpringMVC 上下文中的 ConversionService 组件进行数据类型转换、数据格式化等工作，并将 ServletRequest 对象中的消息填充到参数对象中。

(4) 调用 Validator 组件对已绑定了请求消息数据的参数对象进行数据合法性检验。

(5) 检验完成后会生成数据绑定结果 BindingResult 对象，Spring MVC 会将 BindingResult 对象中的内容赋给处理方法的相应参数。

9.2　数据类型转换

Spring 从 3.0 开始添加了一个位于 org.springframework.core.convert 包中的通用类型转换模块，可以在 Spring MVC 处理方法的参数绑定中使用它进行数据转换。

9.2.1　使用 ConversionService 进行类型转换

org.springframework.core.convert.ConversionService 是 Spring 类型转换体系的核心接口，在该接口中定义了以下 4 个方法。

- boolean canConvert(Class<?> sourceType,Class<?> targetType)：判断是否可以将一个 Java 类转换成另一个 Java 类。
- boolean canConvert(TypeDescriptor sourceType, TypeDescriptor targetType)：需要转换的类将以成员变量的方式出现，TypeDescriptor 描述了需要转换类的信息，还描述了类的上下文信息。
- <T> T convert(Object source, Class<T> targetType)：将源类型对象转换为目标类型对象。
- Object convert(Object source, TypeDescriptor sourceType, TypeDescriptor targetType)：

将对象从源类型对象转换为目标类型对象，通常会用到类中的上下文信息。

可以利用 org.springframework.context.support.ConversionServiceFactoryBean 在 Spring 的上下文中定义一个 ConversionService，Spring 会自动识别出上下文中的 ConversionService，并在 Spring MVC 处理 Bean 属性配置和处理方法入参绑定时，使用它进行数据转换。例如，对于 Spring MVC 中的前台 form 表单中的时间字符串到后台 Date 数据类型的转换问题，就可以通过 ConversionService 来解决。

在 ConversionServiceFactoryBean 中可以内置很多类型转换器，使用它们可以完成大多数的 Java 类型转换工作，可以通过在 ConversionServiceFactoryBean 的 converters 属性注册自定义的类型转换器，配置的实例代码如下：

```xml
<bean id="conversionService" class="org.springframework.context.
support.ConversionServiceFactoryBean">
    <property name="converters">
        <list>
            <bean class="com.springmvc.converter.
StringToDateConverter" />
        </list>
    </property>
</bean>
```

下面通过示例来讲解如何利用 ConversionService 进行数据类型转换，具体步骤如下。

(1) 将项目 springmvc-3 复制并重命名为 springmvc-4，再导入到 Eclipse 开发环境中。选中 springmvc-4 项目并右击，在快捷菜单中选择 Properties 命令，进入属性设置界面，找到 Web Project Settings 选项，将右侧 Context Root 文本框中的 springmvc-3 改为 springmvc-4，单击 Apply and Close 按钮。删除 src 目录下相关包下的 User 类，并删除 UserController 类中的方法，在 WebContent 下新建一个 ch09 文件夹，并删除 ch08 文件夹，保留相关 jar 包以及 springmvc.xml 配置文件。

(2) 在 ch09 文件夹中，新建一个简单的注册页面 register.jsp，用其传递一个姓名和一个用户的生日信息，代码如下：

```html
<body>
    <h3>注册页面</h3>
    <form action="../register" method="post">
      姓名：<input type="text" id="name" name="name"><br><br>
      生日：<input type="text" id="birthday" name="birthday"><br><br>
        <input id="submit" type="submit" value="提交">
    </form>
</body>
```

(3) 在 com.springmvc.entity 包中新建 User 类，代码如下：

```java
package com.springmvc.entity;
import java.util.Date;
public class User {
    private String name;
    private Date birthday;
    // 省略属性的 setter、getter 方法
    // 省略构造方法
}
```

User 类提供了 name 和 birthday 属性，用于接收 jsp 页面传入的数据。注意，birthday 属性的类型是一个 java.util.Date，而 jsp 页面传入的数据类型都是 String，这里就需要将 String 转换成 Date 对象。

（4）在 UserController 类中添加一个 register 方法，并通过@RequestMapping 注解进行映射，代码如下：

```java
package com.springmvc.controller;
import org.springframework.stereotype.Controller;
import org.springframework.ui.Model;
import org.springframework.web.bind.annotation.ModelAttribute;
import org.springframework.web.bind.annotation.RequestMapping;
import org.springframework.web.bind.annotation.RequestMethod;
import com.springmvc.entity.User;
@Controller
public class UserController {
    @RequestMapping(value="/register",method = RequestMethod.POST)
    public String register(@ModelAttribute User user,Model model) {
        System.out.println(user.getBirthday());
        model.addAttribute("user",user);
        return "success";
    }
}
```

UserController 类中的 register 方法只是简单地接收请求数据，并将其设置到 User 对象中。

（5）新建 com.springmvc.converter 包，并在其中新建 StringToDateConverter 类，实现 Converter<S,T>接口，代码如下：

```java
package com.springmvc.converter;
import java.text.SimpleDateFormat;
import java.util.Date;
import org.springframework.core.convert.converter.Converter;
public class StringToDateConverter implements Converter<String, Date>{
    private String datePattern;   // 日期类型模板：如 yyyy-MM-dd
    public void setDatePattern(String datePattern) {
        this.datePattern = datePattern;
    }
    @Override
    public Date convert(String date) {
        try {
            SimpleDateFormat dateFormat=new SimpleDateFormat(this.datePattern);
            return dateFormat.parse(date);//将字符串转换成 Date 类型
        }catch (Exception e) {
            e.printStackTrace();
            System.out.println("日期转换失败！");
            return null;
        }
    }
}
```

(6) 在 springmvc.xml 文件中加入将自定义字符转换器，并添加 p 引用，将 prefix 前缀改为 ch09，代码如下：

```xml
<mvc:annotation-driven conversion-service="conversionService"/>
……
<bean id="conversionService" class="org.springframework.context.support.ConversionServiceFactoryBean">
    <property name="converters">
        <list>
            <bean class="com.springmvc.converter.StringToDateConverter" p:datePattern="yyyy-MM-dd"></bean>
        </list>
    </property>
</bean>
```

在该配置文件中，使用了<mvc:annotation-driven/>标签，该标签可简化 Spring MVC 的相关配置，自动注册 RequestMappingHandlerMapping 与 RequestMappingHandlerAdapter 两个 bean。<mvc:annotation-driven/>标签没有显式定义时，会注册一个默认的 ConversionService，即 FormattingconversionServiceFactoryBean，以满足大部分类型转换的要求。现在需要注册一个自定义的 StringToDateConverter 转换类，因此需要显式定义一个 ConversionService 覆盖<mvc:annotation-driven/>中的默认实现类，而这时需要通过设置 converters 属性来完成。

在 StringToDateConverter 的 bean 装配中，将属性 datePattern 设置为"yyyy-MM-dd"，即日期格式，在装配好这个 ConversionService 之后，就可在任何控制器的处理方法中使用这个转换器了。

(7) 在 ch09 文件夹中，新建 success.jsp 页面，实现输出信息，代码如下：

```
姓名：${requestScope.user.name}<br><br>
生日：${requestScope.user.birthday}
```

(8) 重启 Tomcat，通过浏览器访问 http://localhost:8080/springmvc-4/ch09/register.jsp 网址进行浏览，如图 9-2 所示。输入姓名和生日，单击"提交"按钮，转换器会自动将输入的日期字符串转换成 Date 类型，查看控制台可以看到时间格式的输出信息。User 对象的 birthday 属性已经获得 jsp 页面传入的日期值，跳转到 success.jsp 页面，如图 9-3 所示。

图 9-2　测试 ConversionService(一)

图 9-3　测试 ConversionService(二)

9.2.2 使用@InitBinder 注解进行类型转换

Spring MVC 默认不支持自动转换表单中的日期字符串和实体类中的日期类型的属性，必须要手动配置，上一小节讲解了使用 ConversionService 进行类型转换，通过自定义数据类型的绑定实现这个功能。本节通过 Spring MVC 的注解@InitBinder 和 Spring 自带的 WebDataBinder 类来实现这一类型转换，具体步骤如下。

（1）在实体类 User.java 不进行修改，还是使用日期型属性 birthday。

（2）在 UserController 类中添加 initBinder 方法，并用@InitBinder 注解标识，将从表单获取的字符串类型的日期转换成 Date 类型。

```java
@InitBinder
public void initBinder(WebDataBinder binder) {
    SimpleDateFormat dateFormat = new SimpleDateFormat("yyyy-MM-dd");
    binder.registerCustomEditor(Date.class, new CustomDateEditor(dateFormat, true));
}
```

@InitBinder 注解标识的方法可以对 WebDataBinder 对象进行初始化，用于完成从表单文本域到实体类属性的绑定。@InitBinder 标识的方法不能有返回值，必须声明为 void。@InitBinder 标识的方法的参数为 WebDataBinder，是 DataBinder 的子类。DataBinder 是数据绑定的核心部件，可用来进行数据类型转换、格式化以及数据校验。

（3）在 UserController 类中添加方法 testInitBinder，用于测试日期类型转换，并在控制台打印输入姓名和生日，如下所示：

```java
@RequestMapping(value="/testInitBinder")
public String testInitBinder(User user1,Model model) {
    System.out.println("姓名："+user1.getName());
    System.out.println("生日："+user1.getBirthday());
    model.addAttribute("user1",user1);
    return "success1";
}
```

（4）新建 register1.jsp 页面，与前面的 register.jsp 页面类似，如下所示：

```html
<form action="../testInitBinder" method="post">
    姓名：<input type="text" id="name" name="name"><br><br>
    生日：<input type="text" id="birthday" name="birthday"><br><br>
    <input id="submit" type="submit" value="提交"><br>
</form>
```

（5）在 ch09 文件夹中，新建 success1.jsp 页面，实现信息输出，如下所示：

```
使用@InitBinder注解进行类型转换之后的跳转页面<br><br>
姓名：${requestScope.user1.name }<br><br>
生日：${requestScope.user1.birthday }<br><br>
```

（6）重启 Tomcat，访问 http://localhost:8080/springmvc-4/ch09/register1.jsp 地址，在表单中输入姓名"nanjing"和字符串类型日期"2018-07-11"，单击"提交"按钮，控制台输出实体对象 user1 中的 name 属性和 birthday 属性值，如下所示：

姓名：nanjing
生日：Wed Jul 11 00:00:00 CST 2018

浏览器的地址栏变为 http://localhost:8080/springmvc-4/testInitBinder，页面显示 success1.jsp 中的内容，如图 9-4 所示。

图 9-4　测试@InitBinder 注解

每次请求都会先调用@InitBinder 注解标识的方法，然后再调用控制器类中处理请求的方法。

对于同一个类型的对象来说，如果既在 ConversionService 中装配了自定义的转换器，同时还在控制器中通过@InitBinder 装配了自定义的编辑器，则 Spring MVC 将先查询通过@InitBinder 装配的自定义编辑器，然后再查询通过 ConversionService 装配的自定义转换器。

9.3　数据格式化

除了可以使用 ConversionService 和@InitBinder 注解实现数据类型的转换外，还可通过在实体类的属性上添加相应的注解来实现数据的格式化。在实体类 User 的 birthday 属性上标识@DateTimeFormat 注解，如下所示：

```
import org.springframework.format.annotation.DateTimeFormat;
……
@DateTimeFormat(pattern="yyyy-MM-dd")
private Date birthday;
```

@DateTimeFormat 可将表单中输入的形如"yyyy-MM-dd"的日期字符串格式化为 Date 类型的数据。

将 UserController 类中用@InitBinder 注解标识的 initBinder()方法注释掉，重启 Tomcat，浏览页面 ch09/register1.jsp，在表单中输入姓名和字符串类型日期"2018-07-11"，单击"提交"按钮，控制台依然会成功输出实体对象 user1 中的 name 和 birthday 属性值。

如果在 Float 类型属性上使用@NumberFormat(pattern="#,###,###.#")注解，则将表单中输入的形如"1,234,567.8"的字符串格式化为 Float 类型的数据。

9.4　数据校验

在实际工作中，得到数据后的第一步就是校验数据的正确性，如果存在录入上的问题，一般会通过注解校验，发现错误后返回给用户，但是对于一些逻辑上的错误，比如购买金额=购买数量×单价，这样的规则就很难使用注解方式进行验证了，此时可以使用 Spring 提供的验证器(Validator)规则去验证，在上一章的 Spring MVC 的 errors 标签中已经演示过 Validator 验证。由于 Validator 框架通过硬编码完成数据校验，在实际开发中会显得比较麻烦，因此现在开发更加推荐使用 JSR 303 完成服务器端的数据校验。

Spring 3 开始支持 JSR 303 验证框架，JSR 303 是 Java 为 Bean 数据合法性校验所提供的标准框架。JSR 303 支持 XML 风格的和注解风格的验证，通过在 Bean 属性上标注类似于 @NotNull、@Max 等的标准注解指定校验规则，并通过标准的验证接口对 Bean 进行验证。访问 http://jcp.org/en/jsr/detail?id=303 可以查看详细内容并下载 JSR 303 Bean Validation。JSR 303 不需要编写验证器，它定义了一套可标注在成员变量、属性方法上的校验注解，如表 9-1 所示。

表 9-1　JSR 303 注解约束

约　　束	说　　明
@Null	被注解的元素必须为 Null
@NotNull	被注解的元素必须不为 Null
@AssertTrue	被注解的元素必须为 true
@AssertFalse	被注解的元素必须为 false
@Min(value)	被注解的元素必须是一个数字，其值必须大于等于最小值
@Max(value)	被注解的元素必须是一个数字，其值必须小于等于最大值
@DecimalMin(value)	被注解的元素必须是一个数字，其值必须大于等于最小值
@DecimalMax(value)	被注解的元素必须是一个数字，其值必须小于等于最大值
@Size(max,min)	被注解的元素的大小必须在指定的范围内
@Digits(integer,fraction)	被注解的元素必须是一个数字，其值必须在可接受范围内
@Past	被注解的元素必须是一个过去的日期
@Future	被注解的元素必须是一个将来的日期
@Pattern(value)	被注解的元素必须符合指定的正则表达式

Hibernate Validator 是 JSR 303 的一个参考实现，除了支持所有标准的校验注解之外，还扩展了如表 9-2 所示的注解。

表 9-2　Hibernate Validator 扩展的注解

约　　束	说　　明
@NotBlank	检查被注解的元素是不是 Null，以及被去掉前后空格的长度是否大于 0
@Email	被注解的元素必须是电子邮件格式
@URL	被注解的元素必须是合法的 URL 地址
@length	被注解的字符串的大小必须在指定的范围内
@NotEmpty	检查被注解的字符串必须非空
@Range	被注解的元素必须在合适的范围内

Spring MVC 支持 JSR 303 标准的验证框架,Spring 的 DateBinder 在进行数据绑定时,可同时调用验证框架来完成数据校验工作,非常方便。在 Spring MVC 中,可以直接通过注解驱动的方式来进行数据校验。

下面通过在注册页面中增加用户时,添加 JSR 303 验证来介绍,本例使用的是 Hibernate Validator 的实现,目前最高版本是 6,步骤如下。

(1) 下载添加验证 jar 包。

访问网址 http://hibernate.org/validator/,即可在页面上看到一个绿色的 Latest stable(6.0) 按钮,单击该按钮跳转到 http://hibernate.org/validator/releases/6.0/页面,在 Releases in this series 下方看到绿色底纹的 "6.0.10.Final",单击右侧的 Download 下载链接,跳转到 https://sourceforge.net/网站下载 hibernate-validator-6.0.10.Final-dist.zip 压缩包。解压该压缩包,将 dist 文件夹中的 hibernate-validator-6.0.10.Final.jar,以及\dist\lib\required\路径下的 validation-api-2.0.1.Final.jar、jboss-logging-3.3.2.Final.jar 和 classmate-1.3.4.jar 共四个 jar 包,复制到 springmvc-4 项目的 WebContent\WEB-INF\lib 目录下,刷新该项目。

(2) 在 Spring MVC 配置文件中添加对 JSR 303 验证框架的支持。

由于在 Spring MVC 配置文件 springmvc.xml 中已经使用了<mvc:annotation-driven>,因此会自动注册 JSR 303 验证框架。

(3) 使用 JSR 303 验证框架注解为模型对象指定验证信息。

修改实体类 User.java,添加 email 属性及其 getter 和 setter 方法,再使用 JSR 303 验证框架注解为这些属性指定验证信息,如下所示:

```java
package com.springmvc.entity;
import javax.validation.constraints.Email;
import javax.validation.constraints.NotEmpty;
import javax.validation.constraints.Size;
import org.hibernate.validator.constraints.Range;
public class User {
    @NotEmpty
    @Size(min=6,max=20)
    private String name;
    @Range(min = 18, max = 45)
    private int age;
    @Email
    @NotEmpty
    private String email;
    // 省略属性的 setter、getter 方法
}
```

对于 name 属性,要求不为空,其长度不小于 6,且不大于 20;对于 email 属性,要求不为空,且格式为 email;对于 age 属性,要求输入的年龄范围在 18 岁到 45 岁之间。

(4) 在 UserController 类中添加方法 testValidate(),测试表单数据校验,如下所示:

```java
@RequestMapping("/testValidate")
public String testValidate(@Valid User user, BindingResult result) {
    if (result.getErrorCount() > 0) {
        for (FieldError error : result.getFieldErrors()) {
```

```
                System.out.println(error.getField() + ":"+
error.getDefaultMessage());
            }
        }
        return "index";
}
```

在 testValidate 方法中，通过@Valid 注解告诉 Spring MVC，User 类的对象 user 在绑定表单数据后需要进行 JSR 303 验证，绑定的结果保存到 BindingResult 类型的对象 result 中。通过判断 result 就可以知道绑定过程是否出现错误，如果出现错误则输出。

（5）在 ch09 文件夹中，新建 index.jsp 页面，在页面中创建一个表单，如下所示：

```
<form action="../testValidate" method="post">
    姓名：<input type="text" name="name"><br><br>
    年龄：<input type="text" name="age"><br><br>
    邮箱：<input type="text" name="email"><br><br>
    <input type="submit" value="提交" />
</form>
```

（6）重启 Tomcat，浏览页面 http://localhost:8080/springmvc-4/ch09/index.jsp，在表单页面中的年龄和邮箱处填入"10"和"yz"，单击"提交"按钮，如图 9-5 所示。

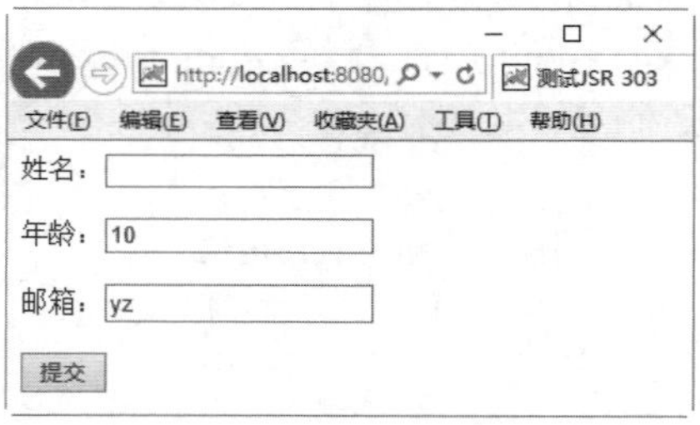

图 9-5　测试 JSR 303

控制台输出如下校验错误信息：

```
name:不能为空
age:需要在 18 和 45 之间
email:不是一个合法的电子邮件地址
name:个数必须在 6 和 20 之间
```

只有表单输入信息满足所有校验要求，控制台才不会输出错误信息。

9.5　小　　结

本章介绍了 Spring MVC 的数据类型转换、数据格式化和数据校验。对于类型转换，讲解了使用 ConversionService 进行类型转换和使用@InitBinder 注解进行类型转换；对于数据格式化，讲解了使用注解来实现数据的格式化。对于数据校验，讲解了通过注解来实现信息输出，现阶段更多的是使用 JSR 303 验证规范。

第 10 章　Spring MVC 文件上传和下载

在当前互联网应用中，上传头像、图片、证件和相关文件等是十分常见的，这就涉及到文件的上传功能，当然我们有时也需要在网上下载相关资源。文件上传和下载是项目中常用的功能。Spring MVC 为文件的上传和下载提供了良好的支持，本章将对 Spring MVC 环境中文件的上传和下载进行讲解。

10.1　文 件 上 传

在 Spring MVC 中实现文件上传十分方便，它为文件上传提供了直接的支持，即 MultipartResolver(多部件解析器)接口。MultipartResolver 用于处理上传请求，将上传请求包装成可以直接获取文件的数据，从而方便操作。它有两个实现类：

- StandardServletMultipartResolver：它是 Spring 3.1 版本后的产物，使用 Servlet 3.0 标准的上传方式，不用依赖于第三方包。
- CommonsMultipartResolver：使用了 Apache 的 commons-fileupload 完成具体的上传操作，可以在 Spring 的各个版本中使用，需要依赖第三方包才能实现。

默认情况下，Spring 不会处理 multipart 的 form 信息，因为默认用户会自己去处理这部分信息，当然可以随时打开这个支持。这样对于每一个请求，都会查看它是否包含 multipart 信息，如果没有则按流程继续执行。如果有，就会交给已经被声明的 MultipartResolver 进行处理，然后就能像处理其他普通属性一样处理文件上传了。

10.1.1　单文件上传

下面使用 CommonsMultipartResolver 实现单文件的上传功能，具体流程如下。

（1）将项目 springmvc-4 复制并重命名为 springmvc-5，再导入到 Eclipse 开发环境中。选中 springmvc-5 项目并右击，在快捷菜单中选择 Properties 命令，进入属性设置界面，找到 Web Project Settings 选项，将右侧 Context Root 文本框中的 springmvc-4 改为 springmvc-5，单击 Apply and Close 按钮。删除 src 目录下相关包下的类，在 WebContent 下新建一个 ch10 文件夹，并删除 ch09 文件夹，保留相关 jar 包以及 springmvc.xml 配置文件。

（2）下载并添加 jar 包，访问 http://commons.apache.org/proper/ 网址，进入后找到 commons-fileupload 和 commons-io 进行下载，得到 commons-fileupload-1.3.3.jar 和 commons-io-2.6.jar(也可直接使用提供的 jar 包)两个 jar 包，复制到项目 springmvc-5 的 WebContent\WEB-INF\lib 目录下，并刷新项目发布到 Libraries 路径下。

（3）在 Spring MVC 配置文件中配置 CommonsMultipartResolver 类，如下所示：

```
<bean id="multipartResolver" class="org.springframework.web
.multipart.commons.CommonsMultipartResolver">
    <!-- 设置上传文件的最大尺寸为1MB -->
    <property name="maxUploadSize" value="1048576" />
    <!-- 字符编码 -->
    <property name="defaultEncoding" value="UTF-8" />
</bean>
```

Spring MVC 环境所需要的组件扫描器、注解驱动和视图解析器以前已经配置。MultipartResolver 接口的实现类 CommonsMultipartResolver 是引用 multipartResolver 字符串获取该实现类对象并完成文件解析的，所以在配置 CommonsMultipartResolver 时必须指定该 Bean 的 id 为 multipartResolver。

在上述配置代码中，除配置了 CommonsMultipartResolver 类外，还通过<property>元素配置了允许上传文件的大小和编码格式。通过<property>元素可以对文件解析器类 CommonsMultipartResolver 的如下属性进行配置。

- maxUploadSize：上传文件的最大长度，长度以字节为单位。
- maxInMemorySize：缓存中的最大尺寸。
- defaultEncoding：默认的字符编码格式。
- resolveLazily：延迟文件解析，以便在 Controller 中捕获文件大小异常。

（4）在 ch10 文件夹中新建 index.jsp 页面，并在该页面上创建一个表单，用于上传文件。提交表单后，以 POST 方式提交到一个名为 "/fileUpload" 的请求中，如下所示：

```
<form action="../fileUpload" method="post" enctype="multipart/form-data">
    <input type="file" name="file" /><br><br>
    <input type="submit" value="上传" /><br>
</form>
```

新建 success.jsp 页面，如下所示：

文件上传成功！

上传文件路径：${requestScope.fileUrl }

(5) 在 com.springmvc.controller 包中，新建 FileUploadController 类，并在其中添加方法 fileUpload，处理文件上传，如下所示：

```java
package com.springmvc.controller;
import java.io.File;
import javax.servlet.http.HttpServletRequest;
import org.springframework.stereotype.Controller;
import org.springframework.ui.ModelMap;
import org.springframework.web.bind.annotation.RequestMapping;
import org.springframework.web.bind.annotation.RequestParam;
import org.springframework.web.multipart.MultipartFile;
@Controller
public class FileUploadController {
    @RequestMapping(value = "/fileUpload")
    public String fileUpload(@RequestParam(value="file", required=false) MultipartFile file, HttpServletRequest request, ModelMap model) {
        //服务器端 upload 文件夹物理路径
        String path = request.getSession().getServletContext().getRealPath("upload");
        //获取文件名
        String fileName = file.getOriginalFilename();
        //实例化一个 File 对象，表示目标文件(含物理路径)
        File targetFile = new File(path, fileName);
        if(!targetFile.exists()){
            targetFile.mkdirs();
        }
        try {
            //将上传文件保存到服务器上指定位置
            file.transferTo(targetFile);
        } catch (Exception e) {
            e.printStackTrace();
        }
        model.put("fileUrl",request.getContextPath()+"/upload/"+fileName);
        return "success";
    }
}
```

在 FileUpload 方法参数中，@RequestParam 注解用于在控制器 FileUploadController 中绑定请求参数到方法参数。请求参数为 file，将 index.jsp 页面文件上传表单中名为 file 的 value 值赋给 MultipartFile 类型的 file 属性；"required=false"表示使用@RequestParam 注解可以不传 file 参数，如果"required=true"时则必须传递该参数，required 的默认值是 true。

Spring MVC 会将上传的文件绑定到 MultipartFile 对象中。MultipartFile 提供了获取上传文件内容、文件名等方法。通过 transferTo()方法还可以将文件存储到硬件中。MultipartFile 对象中的常用方法如下。

- byte[] getBytes()：获取文件数据。
- String getContentType[]：获取文件 MIME 类型，如 image/jpeg 等。
- InputStream getInputStream()：获取文件流。

- String getName()：获取表单中文件组件的名字。
- String getOriginalFilename()：获取上传文件的原名。
- Long getSize()：获取文件的字节大小，单位为 byte。
- boolean isEmpty()：确定是否有上传文件。
- void transferTo(File dest)：将上传文件保存到一个目录文件中。

（6）部署 springmvc-5，启动 Tomcat，浏览 http://localhost:8080/springmvc-5/ch10/index.jsp 页面，在文件上传表单中先通过"浏览"按钮选择一个文件，然后单击"上传"按钮，如图 10-1 所示。

文件成功上传后，在 Tomcat 的根路径\webapps\springmvc-5\upload 下就能看到上传的文件。success.jsp 页面显示的文件路径如图 10-2 所示。

图 10-1　文件上传

图 10-2　文件上传后的路径

> **注意**：upload 文件夹在项目的发布路径中，而不是创建的项目所在目录，本书在第 1 章已经将项目的发布路径修改为 Tomcat 的 webapps 目录。如果未更改项目的发布路径，则要到工作空间的 .metadata 目录中寻找项目发布目录（路径为 workspace\.metadata\.plugins\org.eclipse.wst.server.core\tmp1\wtpwebapps\）下找到相应项目中的 upload 文件夹。

10.1.2　多文件上传

如果想通过浏览文件一次选择多个文件上传，可以将方法参数中的参数设置为 @RequestParam("files") MultipartFile[] files 数组形式，这样可提交多个文件，如下所示：

```
@RequestMapping(value = "/fileUpload")
public String fileUpload(@RequestParam("files") MultipartFile[] files,
HttpServletRequest request, ModelMap model) {
    // 利用数组的方式来处理多个文件，处理过程略
}
```

随着 HTML 5 的广泛应用，可利用 HTML 5 的特性，一次性选择多张图片的选择方式以前端设计为主。下面来介绍用 HTML 5 的方式实现多文件上传，过程如下。

（1）在 springmvc-5 项目的 ch10 文件夹中，新建 fileUpload.jsp 页面，代码如下：

```
<%@ page language="java" contentType="text/html; charset=UTF-8"
pageEncoding="UTF-8"%>
<!DOCTYPE HTML>
<html>
```

```
<head>
<meta http-equiv="Content-Type" content="text/html; charset=UTF-8">
<title>文件上传</title>
</head>
<body>
    <form action="../upload" method="post" enctype="multipart/form-data">
        文件描述：<input type="text" name="description"><br><br>
        请选择文件<input type="file" name="files" multiple="multiple"/>
<br><br>
        <input type="submit" value="上传" /><br>
    </form>
</body>
</html>
```

上述代码中，首先使用<!DOCTYPE HTML>声明 HTML 5 标准网页。在表单中，除了满足上传表单所必需的条件外，在类型为 file 的<input>元素中还增加了一个 multiple 属性，该属性为 HTML 中的新属性，使用该属性，就可以同时选择多个文件进行上传。

新建 error.jsp 页面，在<body>元素内编写"文件上传失败，请重新上传！"的提示信息。

(2) 在控制器类 FileUploadController 中添加 upload()方法，如下所示：

```
import java.util.List;
import java.util.UUID;
……
@RequestMapping(value = "/upload")
public String upload(@RequestParam("description") String description,
@RequestParam(value="files", required=false) List<MultipartFile>
files,HttpServletRequest request) {
    //判断上传文件是否存在
    if (!files.isEmpty() && files.size()>0 ) {
        //循环输出上传的文件
        for(MultipartFile file : files) {
            //获取上传文件的原始名称
            String originalFilename=file.getOriginalFilename();
            //设置上传文件的保存地址目录
            String dirPath=request.getServletContext().getRealPath("/upload/");
            File filePath=new File(dirPath);
            //如果保存文件的地址不存在，就先创建目录
            if (!filePath.exists()) {
                filePath.mkdirs();
            }
            //使用 UUID 重新命名上传的文件名称(文件描述_uuid_原始文件名称)
            String newFileName= description+"_"+UUID.randomUUID()+"_"
+originalFilename;
            try {
                //使用 MultipartFile 接口的方法将文件上传到指定位置
                file.transferTo(new File(dirPath+newFileName));
            } catch (Exception e) {
                e.printStackTrace();
                return "error";
```

```
            }
        }
        //跳转到成功页面
        return "success";
    }else {
        return "error";
    }
}
```

在 upload 方法参数中,使用 List<MultipartFile>集合类型接收用户上传的文件,然后判断上传文件是否存在。如果存在,则继续执行上传操作,用 MultipartFile 接口的 transferTo()方法将上传文件保存到用户指定的目录位置后,会跳转到 success.jsp 页面,如果文件不存在或上传失败,则跳转到 error.jsp 页面。

(3) 重启 Tomcat,浏览 http://localhost:8080/springmvc-5/ch10/fileUpload.jsp 网址,其显示效果如图 10-3 所示。

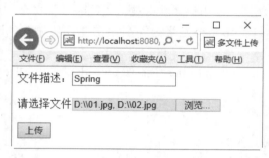

图 10-3 upload.jsp 多文件上传页面

在文件上传页面中,填入文件描述"Spring",单击"浏览"按钮,选择所要上传的文件。单击"上传"按钮,程序正确执行后浏览器就会跳转到 success.jsp 页面,查看项目发布目录,即可在 springmvc-5 项目中出现一个 upload 文件夹,该文件夹的内容如图 10-4 所示。

从图 10-4 可以看出,已经成功上传了两张图片,图片文件的命名规则为"文件描述_UUID_原始文件名称"的形式。

图 10-4 上传后的 upload 文件夹

10.2 文 件 下 载

文件下载就是将文件服务器中的文件下载到本机,操作相对比较简单,直接在页面给出一个超链接,该链接的 href 属性等于要下载文件的文件名,就可以实现文件下载了。但是如果该文件的文件名为中文,在某些早期的浏览器上就会导致下载失败;如果使用最新的Firefox、Chrome等浏览器则可以正常下载文件名为中文的文件。

修改10.1.1小节中单文件上传的示例，实现文件下载，操作步骤如下。

(1) 修改FileUploadController类中的fileUpload()方法，将获得的fileName文件名封装到model中，这样便于在success.jsp页面得到文件名，代码如下：

```
model.put("fileName", fileName);
```

(2) 修改success.jsp页面，添加一个文件下载的超链接，该链接的href属性要指定后台文件下载的方法及文件名(这里就用我们上传的文件名)，代码如下：

```
文件上传成功！<br>
上传文件路径：${requestScope.fileUrl}<br>
<a href="fileDownload?fileName=${requestScope.fileName}">
    ${requestScope.fileName}
</a>
```

(3) 在后台FileUploadController类中，添加一个fileDownload()方法，并进行相应的映射，使用Spring MVC提供的文件下载方法进行文件下载。Spring MVC提供了一个ResponseEntity类型的对象，使用它可以很方便地定义返回的HttpHeaders对象和HttpStatus对象，通过对这两个对象的设置，即可完成下载文件时所需要的配置信息。代码如下：

```java
import org.apache.commons.io.FileUtils;
import org.springframework.http.HttpHeaders;
import org.springframework.http.HttpStatus;
import org.springframework.http.MediaType;
import org.springframework.http.ResponseEntity;
import org.springframework.ui.Model;
……
@RequestMapping(value="/fileDownload")
public ResponseEntity<byte[]> fileDownload(HttpServletRequest request,
@RequestParam("fileName") String fileName,Model model)throws Exception {
    //下载文件路径
    String path = request.getServletContext().getRealPath("/upload/");
    //创建文件对象
    File file = new File(path + File.separator + fileName);
    //设置响应头
    HttpHeaders headers = new HttpHeaders();
    //下载显示的文件名，解决中文名称乱码问题
    String downloadFileName = new String(fileName.getBytes("UTF-8"),"ISO-8859-1");
    //通知浏览器以下载方式(attachment)打开文件
    headers.setContentDispositionFormData("attachment", downloadFileName);
    //定义以二进制流数据(最常见的文件下载)的形式下载返回文件数据
    headers.setContentType(MediaType.APPLICATION_OCTET_STREAM);
    //使用Spring MVC框架的ResponseEntity对象封装返回下载数据
    return new ResponseEntity<byte[]>
(FileUtils.readFileToByteArray(file),headers,HttpStatus.CREATED);
}
```

在 fileDownload()方法中，首先根据文件路径和需要下载的文件名来创建文件对象，然后对响应头中文件下载时的打开方式即下载方式进行设置，最后返回 ResponseEntity 封装的下载结果对象。

使用 ResponseEntity 对象，可以很方便地定义返回的 HttpHeaders 和 HttpStatus。上面代码中 MediaType 代表的是 Internet Media Type，即互联网媒体类型，也叫做 MIME 类型。在 Http 协议消息头中，使用 Content-Type 来表示具体请求中的媒体类型信息。HttpStatus 类型代表的是 Http 协议中的状态。

(4) 重启 Tomcat，浏览 http://localhost:8080/springmvc-5/ch10/index.jsp 网址，如图 10-5 所示。浏览一个文件并上传该文件，上传成功后的效果如图 10-6 所示。

图 10-5　文件上传表单页面

图 10-6　文件上传成功后的页面

可以看到在图 10-6 所示页面上有一个上传文件名的超链接，单击该超链接，会出现下载提示框，如图 10-7 所示(这里以 IE 浏览器为例进行演示)。

单击"打开"选项将直接打开该文件，如果选择"保存"或"另存为"选项，将弹出"另存为"对话框，选择保存路径。

如果上传的文件名称中带有中文，使用 IE 浏览器下载就会出现问题，而使用搜狗、Chrome 浏览器下载，则不会出现问题，其效果如图 10-8 所示。

图 10-7　文件下载弹出窗口

图 10-8　搜狗浏览器下载带中文名称的效果

10.3　小　　结

本章介绍了 Spring MVC 环境下的文件上传和文件下载操作。首先讲解了如何实现文件上传，并通过案例演示了单文件上传和在 HTML 5 模式下的多文件上传功能的实现。在文件下载中，介绍了 Spring MVC 中专门提供的 ResponseEntity 类型，用于文件下载功能的实现。

第 11 章　Spring MVC 的国际化和拦截器

"国际化"是指一个应用程序在运行时能够根据客户端请求所在国家或地区语言的不同而显示不同的用户界面。程序国际化是商业系统的一个基本要求，当前的系统不再是简单的单机系统，而是一个开放的系统，需要面对来自全世界各个地方的访问者，因此，国际化成为当前商业系统不可少的一部分。Spring MVC 的国际化建立在 Java 国际化的基础上。在实际项目中，拦截器的使用非常普遍，如在某购物网站中通过拦截器可拦截未登录的用户，禁止其添加购物车、购买商品等操作。在 Struts 2 框架中，拦截器是重要的组成部分，Spring MVC 也提供了 Interceptor 拦截器机制，通过配置即可对请求进行拦截处理。本章针对 Spring MVC 的国际化和拦截器进行讲解。

11.1　Spring MVC 国际化

全球化的 Internet 需要全球化的软件。全球化软件，意味着一个软件能够很容易地适应不同地区的市场。当一个软件需要在全球范围内使用时，就必须考虑在不同的地域和语言环境下的使用情况，最简单的要求就是在用户界面上显示的信息可以使用本地化语言来表示。软件的全球化意味着国际化和本地化。

11.1.1　Spring MVC 国际化概述

国际化(internationalization)是指程序在不做任何修改的情况下，就可以在不同的国家或地区和不同的语言环境下,按照当地的语言和格式习惯显示字符。例如,对于中国大陆的用户,会自动显示中文简体的提示信息、错误信息等;而对于美国的用户,会自动显示英文的提示信息、错误信息等。

国际化的程序运行在本地机器上时,能够根据本地机器的语言和地区设置显示相应的字符,这个过程叫做本地化(Localization)。

目前,国内很多大型的公司网站主页上都有简体中文、繁体中文和英文可以选择。例如,我们访问中国建设银行的网站,默认进入的是中文网站,在网页的右下角有"繁体/ENGLISH"超链接,单击 ENGLISH 就切换到英文网页,单击"繁体"就切换到繁体中文网页。当内地用户在内地使用 Google 时,默认显示的是简体中文,而香港用户在香港打开 Google 时,则默认显示的是繁体中文。用户也可以自定义选择语言资源文件,包括英文。

Spring MVC 的国际化是建立在 Java 国际化的基础之上的,其一样也是通过提供不同国家/语言环境的消息资源,然后通过 ResourceBundle 加载指定 Locale 对应的资源文件,再取得该资源文件中指定的 key 对应的消息。过程与 Java 程序的国际化完全相同,只不过 Spring MVC 框架对 Java 程序的国际化进行了进一步的封装,从而简化了应用。

首先熟悉一下 Spring MVC 的国际化结构,DispatcherServlet 会解析一个 LocaleResolver 接口对象,通过它来决定用户区域,读出对应用户系统设定的语言或者用户选择的语言,确定其国际化。对于 DispatcherServlet 而言,只能够注册一个 LocaleResolver 接口对象,LocaleResolver 接口的实现类在 Spring MVC 中也提供了多个实现类。

Spring MVC 也支持国际化的操作,主要是前端控制器内部拥有国际化解析器。在 Spring MVC 中选择语言区域,可以使用 Spring MVC 提供的语言区域解析器接口 LocaleResolver,该接口的常用实现类都在 org.springframework.web.servlet.i18n 包下,包括如下实现类。

- AcceptLanguageLocaleResolver:控制器无需写额外的内容,可以不用显式配置。
- SessionLocaleResolver:使用 Session 传输语言环境,根据用户 Session 的变量读取区域设置,它是可变的。如果 Session 没有设置,那么它也会使用开发者设置的默认值。
- CookieLocaleResolver:使用 Cookie 传送语言环境,根据 Cookie 数据获取国际化信息,如果用户禁止 Cookie 或者没有设置,它会根据 accept-language HTTP 头部确定默认区域。

由于 AcceptLanguageLocaleResolver 是固定的,所以现实中使用比较多的是可以手动显式配置的 HttpSessionLocaleResolver 和 CookieLocaleResolver。

在 Spring MVC 中,不直接使用 java.util.ResourceBundle 的抽象类,而是使用 ResourceBundleMessageSource 类作为 messageResource 的 bean,告知 Spring MVC 国际化的属性文件保存在哪里,配置信息代码如下:

```
<bean id="messageSource" class="org.springframework.context.support
.ResourceBundleMessageSource">
    <property name="basename">
        <list>
            <value>message</value>
            <value>yzpc</value>
        <list>
```

```
        </property>
    </bean>
```

如果项目中只有一组属性文件,则可以使用 basename 来指定国际化属性文件的名称,代码如下:

```
<bean id="messageSource" class="org.springframework.context.support
.ResourceBundleMessageSource">
    <property name="basename" value="message" />
</bean>
```

11.1.2 基于浏览器请求的国际化实现

基于浏览器请求的国际化使用的是 AcceptLanguageLocaleResolver 类,该类是默认的实现类,也是最容易使用的语言区域解析器。Spring MVC 会读取浏览器的 accept-language 标题,根据请求消息头自动获取语言区域。AcceptLanguageLocaleResolver 可以不用显式配置,也可以显式配置。

下面通过一个注册示例来讲解基于浏览器请求的国际化实现,操作步骤如下。

(1) 将项目 springmvc-5 复制并重命名为 springmvc-6,再导入到 Eclipse 开发环境中。选中 springmvc-6 项目并右击,在快捷菜单中选择 Properties 命令,进入属性设置界面,找到 Web Project Settings 选项,将右侧 Context Root 文本框中的 springmvc-5 改为 springmvc-6,单击 Apply and Close 按钮。删除 src 目录下相关包下的类,在 WebContent 下新建一个 ch11 文件夹,并删除 ch10 文件夹,保留相关 jar 包以及 springmvc.xml 配置文件。

(2) 在 com.springmvc.entity 包中,新建 User 的实体类,代码如下:

```
package com.springmvc.entity;
public class User {
    private String loginName;
    private String password;
    private int age;
    private String email;
    private String phone;
    // 省略属性的 getter 和 setter 方法
    // 省略构造方法
    ......
}
```

(3) 在 src 根路径下,新建 message_en_US.properties 和 message_zh_CN.properties 两个资源文件。message_en_US.properties 的内容如下:

```
loginName=LoginName
password=Password
age=Age
email=Email
phone=Phone
submit=Submit
welcome=Welcome {0} , Congratulations on your registration.
title=Register Page
```

```
userName=Administrator
info=Your registration information is as follows
```

message_zh_CN.properties 的内容如下所示:

```
loginName=\u540D\u79F0
password=\u5BC6\u7801
age=\u5E74\u9F84
email=\u90AE\u7BB1
phone=\u7535\u8BDD
submit=\u6CE8\u518C
welcome=\u6B22\u8FCE {0} \uFF0C
\u606D\u559C\u60A8\u6CE8\u518C\u6210\u529F\u3002
title=\u6CE8\u518C\u9875\u9762
userName=\u7BA1\u7406\u5458
info=\u60A8\u7684\u6CE8\u518C\u4FE1\u606F\u5982\u4E0B
```

在最新版的 Eclipse 中,输入中文就会自动转换为相应编码。如果没有转换,可通过 native2ascii.exe 工具进行转换。

(4) 在 springmvc.xml 配置文件中加载国际化资源文件,代码如下:

```
<bean id="messageSource" class="org.springframework.context.support
.ResourceBundleMessageSource">
    <!-- 国际化资源文件名 -->
    <property name="basename" value="message" />
</bean>
<!-- AcceptHeaderLocaleResolver 因为是默认语言区域解析,可不配置 -->
<bean id="localeResolver" class="org.springframework.web.servlet
.i18n.AcceptHeaderLocaleResolver"/>
```

在 web.xml 文件中配置 Spring MVC 的前端控制 DispatcherServlet,前面已经配置。

(5) 在 ch11 文件夹中,新建 registerForm.jsp 页面,代码如下:

```
<%@ page language="java" contentType="text/html; charset=UTF-8"
pageEncoding="UTF-8"%>
<%@ taglib prefix="spring" uri="http://www.springframework.org/tags" %>
<%@ taglib prefix="fm" uri="http://www.springframework.org/tags/form" %>
……
<body>
    <h3><spring:message code="title"/></h3>
    <fm:form modelAttribute="user" method="post" action="register">
        <spring:message code="loginName"/>
        <fm:input path="loginName"/><br>
        <spring:message code="password"/>
        <fm:input path="password"/><br>
        <spring:message code="age"/>
        <fm:input path="age"/><br>
        <spring:message code="email"/>
        <fm:input path="email"/><br>
        <spring:message code="phone"/>
        <fm:input path="phone"/><br>
        <input type="submit" value="<spring:message code="submit"/>"/><br>
```

```
        </fm:form>
    </body>
```

在注册页面中,通过<spring:message />标签输出国际化信息,通过 Spring MVC 的表单标签显示文本框。表单标签不进行数据绑定是无法操作的,在运行程序的时候会报错,因为表单标签是依赖于数据绑定操作的。在控制器中首先需要新建一个 User 的引用,也就是说要有一个 User 对象才能使 User 对象的相应属性绑定到表单 input 标签。

(6) 在 com.springmvc.controller 包下的 UserController 类中,添加动态跳转的 registerForm()方法、注册的 register()方法,如下所示:

```
package com.springmvc.controller;
import javax.servlet.http.HttpServletRequest;
import org.springframework.stereotype.Controller;
import org.springframework.ui.Model;
import org.springframework.validation.annotation.Validated;
import org.springframework.web.bind.annotation.*;
import org.springframework.web.servlet.support.RequestContext;
import com.springmvc.entity.User;
@Controller
public class UserController {
    @RequestMapping(value="/{formName}")
    public String registerForm(@PathVariable String formName, Model model) {
        User user=new User();
        model.addAttribute("user", user);
        return formName;    //动态跳转到页面
    }
    @RequestMapping(value="/register",method=RequestMethod.POST)
    public String register(@ModelAttribute @Validated User user,Model model,HttpServletRequest request) {
        //从后台代码获取国际化资源文件中的信息 userName
        RequestContext requestContext = new RequestContext(request);
        String username = requestContext.getMessage("userName");
        System.out.println(userName);
        model.addAttribute("user",user);
        return "success";
    }
}
```

register()方法接收请求,可通过 RequestContext 对象的 getMessage()方法来获取国际化消息,跳转到 success.jsp 页面。

(7) 在 ch11 文件夹中,新建 success.jsp 页面,并添加 spring 标签,代码如下:

```
<font color="blue"><h4><spring:message code="welcome" arguments="${requestScope.user.loginName }"/></h4></font>
<spring:message code="info"/><br>
<spring:message code="password"/>:${requestScope.user.password }<br>
<spring:message code="age"/>:${requestScope.user.age }<br>
<spring:message code="email"/>:${requestScope.user.email }<br>
<spring:message code="phone"/>:${requestScope.user.phone }<br>
```

在该页面中，使用 spring 的 message 标签读取资源文件中名为 welcome 的消息，并设置一个参数，参数的值为 user 对象的 loginName 属性。其他几个属性也在页面上输出。

(8) 部署 springmvc-6 这个 Web 项目，在浏览器的地址栏中输入 URL 进行测试：http://localhost:8080/springmvc-6/registerForm，运行界面如图 11-1 所示。在注册页面上，输入名称、密码等相关信息，单击"注册"按钮，请求将被提交到 Controller，可从后台代码获取国际化资源文件中的信息 userName，并在控制台打印出该信息，然后跳转到 success.jsp 页面，如图 11-2 所示。

图 11-1　基于浏览器请求的注册页面　　　　图 11-2　基于浏览器请求的成功页面

(9) 为测试 springmvc-6 项目的国际化，需要修改浏览器的语言顺序(以 Windows 10 系统为例，不同的系统会有所区别)。在浏览器中依次单击"工具"→"Internet 选项"→"常规"选项卡→"语言"，出现"语言首选项"对话框，单击"设置语言首选项"按钮进入"语言"窗口。在"更改语言首选项"下，如果有 English(United States)选项，则将其上移到最上方；如果没有 English(United States)选项，则需要单击"添加语言"按钮，找到"英语"→"英语(美国)"，双击即可添加，再调整语言顺序，如图 11-3 所示。

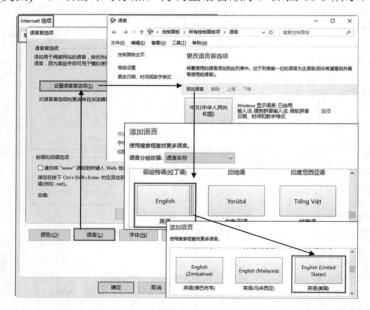

图 11-3　修改语言设置

再次在浏览器中输入 http://localhost:8080/springmvc-6/registerForm 进行测试，会看到如图 11-4 所示的界面。输入相关信息，单击 Submit 按钮，请求将被提交到 Controller，而后跳转到 success.jsp 页面，如图 11-5 所示。

图 11-4　测试英文注册页面

图 11-5　测试英文注册成功页面

可以看到，页面显示和注册完后的信息都变成了英文版，实现了国际化的功能。

11.1.3　基于 HttpSession 的国际化实现

基于 HttpSession 的国际化实现使用的是 LocaleResolver 接口的 SessionLocaleResolver 实现类，SessionLocaleResolver 不是默认的语言区域解析器，需要对其进行显式配置。如果使用它，Spring MVC 会从 HttpSession 作用域中获取用户所设置的语言区域，来确定使用哪个语言区域。通过请求参数改变国际化的值时，可使用 Spring 提供的国际化拦截器 LocaleChangeInterceptor，拦截器的知识在下节讲解。这里我们来看一下 SessionLocaleResolver 实现国际化时的工作原理，如图 11-6 所示。

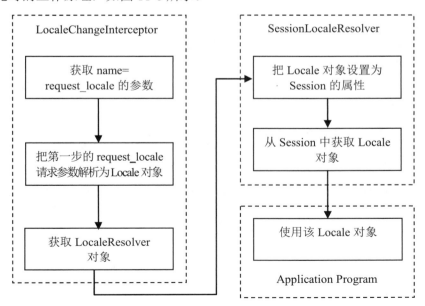

图 11-6　SessionLocaleResolver 实现国际化的工作原理

下面通过修改注册示例来讲解 SessionLocaleResolver 的实现，步骤如下。

(1) 修改 springmvc.xml 配置文件，注释默认的 AcceptLanguageLocaleResolver 类型的 Bean localeResolver，添加 SessionLocaleResolver 类型的 Bean，并添加国际化操作的拦截器，配置如下：

```xml
<!-- SessionLocaleResolver 配置 -->
<bean id="localeResolver"
class="org.springframework.web.servlet.i18n.SessionLocaleResolver"/>
<mvc:interceptors>
    <!--如果采用基于 Session/Cookie 的国际化，必须配置国际化操作拦截器-->
    <bean class=
"org.springframework.web.servlet.i18n.LocaleChangeInterceptor"/>
</mvc:interceptors>
```

(2) 为了方便切换，在 registerForm.jsp 的注册页面上，添加中文和英文的超链接，分别用于切换中文和英文语言环境，代码如下：

```html
<body>
    <a href="registerForm?request_locale=zh_CN">中文</a> |
    <a href="registerForm?request_locale=en_US">英文</a> <br/>
    <h3><spring:message code="title"/></h3>
    ……<!-- 省略未修改的表单 -->
    </fm:form>
</body>
```

(3) 修改动态获取跳转页面的 registerForm() 方法，代码如下：

```java
@RequestMapping(value="/{formName}")
public String registerForm(@PathVariable String formName,String request_locale,Model model,HttpServletRequest request) {
    System.out.println("request_locale="+request_locale);
    if (request_locale!=null) {
        if (request_locale.equals("zh_CN")) {//设置中文环境
            Locale locale=new Locale("zh","CN");
            request.getSession().setAttribute(SessionLocaleResolver
                    .LOCALE_SESSION_ATTRIBUTE_NAME,locale);
        }else if(request_locale.equals("en_US")){//设置英文环境
            Locale locale=new Locale("en","US");
            request.getSession().setAttribute(SessionLocaleResolver
                    .LOCALE_SESSION_ATTRIBUTE_NAME,locale);
        }else { //使用之前的语言环境
            request.getSession().setAttribute(SessionLocaleResolver
    .LOCALE_SESSION_ATTRIBUTE_NAME,LocaleContextHolder.getLocale());
        }
    }
    User user=new User();
    model.addAttribute("user", user);
    return formName;    //动态跳转页面
}
```

registerForm 根据提交的 request_locale 参数值，获取 Session 对象，并调用 setAttribute() 方法进行语言切换。

(4) 重新启动 Tomcat，访问 http://localhost:8080/springmvc-6/registerForm 网址，会看到如图 11-7 所示的界面。单击 "英文" 超链接，页面会切换为英文语言环境，如图 11-8 所示。

图 11-7　基于 Session 的国际化中文页面　　图 11-8　基于 Session 的国际化英文页面

页面最上方有两个超链接，用于切换中文和英文语言环境，单击超链接，控制台窗口就会输出相应语言的 request_locale 参数值，UserController 的 registerForm()方法根据提交的参数值，进行语言切换。输入相应内容，单击 "注册" 按钮，请求被提交到 Controller，而后跳转到 success.jsp 页面，该页面将会根据语言环境显示欢迎语句。

> **注意：** 有些版本的 IE 浏览器，不管中文还是英文页面提交后，success.jsp 显示的都是同一种语言，换成 FireFox、搜狗等其他类型浏览器就没有此问题。

11.1.4　基于 Cookie 的国际化实现

基于 Cookie 的国际化实现使用的是 LocaleResolver 接口的 CookieLocaleResolver 实现类，CookieLocaleResolver 不是默认的语言区域解析器，需要对其进行显式配置。如果使用它，Spring MVC 会从 Cookie 域中获取用户所设置的语言区域，来确定使用哪个语言区域。

下面继续修改注册示例来讲解 CookieLocaleResolver 的实现，步骤如下：

(1) 修改 springmvc.xml 配置文件，注释前面配置的 id 为 localeResolver 的 SessionLocaleResolver 类型的 Bean，添加 CookieLocaleResolver 类型的 Bean，配置如下：

```xml
<!-- CookieLocaleResolver 配置 -->
<bean id="localeResolver"
    class="org.springframework.web.servlet.i18n.CookieLocaleResolver"/>
```

(2) 修改 UserController 类中动态获取跳转页面的 registerForm()方法，代码如下：

```java
@RequestMapping(value="/{formName}")
public String registerForm(@PathVariable String formName,String request_locale,Model model,HttpServletRequest request, HttpServletResponse response) {
    System.out.println("request_locale="+request_locale);
    if (request_locale!=null) {
        if (request_locale.equals("zh_CN")) { //设置中文环境
```

```
            Locale locale=new Locale("zh","CN");
            (new CookieLocaleResolver()).setLocale(request, response, locale);
        }else if(request_locale.equals("en_US")){  //设置英文环境
            Locale locale=new Locale("en","US");
            (new CookieLocaleResolver()).setLocale(request, response, locale);
        }else {   //使用之前的语言环境
            (new CookieLocaleResolver()).setLocale(request, response,LocaleContextHolder.getLocale());
        }
    }
    User user=new User();
    model.addAttribute("user", user);
    return formName;        //动态跳转页面

}
```

registerForm()方法根据提交的 request_locale 参数值，创建 CookieLocaleResolver 对象，并调用 setLocale 方法将语言环境设置在 Cookie 中，从而进行了语言环境切换。

（3）重启 Tomcat，使用 Firefox 浏览器访问 http://localhost:8080/springmvc-6/registerForm 地址，会看到中文界面。单击"英文"超链接，页面切换为英文语言环境。

按 F12 键进入 Firefox 火狐浏览器的开发者模式的调试窗口，选择"网络"，单击相应的状态请求，右侧就会出现相应的请求信息，切换到"消息头"选项卡，可以看到，请求头中传递的是"zh_CN"，而响应头中则是"en_US"，这说明程序通过 Cookie 进行了语言环境切换，如图 11-9 所示。

图 11-9　基于 Cookie 的国际化调试窗口

在调试窗口中，切换到 Cookie 选项卡，可以明显看到响应 Cookie 为 en_US 和请求 Cookie 为 zh_CN 是不一样的，如图 11-10 所示。

第 11 章 Spring MVC 的国际化和拦截器

图 11-10 测试基于 Cookie 的国际化

输入相应内容，单击"注册"按钮，请求被提交到 Controller，而后跳转到 success.jsp 页面，该页面将会根据语言环境显示欢迎语句。

11.2 Spring MVC 拦截器

拦截器是 Spring MVC 中强大的控件，它可以在进入处理器之前做一些操作，或者在处理器完成后进行操作，甚至是在渲染视图后进行操作。

11.2.1 拦截器概述

对于任何优秀的 MVC 框架，都会提供一些通用的操作，如请求数据的封装、类型转换、数据校验、解析上传的文件、防止表单的多次提交等。早期的 MVC 框架将这些操作都写在核心控制器中，而这些常用的操作又不是所有的请求都需要实现的，这就导致了框架的灵活性不足，可扩展性降低。

Spring MVC 提供了 Interceptor 拦截器机制，类似于 Servlet 中的 Filter 过滤器，用于拦截用户的请求并做出相应的处理。比如通过拦截器来进行用户权限验证，或者用来判断用户是否已经登录。Spring MVC 拦截器是可插拔式的设计，需要某一功能拦截器，只需在配置文件中应用该拦截器即可；如果不需要这个功能拦截器，只需在配置文件中取消应用该拦截器。

要在 Spring MVC 中使用拦截器，就需要对拦截器进行定义和配置。在 Spring MVC 中定义拦截器有两种方法：

- 实现 HandlerInterceptor 接口，或者继承实现 HandlerInterceptor 接口的实现类(例如 HandlerInterceptorAdapter)。
- 实现 WebRequestInterceptor 接口，或者继承实现 WebRequestInterceptor 接口的实现类。

1. 实现 HandlerInterceptor 接口

首先来看看 HandlerInterceptor 接口的源码，该接口位于 org.springframework.web.servlet 的包中，定义了三个方法，代码如下：

```
package org.springframework.web.servlet;
……
public interface HandlerInterceptor {
    boolean preHandle(HttpServletRequest request, HttpServletResponse
response, Object handler)   throws Exception;
    void postHandle(HttpServletRequest request, HttpServletResponse
response, Object handler,ModelAndView modelAndView) throws Exception;
    void afterCompletion(HttpServletRequest request, HttpServletResponse
response,Object handler,Exception ex) throws Exception;
}
```

如果要实现 HandlerInterceptor 接口，就要实现三个方法，分别是 preHandle、postHandle 和 afterCompletion，关于这三个方法的具体描述如下。

- preHandle 方法：该方法在执行控制器方法之前执行。返回值为 Boolean 类型，如果返回 false，表示拦截请求，不再向下执行；如果返回 true，表示放行，程序继续向下执行(如果后面没有其他 Interceptor，就会执行 Controller 方法)。所以，此方法可对请求进行判断，决定程序是否继续执行，或者进行一些初始化操作及对请求进行预处理。
- postHandle 方法：该方法在执行控制器方法调用之后，且在返回 ModelAndView 之前执行。由于该方法会在 DispatcherServlet 进行返回视图渲染之前被调用，所以此方法多被用于处理返回的视图，可通过此方法对请求域中的模型和视图做进一步的修改。
- afterCompletion 方法：该方法在执行完控制器之后执行。由于是在 Controller 方法执行完毕后执行该方法，所以该方法适合进行一些资源清理、记录日志信息等处理操作。

这里需要注意的是，由于 preHandle 方法决定了程序是否继续执行，所以 postHandle 及 afterCompletion 方法只能在当前 Interceptor 的 pretHandle 方法的返回值为 true 时才会执行。

在实现了 HandlerInterceptor 接口之后，需要在 Spring 的类加载配置文件中配置拦截器实现类，才能使拦截器起到拦截的效果。HandlerInterceptor 类加载配置有两种方式，分别是"针对 HandlerMapping 配置"和"全局配置"。

(1) 针对 HandlerMapping 配置，样例代码如下：

```
<bean class="org.springframework.web.servlet.handler
.BeanNameUrlHandlerMapping">
    <property name="interceptors">
        <list>
            <ref bean="myInterceptor1"/>
            <ref bean="myInterceptor2"/>
        </list>
    </property>
</bean>
<bean id="myInterceptor1" class="com.interceptor.MyInterceptor1"/>
<bean id="myInterceptor2" class="com.interceptor.MyInterceptor2"/>
```

这里为 BeanNameUrlHandlerMapping 处理映射器配置了一个 interceptors 拦截器链，该拦截器链中包含 myInterceptor1 和 myInterceptor2 两个拦截器，具体实现分别对应下面 id 为

myInterceptor1 和 myInterceptor2 的 bean 配置。此种配置的优点是针对具体的处理映射器进行拦截操作，缺点是如果使用多个处理映射器，就要在多处添加拦截器的配置信息，比较烦琐。

（2）针对全局配置，只需在 Spring 的类加载配置文件中添加<mvc:interceptors>标签对，在该标签对中配置拦截器，可起到全局拦截器的作用，样例代码如下：

```xml
<!-- 配置拦截器 -->
<mvc:interceptors>
    <!-- 使用 bean 直接定义在<mvc:interceptors>下面的拦截器将拦截所有请求 -->
    <bean class="com.springmvc.interceptor.MyInterceptor"/>
    <!-- 定义多个拦截器，顺序执行 -->
    <mvc:interceptor>    <!-- 拦截器 1 -->
        <mvc:mapping path="/**"/>        <!-- 配置拦截器作用的路径 -->
        <mvc:exclude-mapping path=""/><!-- 配置不需要拦截器作用的路径 -->
        <!-- 定义在<mvc:interceptor>下面的拦截器,表示对匹配路径请求才进行拦截 -->
        <bean class="com.springmvc.interceptor.MyInterceptor1"/>
    </mvc:interceptor>
    <mvc:interceptor>    <!-- 拦截器 2 -->
        <mvc:mapping path="/hello"/>
        <bean class="com.springmvc.interceptor.MyInterceptor2"/>
    </mvc:interceptor>
    ……
</mvc:interceptors>
```

在上面的配置中，可在<mvc:interceptors>标签下配置多个拦截器，其子元素<bean>定义的是全局拦截器，它会拦截所有请求；而<mvc:interceptor>元素中定义的是指定元素的拦截器，它会对指定路径下的请求生效，其子元素必须按照<mvc:mapping …/>→<mvc:exclude-mapping …/>→<bean .../>的顺序，否则文件会报错。<mvc:interceptor>元素的<mvc:mapping>子元素用于配置拦截器作用的路径，该路径在其属性 path 中定义。如上述 path 的属性值为"/**"表示拦截所有路径，"/hello"表示拦截所有以"/hello"结尾的路径。如果在请求路径中包含不需要拦截的内容，可以通过<mvc:exclude-mapping>元素进行配置。

2．实现 WebRequestInterceptor 接口

WebRequestInterceptor 中也定义了三个方法，也是通过这三个方法来实现拦截的。这三个方法都传递同一个参数 WebRequest，那么这个 WebRequest 是什么呢？这个 WebRequest 是 Spring 定义的一个接口，它里面的方法定义都基本与 HttpServletRequest 一样，在 WebRequestInterceptor 中对 WebRequest 进行的所有操作都将同步到 HttpServletRequest 中，然后在当前请求中一直传递。三个方法介绍如下。

（1）preHandle(WebRequest request) 方法。该方法将在请求处理之前进行调用，也就是说会在 Controller 方法调用之前被调用。这个方法与 HandlerInterceptor 中的 preHandle 不同，主要区别在于该方法的返回值是 void，也就是没有返回值，所以我们一般主要用它来进行资源的准备工作，比如我们在使用 Hibernate 的时候可以在这个方法中准备一个 Hibernate 的 Session 对象，然后利用 WebRequest 的 setAttribute(name, value, scope)方法把它放到 WebRequest 的属性中。setAttribute 方法的第三个参数 scope 是 Integer 类型的，在 WebRequest

的父层接口 RequestAttributes 中对它定义了三个常量：
- SCOPE_REQUEST：它的值是 0，代表只有在 request 中可以访问。
- SCOPE_SESSION：它的值是 1，如果环境允许的话，它代表的是一个局部隔离的 session，否则就代表普通的 session，并且在该 session 范围内可以访问。
- SCOPE_GLOBAL_SESSION：它的值是 2，如果环境允许的话，它代表的是一个全局共享的 session，否则就代表普通的 session，并且在该 session 范围内可以访问。

(2) postHandle(WebRequest request, ModelMap model) 方法。该方法将在请求处理之后，也就是在 Controller 方法调用之后被调用，但是会在视图返回被渲染之前被调用，所以可以在这个方法里面通过改变数据模型 ModelMap 来改变数据的展示。该方法有两个参数，WebRequest 对象是用于传递整个请求数据的，比如在 preHandle 中准备的数据都可以通过 WebRequest 来传递和访问；ModelMap 就是 Controller 处理之后返回的 Model 对象，我们可以通过改变它的属性来改变返回的 Model 模型。

(3) afterCompletion(WebRequest request, Exception ex) 方法。该方法会在整个请求处理完成，也就是在视图返回并被渲染之后执行。所以在该方法中可以进行资源的释放操作。而 WebRequest 参数就可以把我们在 preHandle 中准备的资源传递到这里进行释放。Exception 参数表示当前请求的异常对象，如果在 Controller 中抛出的异常已经被 Spring 的异常处理器给处理了，那么这个异常对象就是 null。

11.2.2 拦截器执行流程

1. 单个拦截器的执行流程

在运行程序时，拦截器的执行是有一定顺序的，该顺序与配置文件中所定义的拦截的顺序相关。如果在程序中只定义了一个拦截器，则该单个拦截器在程序中的执行流程如图 11-11 所示。

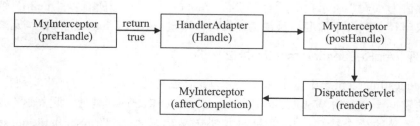

图 11-11　单个拦截器的执行流程

程序首先执行拦截器类中的 preHandle()方法，如果该方法的返回值是 true，则程序会继续向下执行处理器中的方法，否则不再向下执行；在业务控制器类 Controller 处理完请求后，会执行 postHandle() 方法，而后会通过 DispatcherServlet 向客户端返回响应；在 DispatcherServlet 处理完请求后，才会执行 afterCompletion()方法。

下面在 springmvc-6 的项目中通过示例来演示单个拦截器的执行流程，步骤如下：

(1) 在 src 目录下的 com.springmvc.controller 包中的 UserController 类中，注释动态跳转页面的方法。新建一个 hello()方法，并使用@RequestMapping 注解进行映射，代码如下：

```java
// 页面跳转
@RequestMapping("/hello")
public String hello() {
    System.out.println("Hello! Controller 控制器类执行 hello()方法");
    return "hello";
}
```

（2）在 src 目录下，新建一个 com.springmvc.interceptor 包，创建拦截器类 MyInterceptor，该类需要实现 HandlerInterceptor 接口，并重写其中的相应方法，在方法中通过输出语句来输出信息，代码如下：

```java
package com.springmvc.interceptor;
import javax.servlet.http.*;
import org.springframework.web.servlet.*;
public class MyInterceptor implements HandlerInterceptor {
    @Override
    public boolean preHandle(HttpServletRequest request,
HttpServletResponse response, Object handler) throws Exception {
        System.out.println("MyInterceptor 拦截器执行 preHandle()方法");
        return true;
    }
    @Override
    public void postHandle(HttpServletRequest request, HttpServletResponse response, Object handler,
            ModelAndView modelAndView) throws Exception {
        System.out.println("MyInterceptor 拦截器执行 postHandle()方法");
    }
    @Override
    public void afterCompletion(HttpServletRequest request,
HttpServletResponse response, Object handler, Exception ex)
    throws Exception {
        System.out.println("MyInterceptor 拦截器执行 afterCompletion()方法
");
    }
}
```

（3）在 springmvc.xml 的配置文件中，添加拦截器配置，代码如下：

```xml
<mvc:interceptors>    <!-- 配置拦截器 -->
    <!--使用bean 直接定义在<mvc:interceptors>下面的拦截器将拦截所有请求-->
    <bean class="com.springmvc.interceptor.MyInterceptor"/>
</mvc:interceptors>
```

组件扫描器、视图解析器在前面章节已经讲解和配置过，这里不再叙述。

（4）在 ch11 文件夹中，创建一个 hello.jsp 页面文件，在主体部分编写"拦截器执行过程完成！"提示信息。

（5）重启 Tomcat，在浏览器访问网址 http://localhost:8080/springmvc-6/hello，程序正确运行后，浏览器会跳转到 hello.jsp 页面，此时控制台的输出结果如下：

```
MyInterceptor 拦截器执行 preHandle()方法
Hello! Controller 控制器类执行 hello()方法
```

MyInterceptor 拦截器执行 postHandle()方法
MyInterceptor 拦截器执行 afterCompletion()方法

从输出结果可以看出，程序首先执行了拦截器类中的 preHandle()方法，而后执行了控制器中的 hello()方法，最后分别执行了拦截器类中的 postHandle()方法和 afterCompletion()方法。这与前面所描述的单个拦截器的执行顺序是一致的。

2．多个拦截器的执行流程

在一个 Web 工程中，甚至在一个 HandlerMapping 处理器适配器中都可以配置多个拦截器，每个拦截器都按照提前配置好的顺序执行。但需注意的是，它们内部的执行规律并不像多个普通 Java 类一样，它们的设计模式是基于"责任链"的模式。我们知道，拦截器的 preHandle 是有请求放行或拦截的规律的，所以拦截器链的执行首先从 preHandle 方法开始，逐步执行每一个拦截器的 preHandle 方法，若是某一个拦截器的 preHandle 返回 false，则后面拦截器的 preHandle 方法就无法执行了。与此同时，postHandle 与 afterCompletion 的执行也对责任链中的其他拦截器的执行有影响，其实这些方法就是紧紧围绕 Controller 的执行来根据不同的执行周期顺序执行的。

下面通过图例描述多个拦截器的执行流程，假设有两个拦截器 MyInterceptor1 和 MyInterceptor2，将 MyInterceptor1 配置在前，如图 11-12 所示。

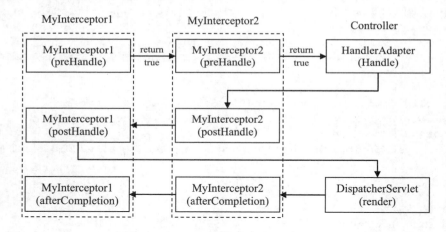

图 11-12 多个拦截器的执行流程

当多个拦截器同时工作时，它们的 preHandle()方法会按照配置文件中拦截器的配置顺序执行，而它们的 postHandle()方法和 afterCompletion()方法则会按照配置顺序的反序执行。

下面通过修改单个拦截器执行流程的实例，来演示多个拦截器的执行，步骤如下：

（1）在 com.springmvc.interceptor 包中，新建两个拦截器类 MyInterceptor1 和 MyInterceptor2，这两个拦截器类均实现了 HandlerInterceptor 接口，其代码与 MyInterceptor 相似，主要就是在输出语句中体现出不同的拦截器。

（2）在 springmvc.xml 的配置文件中，首先注释掉前面配置的 MyInterceptor 拦截器，而后在<mvc:interceptors>元素内配置上面所定义的两个拦截器，代码如下：

```
<mvc:interceptors>
```

```xml
<!-- 定义多个拦截器,顺序执行 -->
<mvc:interceptor>    <!-- 拦截器1 -->
    <mvc:mapping path="/**" />  <!-- 配置拦截器作用的路径 -->
    <!--定义在<mvc:interceptor>下面的拦截器,表示对匹配路径请求才进行拦截-->
    <bean class="com.springmvc.interceptor.MyInterceptor1"/>
</mvc:interceptor>
<mvc:interceptor>    <!-- 拦截器2 -->
    <mvc:mapping path="/hello" />
    <bean class="com.springmvc.interceptor.MyInterceptor2"/>
</mvc:interceptor>
</mvc:interceptors>
```

在上述拦截器的配置代码中,MyInterceptor1 拦截器会作用于所有路径下的请求,MyInterceptor2 拦截器会作用于以"/hello"结尾的请求。

(3) 重启 Tomcat,在浏览器访问网址 http://localhost:8080/springmvc-6/hello,程序正确运行后,浏览器会跳转到 hello.jsp 页面,控制台输出内容如下:

```
MyInterceptor1 拦截器执行 preHandle 方法
MyInterceptor2 拦截器执行 preHandle 方法
Hello! Controller 控制器类执行 hello()方法
MyInterceptor2 拦截器执行 postHandle 方法
MyInterceptor1 拦截器执行 postHandle 方法
MyInterceptor2 拦截器执行 afterCompletion 方法
MyInterceptor1 拦截器执行 afterCompletion 方法
```

通过结果可以观察,两个拦截器的执行顺序并不是完全线性的,而是根据不同的方法功能穿插运行,这也是拦截器链设计模式的一个特点。

11.2.3 使用拦截器实现用户登录权限验证

本节通过一个示例来使用拦截器实现用户登录权限验证。具体为拦截用户的请求,判断用户是否已经登录,如果没有登录,则跳转到 login.jsp 的登录页面,并给出提示信息;如果用户已经登录,则放行;如果账号和密码错误,在登录页面给出相应的提示;当已经登录的用户在系统主页中单击"退出"超链接时,系统同样会回到登录页面。该示例整个流程的执行过程如图 11-13 所示。

在springmvc-6项目中使用拦截器实现用户登录权限验证的步骤如下:

(1) 在com.springmvc.controller包中,在控制器UserController类中,注释以前的方法,并在该类中定义向主页跳转、向登录页跳转、执行用户登录等操作的方法,代码如下:

```java
//向用户登录页面跳转方法
@RequestMapping(value="/login",method=RequestMethod.GET)
public String loginPage() {
    System.out.println("用户从 login 请求到登录跳转 login.jsp 页面");
    return "login";    //跳转到登录页面
}
//用户实现登录方法
@RequestMapping(value="/login",method=RequestMethod.POST)
public String login(User user,Model model,HttpSession session) {
```

```java
        String loginName=user.getLoginName();
        String password=user.getPassword();
        if (loginName!=null && loginName.equals("yzpc") && password!=null && 
password.equals("123456")) {
            System.out.println("用户登录功能实现");
            //将用户添加到session中保存
            session.setAttribute("CURRENT_USER", user);
            return "redirect:index";   //重定向到主页面的跳转方法
        }
        model.addAttribute("message","账号或密码错误,请重新登录!");
        return "login";    //跳转到登录页面
    }
    //向主页跳转方法
    @RequestMapping(value="/index")
    public String indexPage() {
        System.out.println("用户从 index 请求到主页跳转 index.jsp 页面");
        return "index";    //跳转到主页面
    }
    //用户退出登录方法
    @RequestMapping(value="/logout")
    public String logout(HttpSession session) {
        session.invalidate();    // 清除 Session
        System.out.println("退出功能实现,清除 session,重定向到 login 请求");
        return "redirect:login";    //重定向到登录页面的跳转方法
    }
```

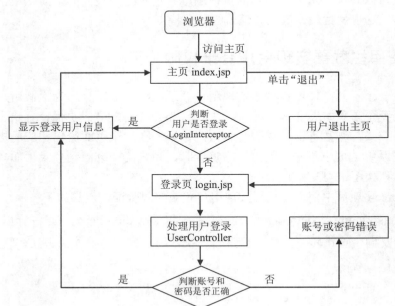

图 11-13　用户登录权限验证的流程

向用户登录页面跳转方法和用户登录方法的@RequestMapping 注解的 value 属性值相同,但 method 属性值不同,因为超链接提交使用 GET 方式,表单提交使用 POST 方式。在用户登录方法中，通过 User 类型参数获取账号和密码，通过 if 语句模拟从数据库获取到账户和密码。如果存在此用户，将信息保存在 Session 中，并重定向到主页，否则跳转到登录页面。

User 实体类已经定义，组件扫描器、视图解析器在前面章节配置过，这里不再叙述。

(2) 在 com.springmvc.interceptor 包中，新建 LoginInterceptor 的拦截器类，代码如下：

```java
// 登录拦截器类
public class LoginInterceptor implements HandlerInterceptor {
    @Override
    public boolean preHandle(HttpServletRequest request,
HttpServletResponse response, Object handler) throws Exception {
        // 获取请求的 URL
        String url=request.getRequestURI();
        if (!(url.contains("Login")||url.contains("login"))) {
            // 非登录请求，获取 Session，判断是否有用户数据
            if (request.getSession().getAttribute ("CURRENT_USER")!=null) {
                return true;          //说明已经登录，放行
            }else {          //没有登录，则跳转到登录页面
                request.setAttribute("message", "您还没有登录，请先登录！");
                request.getRequestDispatcher
("/ch11/login.jsp").forward(request, response);
            }
        }else {
            return true;       // 登录请求，放行
        }
        return false;          // 默认拦截
    }
    // 省略 postHandle()方法和 afterCompletion()方法
}
```

用户登录成功后，会将用户信息封装到 user 对象中，并放置在全局的 session 会话对象中。上面的代码中，在 preHandle()方法中编写了控制用户登录权限的逻辑。首先判断请求是否去往登录页面，如果是则直接返回 true 放行。如果不是，则检测用户的 user 信息是否在 session 中，如果不在，则说明用户没有登录，跳转至 login.jsp 页面。如果 session 中包含 user 对象，则说明用户已经登录，此时直接返回 true 放行。

(3) 在 springmvc.xml 的配置文件中，首先注释前面配置过的拦截器，而后在 <mvc:interceptors>元素内配置上面所定义的 LoginInterceptor 拦截器，代码如下：

```xml
<!-- 配置拦截器 -->
<mvc:interceptors>
    <mvc:interceptor>
        <mvc:mapping path="/**"/>   <!-- /**表示所有 url 包括子 url 路径-->
        <bean class="com.springmvc.interceptor.LoginInterceptor"/>
    </mvc:interceptor>
</mvc:interceptors>
```

在上述拦截器的配置代码中，LoginInterceptor 拦截器会作用于所有路径下的请求。

(4) 在 ch11 文件夹中，新建登录页面 login.jsp 和主页面 index.jsp。

在 login.jsp 登录页面中，编写一个用于实现登录操作的 form 表单，代码如下：

```
<body>
    <font color="red"> ${requestScope.message }</font> <br><br>
    <form action="${pageContext.request.contextPath}/login" method="post">
        账号：<input type="text" name="loginName"><br><br>
        密码：<input type="password" name="password"><br><br>
        <input type="submit" value="登录"><br>
    </form>
</body>
```

在 index.jsp 主页面中，使用 EL 表达式获取用户信息，并通过一个超链接来实现"退出"功能，代码如下：

```
欢迎：${sessionScope.CURRENT_USER.loginName} |
<a href="${pageContext.request.contextPath }/logout">退出</a>
```

(5) 重启 Tomcat，在浏览器访问网址 http://localhost:8080/springmvc-6/index，运行界面如图 11-14 所示。

从图中可以看出，当用户未登录而直接访问主页面时，访问就会被拦截器拦截，从而跳转到登录页面，并提示用户"您还没有登录，请先登录！"。如果在登录页面输入错误的账号和密码，单击"登录"按钮后，浏览器的运行效果如图 11-15 所示。

图 11-14　登录页面　　　　　　　图 11-15　账户或密码错误时登录页面

如果在登录页面输入模拟的"yzpc"账号和"123456"密码，单击"登录"按钮后，浏览器会跳转到主页面，并显示"欢迎：yzpc | 退出"的提示，如图 11-16 所示。

单击"退出"超链接后，即可退出当前系统，系统会从主页面重新定向到登录页面。

图 11-16　系统主页面

11.3 小　　结

　　本章介绍了 Spring MVC 的国际化和拦截器知识。对于 Spring MVC 的国际化，介绍了 Spring MVC 的国际化文件 messageSource、国际化语言区域解析器接口 LocaleResolver 以及该接口的三个常用实现类 AcceptHeaderLocaleResolver、SessionLocaleResolver 和 CookieLocaleResolver 的使用。对于 Spring MVC 的拦截器，介绍了如何在 Spring MVC 项目中定义和配置拦截器，讲解了单个拦截器和多个拦截器的执行流程，最后通过一个用户登录权限验证的示例讲解了拦截器的实际应用。通过应用拦截器机制，Spring MVC 框架可以使用可插拔方式管理各种功能。

第 12 章　MyBatis 入门

MyBatis 作为一个流行的持久层框架，本是 Apache 的一个开源项目 iBatis，2010 年这个项目由 Apache Software Foundation 迁移到了 Google Code，并且改名为 MyBatis。

12.1　MyBatis 概述

MyBatis 是支持普通 SQL 查询、存储过程和高级映射的优秀持久层框架。MyBatis 几乎消除了所有的 JDBC 代码和参数的手工设置以及对结果集的检索。MyBatis 可以使用简单的 XML 或注解来配置和映射基本数据类型，将接口和 Java 的 POJO(Plain Old Java Objects，普通的 Java 对象)映射成数据库中的记录。

对比持久层框架 Hibernate，这两个框架都是 ORM 对象关系映射框架，都是用于将数据持久化的框架技术。

Hiberante 较深度地封装了 JDBC，对开发者编写 SQL 的能力要求不高，只要通过 SQL 语句操作对象即可完成对数据持久化的操作。另外 Hibernate 的可移植性好，如一个项目开始使用的是 MySQL 数据库，现在决定使用 Oracle 数据库，由于不同的数据库使用的 SQL 标准还是有差距的，因此手动修改会存在很大的困难，而使用 Hibernate 只需改变一下数据库方言即可。使用 Hibernate 框架，数据库的移植变得非常方便。但是 Hibernate 也存在诸多的不足，比如在实际开发过程中会生成很多不必要的 SQL 语句耗费程序资源，优化起来也不是很方便，且对存储过程的支持也不够强大。

Mybatis 也是对 JDBC 的封装，但是封装得没有 Hibernate 那么深，通过在配置文件中编写SQL语句，可以根据需求定制SQL语句，数据优化起来比Hibernate容易得多。但Mybatis要求程序员编写SQL的能力要比Hibernate高，且可移植性也不是很好。涉及大数据的系统使用Mybatis比较好，因为优化方便。涉及的数据量不大且对优化要求不高的系统，可以使用Hibernate。

12.2　MyBatis 的下载与安装

在本书编写时，MyBatis的最新版本是3.4.6，本书所讲解的MyBatis框架也是基于这个版本。

读者可以通过官方网站 https://github.com/mybatis/mybatis-3/releases 下载这个版本的MyBatis，MyBatis的下载页面如图12-1所示。

图 12-1　MyBatis 的下载页面

单击 mybatis-3.4.6.zip 链接，就可以下载此版本的 MyBatis 框架压缩包，该压缩包的文件结构如图 12-2 所示。

图 12-2　MyBatis 压缩包的文件结构

lib 文件夹中存放 mybatis-3.4.6 所依赖的 jar 包(如日志包 log4j-1.2.17.jar 等)，mybatis-3.4.6.jar 是 MyBatis 必需的核心包，mybatis-3.4.6.pdf 是 MyBatis 的使用手册。

读者还可以从网站 http://mvnrepository.com/artifact/org.mybatis/mybatis 直接下载MyBatis的核心包。

12.3　MyBatis 的工作原理

在使用 MyBatis 框架进行数据库操作之前，需要先了解 MyBatis 的工作原理，MyBatis框架的执行流程如图 12-3 所示。

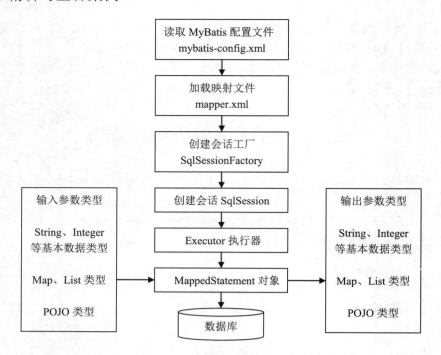

图 12-3　MyBatis 框架的执行流程

从图 12-3 可以看出，使用 MyBatis 操作数据库时，大致经过如下几个步骤。

（1）读取 MyBatis 配置文件 mybatis-config.xml。

mybatis-config.xml 是 MyBatis 的全局配置文件，名称不固定。该文件中配置了数据源、事务等 MyBatis 运行环境。

（2）加载映射文件 mapper.xml。

mapper.xml 是 SQL 映射文件，该文件中定义了数据库操作的 SQL 语句，需要在 mybatis-config.xml 文件中加载。

（3）创建会话工厂。

根据 MyBatis 的配置文件创建会话工厂 SqlSessionFactory。

（4）创建会话。

通过会话工厂 SqlSessionFactory 创建 SqlSession 对象，该对象提供了执行 SQL 的所有方法。

（5）通过 Executor 操作数据库。

Executor 是 Mybatis 的一个核心接口，它与 SqlSession 绑定在一起，每个 SqlSession 都拥有一个新的 Executor 对象，由 Configuration 创建。SqlSession 内部通过执行器操作数据库，增删改语句通过 Executor 接口的 update 方法执行，查询语句通过 query 方法执行。

（6）输入参数和输出结果的映射。

在执行 SQL 语句前，Executor 执行器通过 MappedStatement 对象，将传入的 Java 对象映射到 SQL 语句中。在执行 SQL 语句后，MappedStatement 对象将执行结果映射到 Java 对象。

对于初学 MyBatis 的人来说，此时可能还无法完全理解这些内容。不过没有关系，接下来我们将通过案例来学习如何使用 MyBatis 实现数据的增删改查。

12.4 MyBatis 的增删改查

在 Java 或 Java Web 项目中添加 MyBatis 必需的核心包,就能对数据表进行增删改查操作了。下面以 MySQL 数据库 eshop 中的数据表 user_info 为例,使用 MyBatis 实现数据的增删改查。

12.4.1 查询用户

查询操作通常包括单条记录的精确查询和多条记录的模糊查询,本小节将详细讲解如何使用 MyBatis 框架实现根据用户编号查询用户、根据用户名模糊查询用户的功能。

1. 根据用户编号查询用户

根据用户编号查询用户是通过数据表 user_info 中的主键字段 id 来完成的,具体实现步骤如下。

(1) 新建一个名为 mybatis1 的 Java 项目,在项目中新建文件夹 lib,用于存放项目所需的 jar 包。

(2) 将 MyBatis 必需的核心包 mybatis-3.4.6.jar 和日志包 log4j-1.2.17.jar 复制到项目的 lib 目录中,即完成了 MyBatis 的安装。

(3) 将 MySQL 的驱动包也复制到该项目的 lib 目录中,这里使用的版本为 mysql-connector-java-5.1.18-bin.jar。

(4) 选中项目 lib 目录下的所有 jar 包,右击并选择 Build Path→ Add to Build Path 命令,将这些 jar 包添加到项目的构建路径中。

(5) 创建实体类。

在 src 目录下创建一个 com.mybatis.pojo 包,并在其中创建实体类 UserInfo.java(对应数据库 eshop 中的数据表 user_info)。在 UserInfo 类中声明一些属性(对应数据表 user_info 的部分字段),以及与这些属性对应的 getter 和 setter 方法,还可根据需要添加构造方法和 toString() 方法。

```
package com.mybatis.pojo;
public class UserInfo {
    private int id;
    private String userName;
    private String password;
    private String realName;
    private String sex;
    private String address;
    private String email;
    private String regDate;
    private int status;
    public int getId() {
        return id;
    }
```

```
    public void setId(int id) {
        this.id = id;
    }
    // 此处省略了其他属性的 getter 和 setter 方法
    @Override
    public String toString() {
        return "UserInfo [id=" + id + ", userName=" + userName + ", password="
+ password + "]";
    }
}
```

(6) 创建 SQL 映射的 XML 文件。

在 src 目录下，创建一个 com.mybatis.mapper 包，并在包中创建 SQL 映射的 XML 文件 UserInfoMapper.xml，内容如下：

```
<!DOCTYPE mapper
PUBLIC "-//mybatis.org//DTD Mapper 3.0//EN"
"http://mybatis.org/dtd/mybatis-3-mapper.dtd">
<mapper namespace="com.mybatis.mapper.UserInfoMapper">
    <!-- 根据用户编号查询用户信息 -->
    <select id="findUserInfoById" parameterType="int"
resultType="UserInfo">
        select *
        from user_info where id = #{id}
    </select>
</mapper>
```

映射文件 UserInfoMapper.xml 是一个 XML 格式文件，必须遵循相应的 DTD 文件规范。mybatis-3-mapper.dtd 就是这个规范文件，它规定了映射文件中可以使用的元素。

映射文件以<mapper>作为根结点，其 namespace 属性用于指定<mapper>的唯一命名空间。命名空间设置格式为"包名+SQL 映射文件名"，如 com.mybatis.mapper.UserInfoMapper。

根结点中支持 insert、update、delete、select、cache、cache-ref、resultMap、parameterMap 和 sql 9 个元素。

<select>元素用于映射查询语句，其 id 属性是<select>元素在映射文件中的唯一标识符，这里设置为 findUserInfoById；parameterType 属性用于指定传递给 SQL 语句的参数类型，可以是 int、long、String 等类型，也可以是复杂类型(如对象)，这里设置传入参数类型为 int。在<select>元素映射的查询语句中，#{}表示一个占位符，相当于?，#{id}表示该占位符接收的参数名为 id；resultType 属性用于指定返回结果的类型，这里设置为 UserInfo，表示将查询结果封装到 UserInfo 类型的对象中。

(7) 创建属性文件 db.properties。

在 src 目录下创建属性文件 db.properties，配置 MySQL 数据库的连接信息，如下所示：

```
jdbc.driver=com.mysql.jdbc.Driver
jdbc.url=jdbc:mysql://localhost:3306/eshop
jdbc.username=root
jdbc.password=123456
```

在属性文件 db.properties 中，通过 jdbc.driver 指定数据库驱动名，jdbc.url 指定数据库

URL，jdbc.username 指定数据库用户名，jdbc.password 指定用户密码。

(8) 创建 MyBatis 的核心配置文件。

在 src 目录下，创建 MyBatis 的核心配置文件 mybatis-config.xml，如下所示：

```xml
<?xml version="1.0" encoding="UTF-8" ?>
<!DOCTYPE configuration
PUBLIC "-//mybatis.org//DTD Config 3.0//EN"
"http://mybatis.org/dtd/mybatis-3-config.dtd">
<configuration>
    <!-- 加载属性文件 -->
    <properties resource="db.properties"></properties>
    <!-- 给包中的类注册别名 -->
    <typeAliases>
        <package name="com.mybatis.pojo" />
    </typeAliases>
    <!-- 配置环境 -->
    <environments default="development">
        <!-- 配置一个 id 为 development 的环境 -->
        <environment id="development">
            <!-- 使用 JDBC 事务 -->
            <transactionManager type="JDBC" />
            <!-- 数据库连接池 -->
            <dataSource type="POOLED">
                <property name="driver" value="${jdbc.driver}" />
                <property name="url" value="${jdbc.url}" />
                <property name="username" value="${jdbc.username}" />
                <property name="password" value="${jdbc.password}" />
            </dataSource>
        </environment>
    </environments>
    <!-- 引用映射文件 -->
    <mappers>
        <mapper resource="com/mybatis/mapper/UserInfoMapper.xml" />
    </mappers>
</configuration>
```

在 MyBatis 配置文件中，也必须遵循相应的 DTD 文件规范。mybatis-3-config.dtd 就是这个规范文件，它规定了该配置文件中可以使用的元素。

MyBatis 配置文件以<configuration>元素作为根结点,根结点支持 properties、typeAliases、environments、mappers 等子元素。

<properties>元素用于加载属性文件 db.properties；<typeAliases>元素用于给包中的类注册别名，注册后可以直接使用别名，而不用使用全限定的类名(就是不用包含包名)。

<environments>元素用于环境配置，也就是数据源配置。<environments>元素体中可以配置多个数据库环境，default 属性用于指定默认使用的某种环境的 ID 值。

每个数据库环境通过<environment>子元素配置，其 id 属性用于指定该环境的 ID 值。在<environment>元素体中，使用<transactionManager>子元素配置事务管理，其 type 属性用于指定事务管理类型，type 取值可以是 JDBC(基于 JDBC 的事务)和 MANAGED(托管的事

务)。设置成 JDBC，MyBatis 依赖于从数据源得到的连接来管理事务，即使用 java.sql.Connection 对象完成对事务的提交、回滚。设置成 MANAGED，MyBatis 自身不会去实现事务管理，它会让容器来管理事务的整个生命周期。

在<environments>元素体中，使用<dataSource>子元素声明数据源。其 type 属性用于指定数据源的类型，type 取值可以是 UNPOOLED、POOLED 和 JNDI。设置成 UNPOOLED 时，MyBatis 会为每一次数据库操作创建一个新的连接，并关闭该连接。设置成 POOLED 时，MyBatis 会创建一个数据库连接池，数据库操作时从连接池中获取一个连接，操作完成后再将连接归还给连接池。设置成 JNDI 时，MyBatis 从配置好的 JNDI 数据源获取数据库连接。并发用户规模不大时用 UNPOOLED，测试和开发过程一般用 POOLED，实际运行使用 JNDI。

<mappers>元素用于指定 MyBatis 映射文件的位置，每个映射文件的位置通过<mapper>子元素的 resource 属性来指定。

(9) 创建测试类。

创建 JUnit 测试类 MybatisTest.java，存放在 com.mybatis.test 包中。其代码如下：

```java
package com.mybatis.test;
import java.io.IOException;
import java.io.InputStream;
import org.apache.ibatis.io.Resources;
import org.apache.ibatis.session.SqlSession;
import org.apache.ibatis.session.SqlSessionFactory;
import org.apache.ibatis.session.SqlSessionFactoryBuilder;
import org.junit.After;
import org.junit.Before;
import org.junit.Test;
import com.mybatis.pojo.UserInfo;
public class MybatisTest {
    private SqlSessionFactory sqlSessionFactory;
    private SqlSession sqlSession;
    @Before
    public void init() {
        // 读取mybatis配置文件
        String resource = "mybatis-config.xml";
        InputStream inputStream;
        try {
            // 得到配置文件流
            inputStream = Resources.getResourceAsStream(resource);
            // 根据配置文件信息，创建会话工厂
            sqlSessionFactory = new SqlSessionFactoryBuilder().build(inputStream);
            // 通过工厂得到SqlSession
            sqlSession = sqlSessionFactory.openSession();
        } catch (IOException e) {
            e.printStackTrace();
        }
    }
    // 根据id查询用户
```

```java
    @Test
    public void testFindUserInfoById() {
        // 通过 sqlSession 执行映射文件中定义的 SQL,并返回映射结果
        UserInfo ui = sqlSession.selectOne("findUserInfoById", 1);
        // 打印输出结果
        System.out.println(ui.toString());
    }
    @After
    public void destroy() {
        // 提交事务
        sqlSession.commit();
        // 关闭 sqlSession
        sqlSession.close();
    }
}
```

在测试类 MybatisTest 中，首先添加 init()方法，并使用@Before 注解修饰。JUnit4 使用 Java5 中的@Before 注解，用于初始化，init()方法对于每一个测试方法都要执行一次。在 init() 方法中，先根据 MyBatis 配置文件构建 SqlSessionFactory，再通过 SqlSessionFactory 创建 SqlSession。

在测试类 MybatisTest 中，接着添加 destroy()方法，并使用@After 注解修饰。JUnit4 使用 Java5 中的@After 注解，用于释放资源。destroy()方法对于每一个测试方法都要执行一次。在 destroy()方法中，先执行事务提交，再释放 SqlSession 资源。

在测试类 MybatisTest 中，每一个用@Test 注解修饰的方法称为测试方法，它们的调用顺序为@Before→@Test→@After。

在测试方法 testFindUserInfoById 中，调用 SqlSession 对象的 selectOne 方法执行查询操作。selectOne 方法的第一个参数是映射文件 UserInfoMapper.xml 中定义的<select>元素的 id，这里为 findUserInfoById，表示将执行这个 id 所标识的<select>元素中的 SQL 语句；第二个参数是查询所需的参数，这里查询编号为 1 的用户信息。执行 selectOne 方法后，MyBatis 会将查询结果封装到 UserInfo 对象并返回。

(10) 测试结果。

MyBatis 默认使用 log4j 输出日志信息，为了能在控制台输出 SQL 语句，需要在项目的 src 目录下创建文件 log4j.xml，文件内容可以查阅项目源代码。

执行测试类 MybatisTest 中的 testFindUserInfoById()方法，控制台输出如下：

```
DEBUG [main] - ==>  Preparing: select * from user_info where id = ?
DEBUG [main] - ==> Parameters: 1(Integer)
DEBUG [main] - <==      Total: 1
UserInfo [id=1, userName=tom, password=123456]
```

从控制台输出可以看出，使用 MyBatis 成功查询出了编号为 1 的用户信息。

2．根据用户名模糊查询用户

根据用户名模糊查询用户是通过数据表 user_info 中的字段 userName 来完成的，具体实现步骤如下：

(1) 在映射文件 UserInfoMapper.xml 中，添加根据用户名模糊查询用户的 SQL 语句，如下所示：

```xml
<!-- 根据用户名模糊查询用户 -->
<select id="findUserInfoByUserName" parameterType="String"
    resultType="UserInfo">
    select * from user_info where userName like
    CONCAT(CONCAT('%',#{userName}),'%')
</select>
```

在 id 为 findUserInfoByUserName 的<select>元素中，传入 SQL 查询语句的参数类型为 String，参数值是用户名。SQL 语句执行结果返回的是一个 List 集合类型，resultType 属性设置的返回类型为集合中元素的类型，这里为 UserInfo。在 SQL 查询语句中，使用 CONCAT 函数进行字符串的拼接，这样既能实现模糊查询，又能防止 SQL 注入。

(2) 在测试类 MybatisTest 中，添加一个测试方法 testFindUserInfoByUserName()，代码如下：

```java
// 根据用户名模糊查询用户
@Test
public void testFindUserInfoByUserName() {
    // 执行映射文件中定义的SQL,并返回映射结果
    List<UserInfo> uis = sqlSession.selectList("findUserInfoByUserName",
"m");
    for (UserInfo ui : uis) {
        // 打印输出结果
        System.out.println(ui.toString());
    }
}
```

根据用户名模糊查询的结果可能有多条记录，上述代码调用了 SqlSession 对象的 selectList 方法执行查询操作。如果数据表中没有数据返回，则返回空集合，而不会返回 null；如果有数据返回，则返回集合对象。最后再使用 for 循环输出集合中的对象。

(3) 执行 testFindUserInfoByUserName()方法，控制台输出如下：

```
DEBUG [main] - ==>  Preparing: select * from user_info where userName like CONCAT(CONCAT('%',?),'%')
DEBUG [main] - ==> Parameters: m(String)
DEBUG [main] - <==      Total: 2
UserInfo [id=1, userName=tom, password=123456]
UserInfo [id=3, userName=my, password=123456]
```

从控制台输出可以看出，使用 MyBatis 成功查询出了用户名中包含 m 的两条用户信息。

12.4.2 添加用户

在 MyBatis 映射文件中，插入操作是通过<insert>元素来实现的。在映射文件 UserInfoMapper.xml 中，添加用户的 SQL 语句如下：

```xml
<!-- 添加用户 -->
```

```
<insert id="addUserInfo" parameterType="UserInfo">
    insert into
    user_info(userName,password) values(#{userName},#{password})
</insert>
```

在 id 为 addUserInfo 的<insert>元素中，parameterType 属性设置为 UserInfo，表示传入 SQL 插入语句的参数类型为 UserInfo，将 UserInfo 类型的对象作为参数传递到 SQL 语句中。#{userName}和#{password}占位符接收的参数名分别来自 UserInfo 对象的 userName 和 password 属性。

在测试类 MybatisTest 中，添加一个测试方法 testAddUserInfo()，代码如下：

```
// 添加用户
@Test
public void testAddUserInfo() {
    // 创建 UserInfo 对象 ui
    UserInfo ui = new UserInfo();
    // 向对象 ui 中添加数据
    ui.setUserName("mybatis1");
    ui.setPassword("123456");
    // 执行 sqlSession 的 insert 方法,返回结果是 SQL 语句受影响的行数
    int result = sqlSession.insert("addUserInfo", ui);
    if (result > 0) {
        System.out.println("插入成功");
    } else {
        System.out.println("插入失败");
    }
}
```

在 testAddUserInfo 方法中，首先创建了 UserInfo 对象 ui，并在 ui 中添加了用户名和密码两个数据；然后调用 SqlSession 对象的 insert 方法执行插入操作，返回结果是 SQL 语句受影响的行数。执行 testAddUserInfo()方法，控制台输出如下：

```
DEBUG [main] - ==>  Preparing: insert into user_info(userName,password)
values(?,?)
DEBUG [main] - ==> Parameters: mybatis1(String), 123456(String)
DEBUG [main] - <==    Updates: 1
插入成功
```

打开数据表 user_info，可以看到表中新增了一条用户记录，如图 12-4 所示。

id	userName	password
1	tom	123456
2	john	123456
3	my	123456
4	sj	123456
5	lxf	123456
6	lj	123456
7	mybatis1	123456

图 12-4　数据表 user_info

12.4.3　修改用户

在 MyBatis 映射文件中，更新操作是通过<update>元素来实现的。在映射文件 UserInfoMapper.xml 中，修改用户信息的 SQL 语句如下：

```
<!-- 修改用户信息 -->
<update id="updateUserInfo" parameterType="UserInfo">
    update user_info set
```

```
        userName=#{userName}, password=#{password} where id=#{id}
</update>
```

在 id 为 updateUserInfo 的<update>元素中,parameterType 属性设置为 UserInfo,表示传入 SQL 更新语句的参数类型为 UserInfo,将 UserInfo 类型的对象作为参数传递到 SQL 语句中。#{userName}、#{password} 和#{id}占位符接收的参数名分别来自 UserInfo 对象的 userName、password 和 id 属性。

在测试类 MybatisTest 中,添加一个测试方法 testUpdateUserInfo(),代码如下:

```java
// 修改用户
@Test
public void testUpdateUserInfo() {
    // 加载编号为 7 的用户
    UserInfo ui = sqlSession.selectOne("findUserInfoById", 7);
    // 重新设置用户密码
    ui.setPassword("123123");
    // 执行 sqlSession 的 update 方法,返回结果是 SQL 语句受影响的行数
    int result = sqlSession.update("updateUserInfo", ui);
    if (result > 0) {
        System.out.println("更新成功");
        System.out.println(ui);
    } else {
        System.out.println("更新失败");
    }
}
```

在 testUpdateUserInfo()方法中,首先加载编号为 7 的用户对象 ui,然后重新设置用户密码,最后执行 sqlSession 的 update 方法进行更新操作。

执行 testUpdateUserInfo()方法,控制台输出如下:

```
DEBUG [main] - ==>  Preparing: select * from user_info where id = ?
DEBUG [main] - ==> Parameters: 7(Integer)
DEBUG [main] - <==      Total: 1
DEBUG [main] - ==>  Preparing: update user_info set userName=?, password=? where id=?
DEBUG [main] - ==> Parameters: mybatis1(String), 123123(String), 7(Integer)
DEBUG [main] - <==    Updates: 1
更新成功
UserInfo [id=7, userName=mybatis1, password=123123]
```

从控制台输出可以看出,使用 MyBatis 成功更新了 id 为 7 的用户信息。

12.4.4 删除用户

在 MyBatis 映射文件中,删除操作是通过<delete>元素来实现的。在映射文件 UserInfoMapper.xml 中,删除用户信息的 SQL 语句如下:

```xml
<!-- 根据用户编号删除用户 -->
<delete id="deleteUserInfo" parameterType="int">
    delete from user_info
```

```
        where id=#{id}
</delete>
```

在 id 为 deleteUserInfo 的<delete>元素中，parameterType 属性设置为 int，表示传入 SQL 删除语句的参数类型为 int，这里将用户编号作为参数传递到 SQL 语句中。#{id}占位符接收的参数名为 id。

在测试类 MybatisTest 中，添加一个测试方法 testDeleteUserInfo，该方法用于删除编号为 7 的用户信息，代码如下：

```
// 删除用户
@Test
public void testDeleteUserInfo() {
    // 执行 sqlSession 的 delete 方法,返回结果是 SQL 语句受影响的行数
    int result = sqlSession.delete("deleteUserInfo", 7);
    if (result > 0) {
        System.out.println("成功删除了" + result + "条记录");
    } else {
        System.out.println("插入删除");
    }
}
```

执行 testDeleteUserInfo 方法，控制台输出如下：

```
DEBUG [main] - ==>  Preparing: delete from user_info where id=?
DEBUG [main] - ==> Parameters: 7(Integer)
DEBUG [main] - <==    Updates: 1
成功删除了 1 条记录
```

再次查看数据表 user_info，会看到之前添加的编号为 7 的用户记录被成功删除了。至此，使用 MyBatis 框架实现数据表 user_info 的增删改查操作就讲解完了，读者可以结合之前讲解的 MyBatis 工作原理来理解这个示例。

12.5　使用 resultMap 属性映射查询结果

在上述示例中，实体类 UserInfo 中的属性名与数据表 user_info 中的字段名相同，如果属性名与数据表的字段名不相同，那么就需要使用 resultMap 属性来进行结果集的映射。

将项目 mybatis1 复制并命名为 mybatis2，再导入到 Eclipse 开发环境中。

修改实体类 UserInfo.java，将其属性重新命名，使得属性名与数据表 user_info 的字段名不同。

```
package com.mybatis.pojo;
public class Users {
    private int uid;
    private String uname;
private String upass;
// 省略属性的 getter 和 setter 方法
// 省略 toString()方法
}
```

在映射文件 UserInfoMapper.xml 中，修改 id 为 findUserInfoById 的<select>元素，修改后的映射文件如下：

```xml
<!DOCTYPE mapper
PUBLIC "-//mybatis.org//DTD Mapper 3.0//EN"
"http://mybatis.org/dtd/mybatis-3-mapper.dtd">
<mapper namespace="com.mybatis.mapper.UserInfoMapper">
    <!-- 根据用户编号获取用户信息 -->
    <select id="findUserInfoById" parameterType="int" resultMap="userInfoMap">
        select *
        from user_info where id = #{id}
    </select>
    <resultMap type="UserInfo" id="userInfoMap">
        <id property="uid" column="id" />
        <result property="uname" column="userName" />
        <result property="upass" column="password" />
    </resultMap>
</mapper>
```

在<select>元素中，通过 resultMap 属性引用该映射文件中一个 id 为 userInfoMap 的<resultMap>元素，来完成查询结果的映射。在<resultMap>元素中，type 属性指定映射结果的类型，这里为 UserInfo；<id>和<result>子元素用来将数据表 user_info 的字段映射到 UserInfo 对象的属性，不同的是，<id>元素用来映射标识属性。

修改测试类 MybatisTest，只保留一个测试方法 testFindUserInfoById()。执行该测试方法，控制台输出如下：

```
DEBUG [main] - ==>  Preparing: select * from user_info where id = ?
DEBUG [main] - ==> Parameters: 1(Integer)
DEBUG [main] - <==      Total: 1
UserInfo [id=1, userName=tom, password=123456]
```

可以看出，当数据表字段名与实体类属性名不一致时，使用 resultMap 属性可以实现查询结果的映射。

12.6 使用 Mapper 接口执行 SQL

在测试类 MybatisTest 中，通过 sqlSession 对象调用 selectOne、selectList、delete 和 update 等方法。在这些方法中，需要指定映射文件中执行语句的 id(如 findUserInfoById)。如果 id 的拼写出现错误，只有到运行时才能发现。为此，MyBatis 提供了另一种编程方式，即使用 Mapper 接口执行 SQL，从而避免上述情况的出现，同时也更符合 Java 面向接口编程的习惯。但是，使用 Mapper 接口开发时需要遵循如下规范：

- 映射文件中的 namespace 与 Mapper 接口的类路径相同。
- 在 Mapper 接口中，方法名和映射文件中定义的执行语句的 id 相同。
- 方法的输入参数类型和映射文件中定义的执行语句的 parameterType 的类型相同。
- 方法输出参数类型和映射文件中定义的执行语句的 resultType 的类型相同。

还是以数据表 user_info 的增删改查操作为例,使用 Mapper 接口编程的步骤如下。
(1) 将项目 mybatis1 复制并命名为 mybatis3,再导入到 Eclipse 开发环境中。
(2) 在 com.mybatis.mapper 包中,创建接口 UserInfoMapper.java,并声明方法,如下所示:

```java
package com.mybatis.mapper;
import com.mybatis.pojo.UserInfo;
public interface UserInfoMapper {
    UserInfo findUserInfoById(int id);
}
```

由于映射文件中的 namespace 属性设置为 com.mybatis.mapper.UserInfoMapper,因此 Mapper 接口的类路径必须与之相同。这里创建了接口 UserInfoMapper.java,并将其放置在 com.mybatis.mapper 包中。在接口 UserInfoMapper.java 中,声明了一个 findUserInfoById(int id)方法,方法名为映射文件中一个<select>元素的 id;方法的输入参数类型与该<select>元素的 parameterType 的类型相同;方法的输出参数类型与该<select>元素的 resultType 的类型相同。

(3) 在测试类 MybatisTest 中,修改测试方法 testFindUserInfoById(),如下所示:

```java
// 根据id查询用户
@Test
public void testFindUserInfoById() {
    // 获得 UserInfoMapper 接口的代理对象
    UserInfoMapper uim = sqlSession.getMapper(UserInfoMapper.class);
    // 直接调用接口的方法,查询编号为1的UserInfo对象
    UserInfo ui = uim.findUserInfoById(1);
    // 打印输出结果
    System.out.println(ui.toString());
}
```

MyBatis 通过动态代理的方式实现 Mapper 接口,在测试方法 testFindUserInfoById()中,首先调用 sqlSession 对象的 getMapper 方法获得 UserInfoMapper 接口的代理对象;然后直接调用接口的方法,查询编号为 1 的 UserInfo 对象;最后打印输出结果。

执行测试方法 testFindUserInfoById(),控制台输出如下:

```
DEBUG [main] - ==>  Preparing: select * from user_info where id = ?
DEBUG [main] - ==> Parameters: 1(Integer)
DEBUG [main] - <==      Total: 1
UserInfo [id=1, userName=tom, password=123456]
```

从控制台输出可以看出,使用 Mapper 接口成功地获取了用户信息。

12.7 小 结

本章介绍了 MyBatis 框架的概念、下载与安装和工作原理,并结合数据表 user_info 的增删改查操作,详细讲解了 MyBatis 框架的基本用法。通过本章的学习,相信读者已经对 MyBatis 框架有了初步的了解,下一章将学习更多有关 MyBatis 框架的知识。

第 13 章　MyBatis 的关联映射

MyBatis 不仅支持单张表的增删改查,还支持多张表之间的关联操作。表之间的关联关系通常包括一对一关联、一对多关联和多对多关联。表之间的关联反映到对象层面,就是对象间的关联关系。MyBatis 提供了关联映射,可以很好地处理对象间的关联关系,极大简化持久层数据的操作。

13.1　一对一关联映射

在现实生活中,一个人只能有一个身份证,一个身份证只能对应一个人,人和身份证之间就是一对一关系。以人和身份证为例,使用 MyBatis 框架处理它们之间的一对一关联关系的步骤如下。

(1) 创建数据表。

在数据库 eshop 中,创建数据表 idcard 和 person,分别记录身份证和人的信息,创建表和插入测试数据的 SQL 语句如下:

```
USE eshop
#创建数据表idcard
CREATE TABLE idcard(
    id INT PRIMARY KEY AUTO_INCREMENT,
    cno VARCHAR(18)
);
```

```
#插入测试数据
INSERT INTO idcard(cno) VALUES(320100197001010001);
INSERT INTO idcard(cno) VALUES(320100197001010002);
INSERT INTO idcard(cno) VALUES(320100197001010003);
#创建数据表person
CREATE TABLE person(
    id INT PRIMARY KEY AUTO_INCREMENT,
    NAME VARCHAR(20),
    age INT,
    sex VARCHAR(2),
    cid INT,
    FOREIGN KEY(cid) REFERENCES idcard(id)
);
#插入测试数据
INSERT INTO person(NAME,age,sex,cid) VALUES('zhangsan',22,'男',1);
INSERT INTO person(NAME,age,sex,cid) VALUES('lili',21,'女',2);
INSERT INTO person(NAME,age,sex,cid) VALUES('wangwu',22,'男',3);
```

(2) 将项目mybatis1复制并命名为mybatis4，再导入到Eclipse开发环境中。

(3) 创建实体类。

在com.mybatis.pojo包中，创建实体类Idcard和Person。实体类Idcard的代码如下：

```java
package com.mybatis.pojo;
public class Idcard {
    private int id;
    private String cno;
    // 省略属性的getter和setter方法
    @Override
    public String toString() {
        return "Idcard [id=" + id + ", cno=" + cno + "]";
    }
}
```

实体类Person的代码如下：

```java
package com.mybatis.pojo;
public class Person {
    private int id;
    private String name;
    private int age;
    private String sex;
    // 关联属性
    private Idcard idcard;
    // 省略属性的getter和setter方法
    @Override
    public String toString() {
        return "Person [id=" + id + ", name=" + name + ", age=" + age + ", sex=" + sex + ", idcard=" + idcard + "]";
    }
}
```

(4) 创建 MyBatis 映射文件。

在 com.mybatis.mapper 包中，创建映射文件 IdcardMapper.xml 和 PersonMapper.xml。映射文件 IdcardMapper.xml 的内容如下：

```xml
<!DOCTYPE mapper
PUBLIC "-//mybatis.org//DTD Mapper 3.0//EN"
"http://mybatis.org/dtd/mybatis-3-mapper.dtd">
<mapper namespace="com.mybatis.mapper.IdcardMapper">
    <!-- 根据id查询身份证信息 -->
    <select id="findIdcardById" parameterType="int" resultType="Idcard">
        select * from idcard where id=#{id}
    </select>
</mapper>
```

在映射文件 IdcardMapper.xml 中，添加了一个 id 为 findIdcardById 的<select>元素，根据 id 从数据表 idcard 中查询身份证信息，返回 Idcard 类型对象。

映射文件 PersonMapper.xml 的内容如下：

```xml
<!DOCTYPE mapper
PUBLIC "-//mybatis.org//DTD Mapper 3.0//EN"
"http://mybatis.org/dtd/mybatis-3-mapper.dtd">
<mapper namespace="com.mybatis.mapper.PersonMapper">
    <!-- 根据id查询个人信息，返回resultMap -->
    <select id="findPersonById" parameterType="int" resultMap="personMap">
        select * from person where id = #{id}
    </select>
    <!-- 查询语句查询结果映射 -->
    <resultMap type="Person" id="personMap">
        <id property="id" column="id" />
        <result property="name" column="name" />
        <result property="age" column="age" />
        <result property="sex" column="sex" />
        <!-- 一对一关联映射 -->
        <association property="idcard" column="cid"
            select="com.mybatis.mapper.IdcardMapper.findIdcardById"
javaType="Idcard" />
    </resultMap>
</mapper>
```

在 PersonMapper.xml 中，添加了一个 id 为 findPersonById 的<select>元素，由于实体类 Person 中除了简单的属性 id、name、age 和 sex 之外，还有一个关联的对象属性 idcard，所以需要通过映射文件中 id 为 personMap 的<resultMap>元素完成查询结果的映射。

在 id 为 personMap 的<resultMap>元素中，除了使用<id>和<result>子元素映射数据表 person 的基本字段与实体类 Person 的基本属性外，还通过<association>元素来映射 Person 与 Idcard 对象之间一对一的关联关系。

在<association>元素中，select 属性可以让 MyBatis 找到 com.mybatis.mapper.IdcardMapper 命名空间下 id 为 findIdcardById 的元素，执行该元素中的 SQL 语句，传入的参数来自 column 属性指定的值 cid，查询出的关联结果被封装到 property 属性指定的 Idcard 类型的对象

idcard 中。

(5) 创建 Mapper 接口。

在 com.mybatis.mapper 包中,创建接口 PersonMapper,并声明方法,其代码如下:

```
package com.mybatis.mapper;
import com.mybatis.pojo.Person;
public interface PersonMapper {
    Person findPersonById(int id);
}
```

这里使用 Mapper 接口执行 SQL 语句,在接口 PersonMapper 中声明一个 findPersonById 方法,该方法名为映射文件 PersonMapper.xml 中一个<select>元素的 id。

(6) 引用映射文件。

在配置文件 mybatis-config.xml 中,引入映射文件 IdcardMapper.xml 和 PersonMapper.xml,如下所示:

```
<!-- 引用映射文件 -->
<mappers>
    ……
    <mapper resource="com/mybatis/mapper/IdcardMapper.xml" />
    <mapper resource="com/mybatis/mapper/PersonMapper.xml" />
</mappers>
```

(7) 测试一对一关联映射。

在测试类 MybatisTest 中,添加测试方法 testFindPersonById(),并使用@Test 注解修饰,从数据表 person 中获取记录的同时,获取关联的数据表 idcard 中的记录。其代码如下:

```
@Test
public void testFindPersonById() {
    // 获得 PersonMapper 接口的代理对象
    PersonMapper pm = sqlSession.getMapper(PersonMapper.class);
    // 直接调用接口的方法,查询 id=1 的 Person 对象
    Person person = pm.findPersonById(1);
    System.out.println(person.toString());
}
```

在测试方法 testFindPersonById()中,首先调用 sqlSession 对象的 getMapper 方法获得 PersonMapper 接口的代理对象;然后直接调用接口的方法,查询编号为 1 的 Person 对象;最后打印输出结果。

执行 testFindPersonById()方法,控制台输出如下:

```
DEBUG [main] - ==>  Preparing: select * from person where id = ?
DEBUG [main] - ==> Parameters: 1(Integer)
DEBUG [main] - ====>  Preparing: select * from idcard where id=?
DEBUG [main] - ====> Parameters: 1(Integer)
DEBUG [main] - <====      Total: 1
DEBUG [main] - <==      Total: 1
Person [id=1, name=zhangsan, age=22, sex=男, idcard=Idcard [id=1, cno=321001197001010001]]
```

可以看到，查询 Person 对象时关联的 Idcard 对象也查询出来了。

13.2 一对多关联映射

在实际开发中，一对多关联关系要比一对一更为常见。例如，订单与订单明细之间就是一对多关系。一个订单有多条明细信息，一条订单明细信息对应一个订单。同样，商品类别与商品之间也是一对多关系。一个商品类别对应多个商品，一个商品属于一个类别。

以数据库 eshop 中的商品类别表 type 和商品表 product_info 为例，针对数据查询、插入和删除操作，使用 MyBatis 框架处理它们之间的一对多关联关系。

1．数据查询

数据查询是根据商品类型编号从数据表 type 中查询商品类型，并查询关联的商品列表。其具体实现步骤如下。

(1) 将项目 mybatis1 复制并命名为 mybatis5，再导入到 Eclipse 开发环境中。
(2) 创建实体类。

在 com.mybatis.pojo 包中，创建实体类 ProductInfo 和 Type。实体类 ProductInfo 的代码如下：

```java
package com.mybatis.pojo;
public class ProductInfo {
    private int id;
    private String code;
    private String name;
    // 关联属性
    private Type type;
    // 省略属性的 getter 和 setter 方法
    @Override
    public String toString() {
        return "ProductInfo [id=" + id + ", code=" + code + ", name=" + name + "]";
    }
}
```

为了简化代码，实体类 ProductInfo 中只添加了与数据表 product_info 的部分字段对应的属性。

实体类 Type 的代码如下：

```java
package com.mybatis.pojo;
import java.util.List;
public class Type {
    private int id;
    private String name;
    // 关联集合属性
    private List<ProductInfo> pis;
    // 省略属性的 getter 和 setter 方法
    @Override
```

```java
    public String toString() {
        return "Type [id=" + id + ", name=" + name + ", pis=" + pis + "]";
    }
}
```

(3) 创建 MyBatis 映射文件。

在 com.mybatis.mapper 包中，创建映射文件 ProductInfoMapper.xml 和 TypeMapper.xml。映射文件 ProductInfoMapper.xml 的内容如下：

```xml
<?xml version="1.0" encoding="UTF-8" ?>
<!DOCTYPE mapper
PUBLIC "-//mybatis.org//DTD Mapper 3.0//EN"
"http://mybatis.org/dtd/mybatis-3-mapper.dtd">
<mapper namespace="com.mybatis.mapper.ProductInfoMapper">
    <!-- 根据类型编号查询商品信息 -->
    <select id="findProductInfoById" parameterType="int" resultType="ProductInfo">
        select * from product_info where tid = #{id}
    </select>
</mapper>
```

在 ProductInfoMapper.xml 文件中，定义了 id 为 findProductInfoById 的<select>元素，根据类型编号查询商品信息。

映射文件 TypeMapper.xml 的内容如下：

```xml
<?xml version="1.0" encoding="UTF-8" ?>
<!DOCTYPE mapper
PUBLIC "-//mybatis.org//DTD Mapper 3.0//EN"
"http://mybatis.org/dtd/mybatis-3-mapper.dtd">
<mapper namespace="com.mybatis.mapper.TypeMapper">
    <!-- 根据商品类型编号查询商品类型信息 -->
    <select id="findTypeById" parameterType="int" resultMap="typeMap">
        select *
        from type where id = #{id}
    </select>
    <!-- 查询结果映射 -->
    <resultMap type="Type" id="typeMap">
        <id property="id" column="id" />
        <result property="name" column="name" />
        <!-- 一对多关联映射 -->
        <collection property="pis" column="id" select=
"com.mybatis.mapper.ProductInfoMapper.findProductInfoById">
        </collection>
    </resultMap>
</mapper>
```

在映射文件 TypeMapper.xml 中，定义了 id 为 findTypeById 的<select>元素，根据商品类型编号查询商品类型信息，查询结果通过 id 为 typeMap 的<resultMap>元素进行映射。

在<resultMap>元素中，先通过<id>与<result>子元素实现 Type 类型基本属性的映射，再通过<collection>子元素实现一对多关联映射。

在<collection>元素中，property 属性用于指定关联属性，这里设置为 Type 类型关联的 ProductInfo 类型的集合属性 pis；select 属性用于指定获取关联属性值时需要执行的 SQL 语句的 id，这里设置为 com.mybatis.mapper.ProductInfoMapper，表示 MyBatis 将会执行 com.mybatis.mapper 包下 ProductInfoMapper 接口中定义的方法 findProductInfoById，即执行映射文件 ProductInfoMapper.xml 中定义的 id 为 findProductInfoById 的<select>元素中的 SQL 语句；column 属性用于指定传入上述<select>元素中的 SQL 语句的参数名，这里设置为 id，表示商品类型编号。

(4) 创建 Mapper 接口。

在 com.mybatis.mapper 包中，创建接口 TypeMapper，并声明方法，其代码如下：

```java
package com.mybatis.mapper;
import com.mybatis.pojo.Type;
public interface TypeMapper {
    Type findTypeById(int id);
}
```

在接口 TypeMapper 中声明一个 findTypeById 方法，方法名为映射文件 TypeMapper.xml 中定义的<select>元素的 id。

(5) 引用映射文件。

在配置文件 mybatis-config.xml 中，引入映射文件 ProductInfoMapper.xml 和 TypeMapper.xml，如下所示：

```xml
<!-- 引用映射文件 -->
<mappers>
    ……
    <mapper resource="com/mybatis/mapper/ProductInfoMapper.xml" />
    <mapper resource="com/mybatis/mapper/TypeMapper.xml" />
</mappers>
```

(6) 测试一对多关联映射。

在测试类 MybatisTest 中，添加测试方法 testFindTypeById()，并使用@Test 注解修饰，从数据表 type 获取记录的同时，获取关联的数据表 product_info 中的记录。其代码如下：

```java
// 一对多关联映射：查询数据
@Test
public void testFindTypeById() {
    // 获得 TypeMapper 接口的代理对象
    TypeMapper tm = sqlSession.getMapper(TypeMapper.class);
    // 直接调用接口的方法，查询 id=1 的 Type 对象
    Type type = tm.findTypeById(1);
    System.out.println(type.toString());
}
```

执行 testFindTypeById()方法，控制台输出如下所示：

```
DEBUG [main] - ==>  Preparing: select * from type where id = ?
DEBUG [main] - ==> Parameters: 1(Integer)
DEBUG [main] - ====>  Preparing: select * from product_info where tid = ?
DEBUG [main] - ====> Parameters: 1(Integer)
```

```
DEBUG [main] - <====      Total: 6
DEBUG [main] - <==       Total: 1
Type [id=1, name=电脑, pis=[ProductInfo [id=1, code=1378538,
name=AppleMJVE2CH/A], ProductInfo [id=2, code=1309456,
name=ThinkPadE450C(20EH0001CD)], ProductInfo [id=3, code=1999938, name=联
想小新 300 经典版], ProductInfo [id=4, code=1466274, name=华硕 FX50JX],
ProductInfo [id=5, code=1981672, name=华硕 FL5800], ProductInfo [id=6,
code=1904696, name=联想G50-70M]]]
```

从控制台输出可以看出，MyBatis 执行了查询商品类型的 SQL 语句后，又执行了根据商品类型编号查询商品信息列表的 SQL 语句。

在映射文件 TypeMapper.xml 中，一对多关联映射通过执行另外一个 SQL 映射语句来返回预期的复杂类型，即引用外部定义好的 SQL 语句块，这种关联映射查询方式也称为嵌套查询。

此外，MyBatis 还提供了另一种关联映射查询方式——嵌套结果，它通过嵌套结果映射来处理联结结果集的重复子集。

还是以类别表 type 和商品表 product_info 的一对多关联查询为例，使用嵌套结果查询方式的步骤如下。

（1）在映射文件 TypeMapper.xml 中，编写使用嵌套结果查询方式进行一对多关联查询的配置，如下所示：

```
<!-- 使用嵌套结果查询方式实现一对多关联查询 -->
<select id="findTypeById2" parameterType="int" resultMap="typeMap2">
    select
    t.id tid, t.name tname, pi.* from type t,product_info pi where
    pi.tid=t.id and t.id=#{id}
</select>
<resultMap type="Type" id="typeMap2">
    <id property="id" column="tid" />
    <result property="name" column="tname" />
<!-- 一对多关联映射 -->
    <collection property="pis" ofType="ProductInfo">
        <id property="id" column="id" />
        <result property="code" column="code" />
        <result property="name" column="name" />
    </collection>
</resultMap>
```

在映射文件 TypeMapper.xml 中，添加了一个 id 为 findTypeById2 的<select>元素。在这个<select>元素中，使用嵌套结果的方式定义了一个根据商品类型编号查询商品类型及其关联的商品信息列表的 select 语句。当关联查询的字段名相同时，需要使用别名加以区分。这里的 tid 和 tname 是数据表 type 中字段 id 和 name 的别名，从而与 product_info 表的 id 和 name 字段区分。

select 语句的执行结果通过 id 为 typeMap2 的<resultMap>元素进行映射。在<resultMap>元素中，先通过<id>和<result>子元素映射 Type 类型对象的基本属性 id 和 name，再通过<collection>子元素映射关联的 ProductInfo 类型的集合属性 pis。<collection>元素的 ofType

属性用于指定关联的集合 pis 中的元素类型。

(2) 在接口 TypeMapper 中，添加一个方法 findTypeById2，如下所示：

```
Type findTypeById2(int id);
```

(3) 添加测试方法。

在测试类 MybatisTest 中，添加一个测试方法 testFindTypeById2()，并使用@Test 注解修饰，如下所示：

```java
@Test
public void testFindTypeById2() {
    // 获得 TypeMapper 接口的代理对象
    TypeMapper tm = sqlSession.getMapper(TypeMapper.class);
    // 直接调用接口的方法，查询 id=1 的 Type 对象
    Type type = tm.findTypeById2(1);
    System.out.println(type.toString());
}
```

执行测试方法 testFindTypeById2()，控制台输出与 testFindTypeById()方法执行结果相同。至此，以数据库 eshop 中的商品类别表 type 和商品表 product_info 为例，使用嵌套查询和嵌套结果两种查询方式实现了一对多关联查询。

2．数据插入

数据插入操作要实现将数据同时插入数据表 type 和 product_info 中，具体实现步骤如下。

(1) 在映射文件 TypeMapper.xml 中，编写将数据插入数据表 type 的配置，如下所示：

```xml
<!-- 向 type 表插入数据 -->
<insert id="addType" parameterType="Type">
    <!-- 插入数据,并获得刚插入数据表 type 的记录 id -->
    <selectKey keyProperty="id" resultType="int" order="AFTER">
        SELECT
        LAST_INSERT_ID() AS ID
    </selectKey>
    insert into type(name) values(#{name})
</insert>
```

在映射文件 TypeMapper.xml 中，添加了一个 id 为 addType 的<insert>元素。在这个<insert>元素中，定义了一个向数据表 type 中插入记录的 insert 语句，并通过<selectKey>子元素获取刚插入数据表 type 的记录 id。

在<selectKey>元素中，定义了一个获取刚插入的自动增长的 id 值的 SELECT 语句。keyProperty 属性用于指定将这个 SELECT 语句的执行结果赋值给 Type 类型对象的哪个属性，这里为 id；resultType 属性用于指定执行结果类型，这里为 int；order 属性可以被设置为 BEFORE 或 AFTER。如果设置为 BEFORE，则先选择主键并设置 keyProperty，然后执行插入语句；如果设置为 AFTER，则先执行插入语句，然后是设置 keyProperty。像 MySQL 这种支持自增类型的数据库中，order 需要设置为 AFTER 才会取到正确的值。

(2) 在接口 TypeMapper 中，添加一个 addType 方法，如下所示：

```
void addType(Type type);
```

(3) 在映射文件 ProductInfoMapper.xml 中，编写将数据插入数据表 product_info 的配置，如下所示：

```xml
<!-- 向 product_info 表插入数据 -->
<insert id="addProductInfo" parameterMap="addProductInfoPMap">
    insert into
    product_info(code,name,tid) values(#{code},#{name},#{type.id})
</insert>
<parameterMap type="ProductInfo" id="addProductInfoPMap">
    <parameter property="code" />
    <parameter property="name" />
    <parameter property="type.id" />
</parameterMap>
```

在映射文件 ProductInfoMapper.xml 中，添加了一个 id 为 addProductInfo 的<insert>元素。在这个<insert>元素中，定义了一个向数据表 product_info 插入记录的 insert 语句。由于这条 insert 语句有多个参数且实体类中的属性和数据库中的字段不对应，因此通过 parameterMap 属性指定传入 insert 语句的参数，这里设置为 addProductInfoPMap，即引用当前映射文件中一个 id 为 addProductInfoPMap 的<parameterMap>元素。

在<parameterMap>元素中，type 属性指定 SQL 语句中的参数来自于哪个实体对象，<parameter>子元素的 property 属性进一步指定参数来自于实体对象的哪个属性。

(4) 在 com.mybatis.mapper 包中，创建 ProductInfoMapper 接口，并添加一个 addProductInfo 方法，如下所示：

```java
package com.mybatis.mapper;
import com.mybatis.pojo.ProductInfo;
public interface ProductInfoMapper {
    void addProductInfo(ProductInfo pi);
}
```

(5) 添加测试方法。

在测试类 MybatisTest 中，添加一个测试方法 testAddType()，并使用@Test 注解修饰，向数据表 type 中插入记录的同时向 product_info 表插入记录。

```java
// 添加数据
@Test
public void testAddType() {
    // 创建 Type 对象
    Type type = new Type();
    type.setName("打印机");
    // 获得 TypeMapper 接口的代理对象
    TypeMapper tm = sqlSession.getMapper(TypeMapper.class);
    // 直接调用接口的方法，保存 Type 对象
    tm.addType(type);
    // 创建两个 ProductInfo 对象
    ProductInfo pi1 = new ProductInfo();
    ProductInfo pi2 = new ProductInfo();
    pi1.setCode("111111");
```

```
        pi1.setName("HP1306");
        pi2.setCode("222222");
        pi2.setName("HP11103");
        // 设置关联属性
        pi1.setType(type);
        pi2.setType(type);
        // 获得 ProductInfoMapper 接口的代理对象
        ProductInfoMapper pm = sqlSession.getMapper(ProductInfoMapper.class);
        // 直接调用接口的方法，保存两个 ProductInfo 对象
        pm.addProductInfo(pi1);
        pm.addProductInfo(pi2);
    }
```

在测试方法 testAddType()中，首先创建一个 Type 对象，通过获得的 TypeMapper 接口的代理对象，直接调用接口的方法将 Type 对象中的数据保存到数据表 type；然后创建两个 ProductInfo 对象，并设置关联属性 type，通过获得的 ProductInfoMapper 接口的代理对象，直接调用接口的方法将两个 ProductInfo 对象中的数据保存到数据表 product_info。

执行测试方法 testAddType()，控制台输出如下所示：

```
DEBUG [main] - ==>  Preparing: insert into type(name) values(?)
DEBUG [main] - ==> Parameters: 打印机(String)
DEBUG [main] - <==    Updates: 1
DEBUG [main] - ==>  Preparing: SELECT LAST_INSERT_ID() AS ID
DEBUG [main] - ==> Parameters:
DEBUG [main] - <==      Total: 1
DEBUG [main] - ==>  Preparing: insert into product_info(code,name,tid)
values(?, ?,?)
DEBUG [main] - ==> Parameters: 111111(String), HP1306(String), 6(Integer)
DEBUG [main] - <==    Updates: 1
DEBUG [main] - ==>  Preparing: insert into product_info(code,name,tid)
values(?, ?,?)
DEBUG [main] - ==> Parameters: 222222(String), HP11103(String), 6(Integer)
DEBUG [main] - <==    Updates: 1
```

此时，数据表 type 中添加了一条记录，如图 13-1 所示。product_info 表添加了两条关联的记录，如图 13-2 所示。

id	name
1	电脑
2	冰箱
3	电视机
4	洗衣机
5	数码相机
6	打印机
(Auto)	(NULL)

图 13-1　表 type 中插入的记录

id	code	name	tid
9	1143562	海尔BCD-216S	2
10	1560207	海尔BCD-258W	2
11	1721668	海信（Hisens	2
12	823125	海信BCD-211T	2
13	111111	HP1306	6
14	222222	HP11103	6

图 13-2　表 product_info 中插入的记录

3．数据删除

数据删除操作要实现从数据表 type 删除记录时，将关联表 product_info 中的记录也删除，其具体实现步骤如下。

(1) 在映射文件 TypeMapper.xml 中，编写删除数据的配置，如下所示：

```xml
<!-- 删除数据 -->
<delete id="deleteTypeById" parameterType="int">
    delete from product_info where tid = #{id};
    delete from type where id = #{id};
</delete>
```

在 id 为 deleteTypeById 的<delete>元素中，定义了两条 delete 语句，第一条 delete 语句从关联表 product_info 中删除记录，第二条 delete 语句从主表 type 中删除记录。多条 SQL 语句之间用分号隔开。由于 MySQL 默认没有开启批量执行 SQL 语句的开关，因此无法一次执行多条 SQL 语句，那么如何开启呢？只需在设置 MySQL 连接的 url 时，加上 allowMultiQueries 参数，并设置为 true 即可。打开 db.properties 文件，修改 jdbc.url 属性值，如下所示：

```
jdbc.url=jdbc:mysql://localhost:3306/eshop?allowMultiQueries=true
```

(2) 在接口 TypeMapper 中，添加一个 deleteTypeById 方法，如下所示：

```java
int deleteTypeById(int id);
```

(3) 添加测试方法。

在测试类 MybatisTest 中，添加一个测试方法 testDeleteTypeById()，并使用@Test 注解修饰，删除编号为 6 的商品类型，同时删除关联的商品信息。

```java
// 删除数据
@Test
public void testDeleteTypeById() {
    // 获得 TypeMapper 接口的代理对象
    TypeMapper tm = sqlSession.getMapper(TypeMapper.class);
    // 直接调用接口的方法，删除编号为 6 的商品类型,同时删除关联的商品信息
    int result = tm.deleteTypeById(6);
    if (result > 0) {
        System.out.println("删除成功");
    } else {
        System.out.println("删除失败");
    }
}
```

执行测试方法 testDeleteTypeById()，控制台输出如下：

```
DEBUG [main] - ==>  Preparing: delete from product_info where tid = ?; delete from type where id = ?;
DEBUG [main] - ==> Parameters: 6(Integer), 6(Integer)
DEBUG [main] - <==    Updates: 2
删除成功
```

此时，打开数据表 type，可以看到编号为 6 的商品类型信息被成功删除了；同时数据表 product_info 中 tid 为 6 的商品信息也被成功删除。

至此，以数据表 type 和 product_info 之间的一对多关联映射为例，使用 MyBatis 实现了数据查询、插入和删除操作。

13.3 多对多关联映射

在实际开发中，多对多关联关系也比较常见。例如，订单与商品之间就是多对多关系。一个订单可以包含多个商品，一个商品可以属于多个订单。同样，管理员与系统功能之间也是多对多关系。一个管理员可以拥有多个功能权限，一个系统功能可以属于多个用户。

以数据库 eshop 中的管理员表 admin_info 和系统功能表 functions 为例，它们之间通过中间表 powers 来关联，这个中间表分别与 admin_info 和 functions 构成多对一关联。中间表 powers 以 aid 和 fid 作为联合主键，其中，aid 字段作为外键参照 admin_info 表的 id 字段，fid 字段作为外键参照 functions 表的 id 字段。

以数据查询操作为例，使用 MyBatis 框架处理它们之间的多对多关联关系，其具体实现步骤如下：

(1) 将项目 mybatis1 复制并命名为 mybatis6，再导入到 Eclipse 开发环境中。
(2) 创建实体类。

在 com.mybatis.pojo 包中，创建实体类 AdminInfo 和 Functions。其中，实体类 AdminInfo 的代码如下：

```java
package com.mybatis.po;
import java.util.HashSet;
import java.util.Set;
public class AdminInfo {
    private int id;
    private String name;
    // 关联的属性
    private List<Functions> fs;
    // 省略属性的 getter 和 setter 方法
    // 省略有参构造和无参构造方法
// 省略 toString()方法
}
```

实体类 Functions 的代码如下：

```java
package com.mybatis.po;
import java.util.HashSet;
import java.util.Set;
public class Functions {
    private int id;
    private String name;
    // 省略属性的 getter 和 setter 方法
    // 省略有参构造和无参构造方法
    // 省略 toString()方法
}
```

(3) 创建 MyBatis 映射文件。

在 com.mybatis.mapper 包中，创建映射文件 FunctionsMapper.xml 和 AdminInfoMapper.xml。其中，映射文件 FunctionsMapper.xml 的代码如下：

```xml
<!DOCTYPE mapper
PUBLIC "-//mybatis.org//DTD Mapper 3.0//EN"
"http://mybatis.org/dtd/mybatis-3-mapper.dtd">
<mapper namespace="com.mybatis.mapper.FunctionsMapper">
    <!-- 根据管理员id获取其功能权限列表 -->
    <select id="findFunctionsByAid" parameterType="int"
resultType="Functions">
        select * from functions f where id in (
        select fid from powers where aid = #{id}
        )
    </select>
</mapper>
```

在 FunctionsMapper.xml 文件中，定义了一个 id 为 findFunctionsByAid 的<select>元素，根据管理员 id 获取其功能权限列表。由于管理员和系统功能是多对多的关系，数据库中使用了一个中间表 powers 维护多对多关联关系。此处使用了一个子查询，首先根据管理员 id 到中间表 powers 中查询出所有的功能权限 id，然后再根据功能权限 id 到 functions 表中查询出所有的功能权限信息，并将这些信息封装到 Functions 对象中。

创建映射文件 AdminInfoMapper.xml 的代码如下：

```xml
<!DOCTYPE mapper
PUBLIC "-//mybatis.org//DTD Mapper 3.0//EN"
"http://mybatis.org/dtd/mybatis-3-mapper.dtd">
<mapper namespace="com.mybatis.mapper.AdminInfoMapper">
    <!-- 根据管理员id查询管理员信息 -->
    <select id="findAdminInfoById" parameterType="int"
resultMap="adminInfoMap">
        select * from admin_info where id = #{id}
    </select>
    <resultMap type="AdminInfo" id="adminInfoMap">
        <id property="id" column="id" />
        <result property="name" column="name" />
        <!-- 多对多关联映射 -->
        <collection property="fs" ofType="Functions" column="id"
select="com.mybatis.mapper.FunctionsMapper.findFunctionsByAid">
        </collection>
    </resultMap>
</mapper>
```

在映射文件 AdminInfoMapper.xml 中，定义了一个 id 为 findAdminInfoById 的<select>元素，根据管理员 id 获取管理员信息。由于 AdminInfo 类型对象除了基本的属性 id 和 name 之外，还有一个关联的集合属性 fs，所以返回的是一个<resultMap>元素。由于 fs 是一个 List<Functions>类型的集合，所以 id 为 adminInfoMap 的<resultMap>元素中使用了<collection>元素来映射多对多关联关系。<collection>元素的 select 属性表示会找到 com.mybatis.mapper 包下映射文件 FunctionsMapper.xml 中定义的 id 为 findFunctionsByAid 的<select>元素，执行该元素中的 SQL 语句，参数来自于 column 属性指定的 id 值(管理员 id)，查询到的关联结果被封装到 property 属性指定的 List<Functions>类型的对象 fs 中。

(4) 创建 Mapper 接口。

在 com.mybatis.mapper 包中,创建接口 FunctionsMapper,并声明方法,代码如下:

```java
package com.mybatis.mapper;
import java.util.List;
import com.mybatis.pojo.*;
public interface FunctionsMapper {
    Functions findFunctionsByAid(int id);
}
```

创建接口 AdminInfoMapper,并声明方法,代码如下:

```java
package com.mybatis.mapper;
import com.mybatis.pojo.AdminInfo;
public interface AdminInfoMapper {
    AdminInfo findAdminInfoById(int id);
}
```

(5) 引用映射文件。

在配置文件 mybatis-config.xml 中,引入映射文件 FunctionsMapper.xml 和 AdminInfoMapper.xml,代码如下:

```xml
<!-- 引用映射文件 -->
<mappers>
    ……
    <mapper resource="com/mybatis/mapper/FunctionsMapper.xml" />
    <mapper resource="com/mybatis/mapper/AdminInfoMapper.xml" />
</mappers>
```

(6) 测试多对多关联映射。

在测试类 MybatisTest 中,添加测试方法 testFindTypeById(),并使用@Test 注解修饰,查看管理员及其关联的功能权限。

```java
// 测试多对多关联查询
@Test
public void testFindAdminInfoById() {
    // 获得 AdminInfoMapper 接口的代理对象
    AdminInfoMapper aim = sqlSession.getMapper(AdminInfoMapper.class);
    // 查询 id=1 的 AdminInfo 对象及其关联的功能权限
    AdminInfo ai = aim.findAdminInfoById(1);
    System.out.println(ai.toString());
}
```

执行 testFindTypeById()方法,控制台输出如下:

```
DEBUG [main] - ==>  Preparing: select * from admin_info where id = ?
DEBUG [main] - ==> Parameters: 1(Integer)
DEBUG [main] - ====>  Preparing: select * from functions f where id in ( select fid from powers where aid = ? )
DEBUG [main] - ====> Parameters: 1(Integer)
DEBUG [main] - <====      Total: 10
DEBUG [main] - <==      Total: 1
```

```
AdminInfo [id=1, name=admin, fs=[Functions [id=1, name=电子商城管理后台],
Functions [id=2, name=商品管理], Functions [id=3, name=商品列表], Functions
[id=4, name=商品类型列表], Functions [id=5, name=订单管理], Functions [id=6,
name=查询订单], Functions [id=7, name=创建订单], Functions [id=8, name=用户管
理], Functions [id=9, name=用户列表], Functions [id=11, name=退出系统]]]
```

可以看到，MyBatis 执行了两条 select 语句，查询出了管理员及其关联的功能权限信息。

13.4 小　　结

本章以数据查询操作为例，详细讲解了使用 MyBatis 框架处理数据表之间一对一、一对多和多对多三种关联关系的具体过程。

第 14 章 动态 SQL

在 SQL 语句的 where 条件子句中，往往需要进行一些判断。例如，按名称进行模糊查询，如果传入的参数为空，查询结果很可能是空的。而实际上，当参数为空时，通常希望查出全部的信息。使用 MyBatis 框架提供的动态 SQL，就可以轻松解决这一问题。具体来说，就是使用动态 SQL，增加一个判断，当参数不符合时，就不判断此查询条件。

MyBatis 的动态 SQL 是基于 OGNL 表达式的，MyBatis 中用于实现动态 SQL 的元素主要包括 if、choose(when，otherwise)、trim、where、set、foreach 等。

14.1 <if>元素

<if>元素是简单的条件判断，可用来实现某些简单的条件选择。以数据库 eshop 中的数据表 user_info 为例，要求按用户名模糊查询。如果输入用户名 j，则查询用户名包含 j 的用户信息；如果没有输入，则查询所有用户。使用<if>元素实现这个示例的步骤如下。

(1) 将项目 mybatis1 复制并命名为 mybatis7，再导入到 Eclipse 开发环境中。

(2) 在映射文件 UserInfoMapper.xml 中，添加一个 id 为 findUserInfoByUserNameWithIf 的<select>元素，如下所示：

```
<!-- 动态 SQL 之<if>元素 -->
<select id="findUserInfoByUserNameWithIf" parameterType="UserInfo"
    resultType="UserInfo">
    select * from user_info ui
```

```xml
        <if test="userName != null and userName != ''">
            where ui.userName LIKE CONCAT(CONCAT('%',#{userName}),'%')
        </if>
    </select>
```

在上述<select>元素中，使用<if>元素编写了动态 SQL 语句。<if>元素会对 userName 属性进行非空判断，如果传入的查询条件成立，即 userName 非空，就会将 where 子句拼装到 select 语句中；否则就会忽略 where 子句。

(3) 在 com.mybatis.mapper 包中，创建接口 UserInfoMapper 并声明方法，如下所示：

```java
package com.mybatis.mapper;
import java.util.List;
import com.mybatis.pojo.UserInfo;
public interface UserInfoMapper {
    List<UserInfo> findUserInfoByUserNameWithIf(UserInfo ui);
}
```

(4) 在测试类 MybatisTest 中，添加测试方法 testFindUserInfoByUserNameWithIf()，并用@Test 注解修饰，代码如下：

```java
// 测试动态 SQL 之<if>元素
@Test
public void testFindUserInfoByUserNameWithIf() {
    // 获得 UserInfoMapper 接口的代理对象
    UserInfoMapper uim = sqlSession.getMapper(UserInfoMapper.class);
    // 创建 UserInfo 对象,用于封装查询条件
    UserInfo cond = new UserInfo();
    cond.setUserName("j");
    // 直接调用接口的方法,根据条件查询 UserInfo 对象
    List<UserInfo> uis = uim.findUserInfoByUserNameWithIf(cond);
    for (UserInfo ui : uis) {
        System.out.println(ui.toString());
    }
}
```

在 testFindUserInfoByUserNameWithIf 方法中，首先获得 UserInfoMapper 接口的代理对象，然后将用户名 j 封装到 UserInfo 对象作为查询条件，最后调用接口 UserInfoMapper 中的 findUserInfoByUserNameWithIf 方法，根据条件查询 UserInfo 对象列表。

(5) 执行 testFindUserInfoByUserNameWithIf 方法，控制台输出如下：

```
DEBUG [main] - ==>  Preparing: select * from user_info ui where ui.userName LIKE CONCAT(CONCAT('%',?),'%')
DEBUG [main] - ==> Parameters: j(String)
DEBUG [main] - <==      Total: 3
UserInfo [id=2, userName=john, password=123456]
UserInfo [id=4, userName=sj, password=123456]
UserInfo [id=6, userName=lj, password=123456]
```

从输出结果可以看出，使用<if>元素查询出了用户名包含 j 的用户信息。

在 testFindUserInfoByUserNameWithIf 方法中，如果不设置 userName 的值，生成的 SQL

语句中就不会包含 where 子句。此时，查询结果为所有的用户信息，如下所示：

```
EBUG [main] - ==> Preparing: select * from user_info ui
DEBUG [main] - ==> Parameters:
DEBUG [main] - <== Total: 6
UserInfo [id=1, userName=tom, password=123456]
UserInfo [id=2, userName=john, password=123456]
UserInfo [id=3, userName=my, password=123456]
UserInfo [id=4, userName=sj, password=123456]
UserInfo [id=5, userName=lxf, password=123456]
UserInfo [id=6, userName=lj, password=123456]
```

14.2　<where>、<if>元素

当<if>元素较多时，可能会拼装成 where and 或者 where or 之类的关键字多余的错误 SQL 语句，使用<where>元素可以轻松有效地解决这一问题。只有<where>元素内的条件成立时，才会在拼装 SQL 语句时加上 where 关键字。如果出现 where and 或者 where or 时，<where>元素会自动剔除 where 关键字后面多余的 and 或 or。

以数据表 user_info 为例，要求按用户名模糊查询，同时查询指定状态的用户列表，使用<where>和<if>元素实现这一示例的过程如下。

(1)在映射文件 UserInfoMapper.xml 中，添加一个 id 为 findUserInfoByUserNameAndStatus 的<select>元素，如下所示：

```xml
<!-- 动态 SQL 之<where>、<if>元素 -->
<select id="findUserInfoByUserNameAndStatus" parameterType="UserInfo"
    resultType="UserInfo">
    select * from user_info ui
    <where>
        <if test="userName!=null and userName!=''">
            ui.userName LIKE CONCAT(CONCAT('%', #{userName}),'%')
        </if>
        <if test="status>-1">
            and ui.status = #{status}
        </if>
    </where>
</select>
```

在上述<select>元素中，使用<where>和<if>元素编写了动态 SQL 语句。只有当<where>元素内的条件成立时，才会在组装 SQL 语句时加上 where 关键字。假如第一个 if 条件不成立，第二个 if 条件成立，即 userName 为空，而 status 大于-1 时，从表面上看，在拼装 SQL 语句时 where 关键字之后会多余一个 and 关键字。不过不用担心，<where>元素会将其剔除。

(2) 在接口 UserInfoMapper 中，声明一个方法，如下所示：

```
List<UserInfo> findUserInfoByUserNameAndStatus(UserInfo ui)
```

(3) 在测试类 MybatisTest 中，添加一个测试方法，并用@Test 注解修饰，代码如下：

```
// 测试动态 SQL 之<where>、<if>元素
```

```java
@Test
public void testFindUserInfoByUserNameAndStatus() {
    // 获得 UserInfoMapper 接口的代理对象
    UserInfoMapper uim = sqlSession.getMapper(UserInfoMapper.class);
    // 创建 UserInfo 对象,用于封装查询条件
    UserInfo cond = new UserInfo();
    cond.setUserName("j");
    cond.setStatus(1);
    // 直接调用接口的方法,根据条件查询 UserInfo 对象
    List<UserInfo> uis = uim.findUserInfoByUserNameAndStatus(cond);
    for (UserInfo ui : uis) {
        System.out.println(ui.toString());
    }
}
```

在上述测试方法中,首先获得 UserInfoMapper 接口的代理对象,然后将用户名 j 和用户状态 1 封装到 UserInfo 对象中作为查询条件,最后调用接口 UserInfoMapper 中的 findUserInfoByUserNameAndStatus 方法,根据条件查询 UserInfo 对象列表。

(4) 执行上述测试方法,控制台输出如下:

```
DEBUG [main] - ==>  Preparing: select * from user_info ui WHERE ui.userName
LIKE CONCAT(CONCAT('%', ?),'%') and ui.status = ?
DEBUG [main] - ==> Parameters: j(String), 1(Integer)
DEBUG [main] - <==      Total: 3
UserInfo [id=2, userName=john, password=123456]
UserInfo [id=4, userName=sj, password=123456]
UserInfo [id=6, userName=lj, password=123456]
```

在上述测试方法中,如果将"cond.setUserName("j");"这条语句注释掉,则会执行如下 SQL 语句:

```
DEBUG [main] - ==>  Preparing: select * from user_info ui WHERE ui.status = ?
```

可以看出,如果属性 userName 的值不满足测试条件,在拼装 SQL 语句时,<where>元素会将 where 关键字之后多余的 and 关键字剔除,从而避免出现 SQL 语法错误。

14.3 <set>、<if>元素

<set>和<if>元素可用来组装 update 语句,只有当<set>元素内的条件成立时,才会在组装 SQL 语句时加上 set 关键字。<set>元素内包含<if>子元素,每个<if>元素包含的 SQL 语句后面会有一个逗号,拼接好的 SQL 语句中会包含多余的逗号,从而造成 SQL 语法错误。不过不用担心,<set>元素能将 SQL 语句中多余的逗号剔除。

以数据表 user_info 为例,要求更新某个用户的用户名和密码,使用<set>和<if>元素实现这一示例的过程如下:

(1) 在映射文件 UserInfoMapper.xml 中,添加一个 id 为 updateUserInfo2 的<update>元素,如下所示:

```xml
<!-- 动态SQL之<set>、<if>元素 -->
<update id="updateUserInfo2" parameterType="UserInfo">
    update user_info ui
    <set>
        <if test="userName!=null and userName!=''">
            ui.userName=#{userName},
        </if>
        <if test="password!=null and password!=''">
            ui.password=#{password}
        </if>
    </set>
    where ui.id=#{id}
</update>
```

在上述<update>元素中，使用<set>和<if>元素编写了动态SQL语句。只有当<set>元素内的条件成立时，才会在组装SQL语句时加上set关键字。假如第一个if条件成立，第二个if条件不成立，从表面上看，拼装的SQL语句中会包含一个多余的逗号。不过不用担心，<set>元素会自动将其剔除。

(2) 在接口UserInfoMapper中，声明一个方法，如下所示：

```java
void updateUserInfo2(UserInfo ui);
```

(3) 在测试类MybatisTest中，添加一个测试方法testUpdateUserInfo2，并用@Test注解修饰，如下所示：

```java
// 测试动态SQL之<set>、<if>元素
@Test
public void testUpdateUserInfo2() {
    // 获得UserInfoMapper接口的代理对象
    UserInfoMapper uim = sqlSession.getMapper(UserInfoMapper.class);
    // 创建UserInfo对象，并初始化
    UserInfo cond = new UserInfo();
    cond.setId(3);
    cond.setUserName("miaoyong");
    cond.setPassword("123123");
    // 直接调用接口的方法
    uim.updateUserInfo2(cond);
}
```

(4) 执行测试方法，控制台输出如下：

```
DEBUG [main] - ==> Preparing: update user_info ui SET ui.userName=?, ui.password=? where ui.id=?
DEBUG [main] - ==> Parameters: miaoyong(String), 123123(String), 3(Integer)
DEBUG [main] - <== Updates: 1
```

在上述测试方法中，如果没有给password属性设置值，<set>元素会将SQL语句结尾处多余的逗号剔除。再次执行测试方法，控制台输出如下：

```
DEBUG [main] - ==> Preparing: update user_info ui SET ui.userName=? where ui.id=?
```

14.4 <trim>元素

使用<trim>元素,可以通过 prefix 属性在要拼装的 SQL 语句片段前加上前缀,通过 suffix 属性在要拼装的 SQL 语句片段后加上后缀,通过 prefixOverrides 属性把要拼装的 SQL 语句片段首部的某些内容覆盖,通过 suffixOverrides 属性把要拼装的 SQL 语句片段尾部的某些内容覆盖。因此<trim>元素可用来替代<where>元素和<set>元素实现同样的功能。

使用<trim>元素替代<where>元素,从数据表 user_info 中按用户名模糊查询,同时查询指定状态的用户列表,实现步骤如下:

(1) 在 UserInfoMapper.xml 中,添加一个 id 为 findUserInfoByUserNameWithIf_Trim 的 <select>元素,如下所示:

```xml
<!-- 动态 SQL 之<trim>元素 1 -->
<select id="findUserInfoByUserNameWithIf_Trim" parameterType="UserInfo"
    resultType="UserInfo">
    select * from user_info ui
    <trim prefix="where" prefixOverrides="and|or">
        <if test="userName!=null and userName!=''">
            ui.userName LIKE CONCAT(CONCAT('%', #{userName}),'%')
        </if>
        <if test="status>-1">
            and ui.status = #{status}
        </if>
    </trim>
</select>
```

在上述<select>元素中,使用<trim>元素编写了动态 SQL 语句。在<trim>元素中, prefix 属性设置为 where,将要拼装的 SQL 语句的前缀设置为 where,即使用 where 关键字来连接后面的 SQL 语句片段; prefixOverrides 属性设置为 and|or,是将要拼装的 SQL 语句片段首部多余的 "and" 或 "or" 关键字去除。

(2) 在接口 UserInfoMapper 中,声明一个方法,如下所示:

```
List<UserInfo> findUserInfoByUserNameWithIf_Trim(UserInfo ui);
```

(3) 在测试类 MybatisTest 中,添加一个测试方法,并用@Test 注解修饰,如下所示:

```java
// 测试动态 SQL 之<trim>元素替代<where>元素
@Test
public void testFindUserInfoByUserNameWithIf_Trim() {
    // 获得 UserInfoMapper 接口的代理对象
    UserInfoMapper uim = sqlSession.getMapper(UserInfoMapper.class);
    // 创建 UserInfo 对象,并初始化
    UserInfo cond = new UserInfo();
    cond.setUserName("J");
    cond.setStatus(1);
    // 直接调用接口的方法,根据条件查询 UserInfo 对象
    List<UserInfo> uis = uim.findUserInfoByUserNameWithIf_Trim(cond);
```

```
        for (UserInfo ui : uis) {
            System.out.println(ui.toString());
        }
    }
```

(4) 执行该测试方法，控制台输出如下：

```
DEBUG [main] - ==>  Preparing: select * from user_info ui where ui.userName
LIKE CONCAT(CONCAT('%', ?),'%') and ui.status = ?
DEBUG [main] - ==> Parameters: j(String), 1(Integer)
DEBUG [main] - <==      Total: 3
UserInfo [id=2, userName=john, password=123456]
UserInfo [id=4, userName=sj, password=123456]
UserInfo [id=6, userName=lj, password=123456]
```

在上述测试方法中，如果没有设置 userName 的值，<trim>元素会将 status 条件前多余的 and 关键字剔除。再次执行测试方法，控制台输出的 SQL 语句如下：

```
DEBUG [main] - ==>  Preparing: select * from user_info ui where ui.status = ?
```

使用<trim>元素还可以替代<set>元素，以更新数据表 user_info 中某个用户的用户名和密码为例，具体步骤如下。

(1) 在映射文件 UserInfoMapper.xml 中，添加一个 id 为 updateUserInfo2_trim 的<update>元素，如下所示：

```xml
<!-- 动态 SQL 之<trim>元素替代<set>元素 -->
<update id="updateUserInfo2_trim" parameterType="UserInfo">
    update user_info ui
    <trim prefix="set" suffixOverrides=",">
        <if test="userName!=null and userName!=''">
            ui.userName=#{userName},
        </if>
        <if test="password!=null and password!=''">
            ui.password=#{password}
        </if>
    </trim>
    where ui.id=#{id}
</update>
```

在上述<update>元素中，使用<trim>元素编写了动态 SQL 语句。在<trim>元素中，prefix 属性设置为 set，将要拼装的 SQL 语句的前缀设置为 set，即使用 set 关键字来连接后面的 SQL 语句片段；suffixOverrides 属性设置为 "，"，将要拼装的 SQL 语句片段尾部多余的逗号去除。

(2) 在接口 UserInfoMapper 中，声明一个方法，如下所示：

```
void updateUserInfo2_trim(UserInfo ui);
```

(3) 添加测试方法。

在测试类 MybatisTest 中，添加一个测试方法，并用@Test 注解修饰，如下所示：

```
// 测试动态 SQL 之<trim>元素替代<set>元素
```

```java
@Test
public void testUpdateUserInfo2_trim() {
    // 获得UserInfoMapper接口的代理对象
    UserInfoMapper uim = sqlSession.getMapper(UserInfoMapper.class);
    // 创建UserInfo对象,并初始化
    UserInfo cond = new UserInfo();
    cond.setId(3);
    cond.setUserName("mmm");
    cond.setPassword("321321");
    // 直接调用接口的方法
    uim.updateUserInfo2_trim(cond);
}
```

执行该测试方法,控制台输出如下:

```
DEBUG [main] - ==>  Preparing: update user_info ui set ui.userName=?, ui.password=? where ui.id=?
DEBUG [main] - ==> Parameters: mmm(String), 321321(String), 3(Integer)
DEBUG [main] - <==    Updates: 1
```

在上述测试方法中,如果没有给 password 属性指定值,<trim>元素会将要拼装的 SQL 语句片段尾部多余的逗号去除。再次执行测试方法,控制台输出的 SQL 语句如下:

```
DEBUG [main] - ==>  Preparing: update user_info ui set ui.userName=? where ui.id=?
```

14.5 <choose>、<when>和<otherwise>元素

在查询中,如果不想使用所有的条件,而只是想从多个选项中选择一个,可以使用 MyBatis 提供的<choose>、<when>和<otherwise>元素来实现。<choose>元素会按顺序判断<when>元素中的条件是否成立,如果有一个成立,则不再判断后面<when>元素中的条件是否成立,<choose>元素执行结束;如果所有<when>的条件都不满足,则执行<otherwise>元素中的 SQL 语句。

如果想从数据表 user_info 中根据 userName 或 status 进行查询,当 userName 不为空时则只按照 userName 查询,其他条件忽略;否则当 status 大于-1 时,则只按照 status 查询;当 userName 和 status 都为空时,则查询所有用户记录,使用<choose>、<when>和<otherwise>元素实现这个示例的步骤如下:

(1) 在映射文件 UserInfoMapper.xml 中,添加一个 id 为 findUserInfo_Choose 的<select>元素,如下所示:

```xml
<!-- 动态SQL之<choose>、<when>和<otherwise>元素 -->
<select id="findUserInfo_Choose" parameterType="UserInfo"
    resultType="UserInfo">
    select * from user_info ui
    <where>
        <choose>
            <when test="userName!=null and userName!=''">
```

```
                    ui.userName LIKE
                    CONCAT(CONCAT('%',#{userName}),'%')
                </when>
                <when test="status>-1">
                    and ui.status = #{status}
                </when>
                <otherwise>
                </otherwise>
            </choose>
        </where>
</select>
```

在上述<select>元素中，使用<choose>、<when>和<otherwise>元素编写了动态 SQL 语句。当第一个<when>元素中的条件成立时，只动态拼装第一个<when>元素中的 SQL 语句片段。否则，继续判断下一个<when>元素中的条件。当所有的<when>元素中的条件都不成立时，则只拼接<otherwise>元素内的 SQL 语句片段。

(2) 在接口 UserInfoMapper 中，声明一个方法，如下所示：

```
List<UserInfo> findUserInfo_Choose(UserInfo ui);
```

(3) 添加测试方法。

在测试类 MybatisTest 中，添加一个测试方法，并用@Test 注解修饰，如下所示：

```java
// 测试动态 SQL 之<choose>、<when>和<otherwise>元素
@Test
public void testFindUserInfo_Choose() {
    // 获得 UserInfoMapper 接口的代理对象
    UserInfoMapper uim = sqlSession.getMapper(UserInfoMapper.class);
    // 创建 UserInfo 对象,用于封装查询条件
    UserInfo cond = new UserInfo();
    cond.setUserName("j");
    cond.setStatus(1);
// 直接调用接口的方法
    List<UserInfo> uis = uim.findUserInfo_Choose(cond);
    for (UserInfo ui : uis) {
        System.out.println(ui.toString());
    }
}
```

(4) 执行该测试方法，控制台输出如下：

```
DEBUG [main] - ==>  Preparing: select * from user_info ui WHERE ui.userName LIKE CONCAT(CONCAT('%',?),'%')
DEBUG [main] - ==> Parameters: j(String)
DEBUG [main] - <==      Total: 3
UserInfo [id=2, userName=john, password=123456]
UserInfo [id=4, userName=sj, password=123456]
UserInfo [id=6, userName=lj, password=123456]
```

从控制台输出可以看出，虽然用户名和用户状态这两个条件都成立，但 MyBatis 所拼装的 SQL 语句中只包含用户名，用户状态这个条件被忽略了。

14.6 <foreach>元素

<foreach>元素主要是迭代一个集合,在 SQL 语句中通常用在 in 这个关键字的后面。例如,SQL 语句中的条件形如:where id in(id1, id2, …),这时可使用<foreach>元素,而不必去拼接 id 字符串。

<foreach>元素可以向 SQL 语句传递数组、List<E>等实例。List<E>实例使用 list 作为键,数组实例使用 array 作为键。

如果想从数据表 user_info 中查询 id 为 1 和 3 的用户记录,使用<foreach>元素的 List<E>实例的实现步骤如下:

(1) 在映射文件 UserInfoMapper.xml 中,添加一个 id 为 findUserInfoByIds 的<select>元素,如下所示:

```xml
<!-- 动态 SQL 之<foreach>元素 -->
<select id="findUserInfoByIds" resultType="UserInfo">
    select * from user_info ui where ui.id in
    <foreach collection="list" item="ids" open="(" separator=","
        close=")">
        #{ids}
    </foreach>
</select>
```

<foreach>元素的主要属性有 item、index、collection、open、separator 和 close 等。item 属性表示集合中每个元素迭代时的别名;index 属性指定一个变量名称,表示每次迭代到的位置;open 表示该语句的开始符号;separator 属性表示每次迭代之间的分隔符号;close 属性表示该语句的结束符号;collection 属性需要根据具体情况进行设置,通常有以下两种情况。

- 如果向 SQL 语句传递的是单参数且参数类型为 List<E>,collection 属性的值为 list。
- 如果向 SQL 语句传递的是单参数且参数类型为 array 数组,collection 属性的值为 array。

(2) 在接口 UserInfoMapper 中,声明一个方法,如下所示:

```
List<UserInfo> findUserInfoByIds(List<Integer> ids);
```

(3) 添加测试方法。

在测试类 MybatisTest 中,添加一个测试方法,并用@Test 注解修饰,如下所示:

```java
// 测试动态 SQL 之<foreach>元素
@Test
public void testFindUserInfoByIds() {
    // 获得 UserInfoMapper 接口的代理对象
    UserInfoMapper uim = sqlSession.getMapper(UserInfoMapper.class);
    // 创建集合对象 ids,保存用户 id
    List<Integer> ids = new ArrayList<Integer>();
    ids.add(1);
    ids.add(3);
```

```
        // 直接调用接口的方法
        List<UserInfo> uis = uim.findUserInfoByIds(ids);
        for (UserInfo ui : uis) {
            System.out.println(ui);
        }
    }
```

(4) 执行该测试方法,控制台输出如下:

```
DEBUG [main] - ==>  Preparing: select * from user_info ui where ui.id in ( ? , ? )
DEBUG [main] - ==> Parameters: 1(Integer), 3(Integer)
DEBUG [main] - <==      Total: 2
UserInfo [id=1, userName=tom, password=123456]
UserInfo [id=3, userName=mmm, password=321321]
```

从控制台输出可以看出,使用<foreach>元素的 List<E>实例,对传入的用户 id 集合进行动态 SQL 拼装,最终批量查询出了对应的用户信息。

除了 List<E>实例,使用<foreach>元素的 array 实例,也能实现从数据表 user_info 中查询 id 为 1 和 3 的用户信息,具体步骤如下:

(1) 在映射文件 UserInfoMapper.xml 中,添加一个 id 为 findUserInfoByIds2 的<select>元素,相关代码如下:

```xml
<select id="findUserInfoByIds2" resultType="UserInfo">
    select * from user_info ui where ui.id in
    <foreach collection="array" item="ids" open="(" separator=","
        close=")">
        #{ids}
    </foreach>
</select>
```

(2) 在接口 UserInfoMapper 中,声明一个方法,如下所示:

```java
List<UserInfo> findUserInfoByIds2(int[] ids);
```

(3) 添加测试方法。

在测试类 MybatisTest 中,添加一个测试方法,并用@Test 注解修饰,如下所示:

```java
@Test
public void testFindUserInfoByIds2() {
    // 获得 UserInfoMapper 接口的代理对象
    UserInfoMapper uim = sqlSession.getMapper(UserInfoMapper.class);
    // 创建数组对象 ids,保存用户 id
    int[] ids = new int[2];
    ids[0] = 1;
    ids[1] = 3;
    // 直接调用接口的方法
    List<UserInfo> uis = uim.findUserInfoByIds2(ids);
    for (UserInfo ui : uis) {
        System.out.println(ui);
    }
}
```

(4) 执行该测试方法，控制台输出如下：

```
DEBUG [main] - ==>  Preparing: select * from user_info ui where ui.id in ( ? , ? )
DEBUG [main] - ==> Parameters: 1(Integer), 3(Integer)
DEBUG [main] - <==      Total: 2
UserInfo [id=1, userName=tom, password=123456]
UserInfo [id=3, userName=mmm, password=321321]
```

从控制台输出可以看出，使用<foreach>元素的 array 实例对传入的用户 id 集合进行动态 SQL 拼装，最终批量查询出了对应的用户信息。

14.7 小　　结

本章首先介绍了 MyBatis 框架的动态 SQL 语句及动态 SQL 语句的主要元素，然后结合具体示例，对这些元素进行了详细的讲解。在 MyBatis 框架中，动态 SQL 语句比较重要，熟练使用可以提高开发效率。

第 15 章　MyBatis 的注解配置

　　MyBatis 是一个 XML 驱动的框架，前面介绍的 MyBatis 框架的增删改查、关联映射、动态 SQL 语句等知识，其所有的配置都是通过 XML 完成的，编写大量的 XML 配置比较烦琐。到了 MyBatis 3，可以使用 MyBatis 提供的基于注解的配置方式。本章基于注解，详细讲解 MyBatis 框架的增删改查、关联映射、动态 SQL 语句等知识。

15.1　基于注解的单表增删改查

　　MyBatis 提供了@Insert、@Delete、@Update 和@Select 等常用注解，可以实现数据的增、删、改、查等操作。以数据表 user_info 为例，基于注解实现增删改查操作的具体步骤如下：

　　(1) 将项目 mybatis1 复制并命名为 mybatis8，再导入到 Eclipse 开发环境中。
　　(2) 将项目 mybatis8 的 com.mybatis.pojo 包中的映射文件 UserInfoMapper.xml 删除。
　　(3) 在 com.mybatis.mapper 包中，新建一个接口 UserInfoMapper，编写如下代码：

```
package com.mybatis.mapper;
import java.util.List;
import org.apache.ibatis.annotations.Delete;
import org.apache.ibatis.annotations.Insert;
import org.apache.ibatis.annotations.Select;
import org.apache.ibatis.annotations.Update;
```

```java
import com.mybatis.pojo.UserInfo;
public interface UserInfoMapper {
    @Select("select * from user_info where id=#{id}")
    public UserInfo findUserInfoById(int id);
    @Select("select * from user_info where userName like CONCAT(CONCAT('%',#{userName}),'%')")
    public List<UserInfo> findUserInfoByUserName(String userName);
    @Insert("insert into user_info(userName,password) values(#{userName},#{password})")
    public int addUserInfo(UserInfo ui);
    @Update("update user_info set userName=#{userName},password=#{password} where id=#{id}")
    public int updateUserInfo(UserInfo ui);
    @Delete("delete from user_info where id=#{id}")
    public int deleteUserInfo(int id);
}
```

在接口 UserInfoMapper 中，声明了 findUserInfoById、findUserInfoByUserName、addUserInfo、updateUserInfo 和 deleteUserInfo 五个方法。其中，findUserInfoById 方法用于根据用户编号查询用户，findUserInfoByUserName 方法用于根据用户名模糊查询用户，addUserInfo 方法用于添加用户，updateUserInfo 方法用于修改用户，deleteUserInfo 方法用于删除用户。这五个方法分别使用@Select、@Insert、@Update、@Delete 注解替代了之前映射文件中的 XML 配置，@Select 注解用于映射 select 语句，@Insert 注解用于映射 insert 语句，@Update 注解用于映射 update 语句，@Delete 注解用于映射 delete 语句。对@Select 注解来说，会涉及查询结果的映射。如果实体类属性和数据表字段名保持一致，MyBatis 会自动完成结果映射。如果不一致，则需要使用@Results 注解手动完成结果映射。后面的示例将会用到@Results 注解，到时再作具体讲解。

(4) 修改 MyBatis 配置文件。

在 mybatis-config.xml 文件的<mappers>元素中，先将原先对 UserInfoMapper.xml 映射文件引入的配置删除或注释掉，再添加对接口 UserInfoMapper 的引用，如下所示：

```xml
<!-- 引用接口 -->
<mappers>
<!-- <mapper resource="com/mybatis/mapper/UserInfoMapper.xml" /> -->
    <mapper class="com.mybatis.mapper.UserInfoMapper" />
</mappers>
```

(5) 修改测试类 MybatisTest 中的测试方法。

testFindUserInfoById 方法修改如下所示：

```java
// 根据 id 查询用户
@Test
public void testFindUserInfoById() {
    // 获得 UserInfoMapper 接口的代理对象
    UserInfoMapper um = sqlSession.getMapper(UserInfoMapper.class);
    // 直接调用接口的方法
    UserInfo ui = um.findUserInfoById(1);
    // 打印输出结果
```

```java
        System.out.println(ui.toString());
    }
}
```

testFindUserInfoByUserName 方法修改如下所示:

```java
// 根据用户名模糊查询用户
@Test
public void testFindUserInfoByUserName() {
    // 获得 UserInfoMapper 接口的代理对象
    UserInfoMapper um = sqlSession.getMapper(UserInfoMapper.class);
    // 直接调用接口的方法
    List<UserInfo> uis = um.findUserInfoByUserName("j");
    for (UserInfo ui : uis) {
        // 打印输出结果
        System.out.println(ui.toString());
    }
}
```

testAddUserInfo 方法修改如下所示:

```java
// 添加用户
@Test
public void testAddUserInfo() {
    // 获得 UserInfoMapper 接口的代理对象
    UserInfoMapper um = sqlSession.getMapper(UserInfoMapper.class);
    // 创建 UserInfo 对象 ui
    UserInfo ui = new UserInfo();
    // 向对象 ui 中添加数据
    ui.setUserName("mybatis1");
    ui.setPassword("123456");
    // 直接调用接口的方法
    int result = um.addUserInfo(ui);
    if (result > 0) {
        System.out.println("插入成功");
    } else {
        System.out.println("插入失败");
    }
}
```

testUpdateUserInfo 方法修改如下所示:

```java
// 修改用户
@Test
public void testUpdateUserInfo() {
    // 获得 UserInfoMapper 接口的代理对象
    UserInfoMapper um = sqlSession.getMapper(UserInfoMapper.class);
    // 加载编号为 8 的用户
    UserInfo ui = um.findUserInfoById(8);
    // 重新设置用户密码
    ui.setPassword("123123");
    // 直接调用接口的方法
    int result = um.updateUserInfo(ui);
    if (result > 0) {
```

```
            System.out.println("插入成功");
            System.out.println(ui);
        } else {
            System.out.println("插入失败");
        }
    }
```

testDeleteUserInfo 方法修改如下所示：

```
// 删除用户
@Test
public void testDeleteUserInfo() {
    // 获得 UserInfoMapper 接口的代理对象
    UserInfoMapper um = sqlSession.getMapper(UserInfoMapper.class);
    // 直接调用接口的方法
    int result = um.deleteUserInfo(8);
    if (result > 0) {
        System.out.println("成功删除了" + result + "条记录");
    } else {
        System.out.println("插入删除");
    }
}
```

这些测试方法的执行结果与 12.4 小节中使用映射文件映射 SQL 语句时相同。通过这个示例可以看出，使用注解配置无需编写映射文件，只需在接口的方法上使用注解实现对 SQL 语句的映射，从而简化了程序的开发。

15.2　基于注解的一对一关联映射

使用 MyBatis 的注解配置，除了可以实现单表的增删改查操作外，还可以实现多表的关联映射。以 13.1 小节使用的数据表 idcard 和 person 为例，基于注解配置实现这两张表之间的一对一关联映射，具体步骤如下：

（1）将项目 mybatis4 中与数据表 idcard 和 person 对应的实体类 Idcard.java 和 Person.java 复制到项目 mybatis8 的 com.mybatis.pojo 包中。

（2）在 com.mybatis.mapper 包中，创建接口 IdcardMapper 和 PersonMapper。在接口 IdcardMapper 中添加一个 findIdcardById 方法，如下所示：

```
package com.mybatis.mapper;
import org.apache.ibatis.annotations.Select;
import com.mybatis.pojo.Idcard;
public interface IdcardMapper {
    // 根据 id 查询身份证信息
    @Select("select * from idcard where id=#{id}")
    public Idcard findIdcardById(int id);
}
```

在接口 PersonMapper 中添加一个 findPersonById 方法，如下所示：

```java
package com.mybatis.mapper;
import org.apache.ibatis.annotations.One;
import org.apache.ibatis.annotations.Result;
import org.apache.ibatis.annotations.Results;
import org.apache.ibatis.annotations.Select;
import com.mybatis.pojo.Person;
public interface PersonMapper {
    @Select("select * from person where id = #{id} ")
    @Results({
            @Result(column = "cid", property = "idcard", one = @One(select 
= "com.mybatis.mapper.IdcardMapper.findIdcardById")) })
    // 根据id查询个人信息
    public Person findPersonById(int id);
}
```

在 findPersonById 方法中，使用@Select 注解映射根据 id 查询 Person 对象的 SQL 语句，使用@Results 注解映射查询结果。在@Results 注解中，可以包含多个@Result 注解，一个@Result 注解完成实体类中一个属性和数据表中一个字段的映射。

Person 对象中的基本属性可以自动完成结果映射，而关联的对象属性 idcard 需要手工完成映射。这里，在@Results 注解中使用了一个@Result 注解来映射关联结果。在@Result 注解中，property 属性用来指定关联属性，这里为 idcard；one 属性用来指定数据表属于哪种关联关系，通过@One 注解表明数据表 idcard 和 person 之间是一对一关联关系。在@One 注解中，select 属性用于指定关联属性 idcard 的值是通过执行 com.mybatis.mapper 包中 IdcardMapper 接口里定义的 findIdcardById 方法获得的。

@Result 注解的 column 属性用于指定传入 findIdcardById(int id)方法的参数名，这里为 cid，表示从数据表 person 查询出的 cid 字段值。

(3) 添加接口文件的引用。

在 mybatis-config.xml 文件中，添加对接口 IdcardMapper 和 PersonMapper 的引用，如下所示：

```xml
<!-- 引用接口 -->
<mappers>
    ……
    <mapper class="com.mybatis.mapper.IdcardMapper" />
    <mapper class="com.mybatis.mapper.PersonMapper" />
</mappers>
```

(4) 测试一对一关联映射。

在测试类 MybatisTest 中，添加一个测试方法 testOne2One()，使用@Test 注解修饰，从数据表 person 中查询记录的同时，查询关联表 idcard 中的身份证信息。其代码如下：

```java
@Test
public void testOne2One() {
    // 获得PersonMapper接口的代理对象
    PersonMapper pm = sqlSession.getMapper(PersonMapper.class);
    // 直接调用接口中的方法，根据id查询Person对象及关联的Idcard对象
    Person person = pm.findPersonById(1);
    // 查看Person对象及关联的Idcard对象
```

```
        System.out.println(person.toString());
}
```

执行该测试方法，控制台输出如下：

```
DEBUG [main] - ==>  Preparing: select * from person where id = ?
DEBUG [main] - ==> Parameters: 1(Integer)
DEBUG [main] - ====>  Preparing: select * from idcard where id=?
DEBUG [main] - ====> Parameters: 1(Integer)
DEBUG [main] - <====      Total: 1
DEBUG [main] - <==       Total: 1
Person [id=1, name=zhangsan, age=22, sex=男, idcard=Idcard [id=1,
cno=320100197001010001]]
```

使用 MyBatis 注解配置，在查询 Person 对象时，关联的 Idcard 对象也查询出来了。

15.3　基于注解的一对多关联映射

以 13.2 小节使用的商品类别表 type 和商品表 product_info 为例，基于注解配置实现这两张表之间的一对多关联映射，具体步骤如下：

（1）将项目 mybatis5 中与数据表 type 和 product_info 对应的实体类 Type.java 和 ProductInfo.java 复制到项目 mybatis8 的 com.mybatis.pojo 包中。

（2）在 com.mybatis.mapper 包中，创建接口 ProductInfoMapper 和 TypeMapper。在 ProductInfoMapper 接口中，编写如下代码：

```java
package com.mybatis.mapper;
import java.util.List;
import org.apache.ibatis.annotations.One;
import org.apache.ibatis.annotations.Result;
import org.apache.ibatis.annotations.Results;
import org.apache.ibatis.annotations.Select;
import com.mybatis.pojo.ProductInfo;
public interface ProductInfoMapper {
    // 根据类型编号查询所有商品
    @Select("select * from product_info where tid = #{tid} ")
    List<ProductInfo> findProductInfoByTid(int tid);
    // 根据商品编号获取商品信息
    @Select("select * from product_info where id = #{id} ")
    @Results({
            @Result(column = "tid", property = "type", one = @One(select =
"com.mybatis.mapper.TypeMapper.findTypeById")) })
    ProductInfo findProductInfoByid(int id);
}
```

在接口 ProductInfoMapper 中，首先声明了一个 findProductInfoByTid 方法，通过@Select 注解配置，根据商品类型编号查询所有商品信息。

然后声明了一个 findProductInfoByid 方法，通过@Select 注解配置，根据商品编号获取商品信息。如果希望在查询 ProductInfo 对象时，将关联的 Type 对象也查询出来，可以使用

@Results 注解手工完成关联属性的映射。在@Results 注解中，添加一个@Result 注解，property 属性用来指定关联属性，这里为 type；one 属性用来指定数据表属于哪种关联关系，通过@One 注解表明数据表 product_info 和 type 之间是一对一关联关系。在@One 注解中，select 属性用于指定关联属性 type 的值是通过执行 com.mybatis.mapper 包中 TypeMapper 接口里定义的 findTypeById 方法获得的。@Result 注解的 column 属性设置为 tid，表示传入 findTypeById 方法的参数为从数据表 product_info 查询出的 tid 字段值。

在接口 TypeMapper 中，编写如下代码：

```java
package com.mybatis.mapper;
import org.apache.ibatis.annotations.Many;
import org.apache.ibatis.annotations.Result;
import org.apache.ibatis.annotations.Results;
import org.apache.ibatis.annotations.Select;
import com.mybatis.pojo.Type;
public interface TypeMapper {
    // 根据商品类型编号查询商品类型信息
    @Select("select * from type where id = #{id} ")
    @Results({ @Result(id = true, column = "id", property = "id"),
    @Result(column = "name", property = "name"),
               @Result(column = "id", property = "pis", many = @Many(select = "com.mybatis.mapper.ProductInfoMapper.findProductInfoByTid")) })
    Type findTypeById(int id);
}
```

在接口 TypeMapper 中，声明了一个 findTypeById 方法，通过@Select 注解配置，根据商品类型编号查询商品类型信息。如果希望在查询 Type 对象时，将关联的 ProductInfo 对象也查询出来，可以使用@Results 注解手工完成关联属性的映射。在@Results 注解中，添加了三个@Result 注解。前两个@Result 注解用于完成 Type 对象的基本属性 id 和 name 与数据表 type 的 id 和 name 字段的映射，最后一个@Result 注解用于关联属性的映射。在最后一个@Result 注解中，property 属性用来指定关联属性，这里为 pis；many 属性用来指定数据表属于哪种关联关系，通过@Many 注解表明数据表 type 和 product_info 之间是一对多关联关系。在 @Many 注解中，select 属性用于指定关联属性 pis 的值是通过执行 com.mybatis.mapper 包中 ProductInfoMapper.java 接口里定义的方法 findProductInfoByTid 获得的。@Result 注解的 column 属性设置为 id，表示传入 findProductInfoByTid(int tid)方法的参数为从数据表 type 查询出的 id 字段值。

(3) 添加接口文件的引用。

在 mybatis-config.xml 文件中，添加对接口 ProductInfoMapper 和 TypeMapper 的引用，如下所示：

```xml
<!-- 引用接口 -->
<mappers>
    ……
    <mapper class="com.mybatis.mapper.TypeMapper" />
    <mapper class="com.mybatis.mapper.ProductInfoMapper" />
</mappers>
```

(4) 测试一对多关联映射。

在测试类 MybatisTest 中，添加测试方法 testOne2Many，使用@Test 注解修饰，从数据表 type 查询记录的同时，获取关联的数据表 product_info 中的记录。其代码如下：

```
@Test
public void testOne2Many() {
    // 获得 TypeMapper 接口的代理对象
    TypeMapper tm = sqlSession.getMapper(TypeMapper.class);
    // 直接调用接口的方法，查询 Type 对象及关联的 ProductInfo 对象
    Type type = tm.findTypeById(1);
    System.out.println(type.toString());
}
```

执行测试方法 testOne2Many，控制台输出如下所示：

```
DEBUG [main] - ==>  Preparing: select * from type where id = ?
DEBUG [main] - ==> Parameters: 1(Integer)
DEBUG [main] - ====>  Preparing: select * from product_info where tid = ?
DEBUG [main] - ====> Parameters: 1(Integer)
DEBUG [main] - <====      Total: 6
DEBUG [main] - <==      Total: 1
Type [id=1, name=电脑, pis=[ProductInfo [id=1, code=1378538,
name=AppleMJVE2CH/A], ProductInfo [id=2, code=1309456,
name=ThinkPadE450C(20EH0001CD)], ProductInfo [id=3, code=1999938, name=联
想小新 300 经典版], ProductInfo [id=4, code=1466274, name=华硕 FX50JX],
ProductInfo [id=5, code=1981672, name=华硕 FL5800], ProductInfo [id=6,
code=1904696, name=联想 G50-70M]]]
```

可以看到，查询 Type 对象时关联的 ProductInfo 对象也查询出来了。

(5) 测试多对一关联映射。

在测试类 MybatisTest 中，添加测试方法 testMany2One()，使用@Test 注解修饰，从数据表 product_info 查询记录的同时，获取关联的数据表 type 中的记录。

```
public void testMany2One() {
    // 获得 ProductInfoMapper 接口的代理对象
    ProductInfoMapper pim = sqlSession.getMapper(ProductInfoMapper.class);
    // 直接调用接口的方法，查询 ProductInfo 对象
    ProductInfo pi = pim.findProductInfoByid(1);
    System.out.println(pi.toString());
    // 查看关联的 Type 对象
    System.out.println(pi.getType());
}
```

执行测试 testMany2One()方法，控制台输出如下所示：

```
DEBUG [main] - ==>  Preparing: select * from product_info where id = ?
DEBUG [main] - ==> Parameters: 1(Integer)
DEBUG [main] - ====>  Preparing: select * from type where id = ?
DEBUG [main] - ====> Parameters: 1(Integer)
DEBUG [main] - ======>  Preparing: select * from product_info where tid = ?
DEBUG [main] - ======> Parameters: 1(Integer)
DEBUG [main] - <======      Total: 6
```

```
DEBUG [main] - <====       Total: 1
DEBUG [main] - <==        Total: 1
ProductInfo [id=1, code=1378538, name=AppleMJVE2CH/A]
Type [id=1, name=电脑, pis=[……]]
```

可以看到，查询 ProductInfo 对象时关联的 Type 对象也查询出来了。

15.4　基于注解的多对多关联映射

以 13.3 小节使用的数据表 admin_info 和 functions 为例，基于注解配置实现这两张表之间的多对多关联映射，具体步骤如下所示。

（1）将项目 mybatis6 中与数据表 admin_info 和 functions 对应的实体类 AdminInfo.java 和 Functions.java 复制到项目 mybatis8 的 com.mybatis.pojo 包中。

（2）在 com.mybatis.mapper 包中创建接口 AdminInfoMapper 和 FunctionsMapper。在 FunctionsMapper 接口中，编写如下代码：

```java
package com.mybatis.mapper;
import java.util.List;
import org.apache.ibatis.annotations.Select;
import com.mybatis.pojo.Functions;
public interface FunctionsMapper {
    // 根据管理员id获取其功能权限列表
    @Select("select * from functions where id in (select fid from powers where aid = #{id} )")
    List<Functions> findFunctionsByAid(int aid);
}
```

在 FunctionsMapper 接口中，声明了一个 findFunctionsByAid 方法，使用 @Select 注解映射了一个包含子查询的 select 语句，根据管理员编号获取其功能权限列表。

在 AdminInfoMapper 接口中，编写如下代码：

```java
package com.mybatis.mapper;
import org.apache.ibatis.annotations.Many;
import org.apache.ibatis.annotations.Result;
import org.apache.ibatis.annotations.Results;
import org.apache.ibatis.annotations.Select;
import com.mybatis.pojo.AdminInfo;
public interface AdminInfoMapper {
    // 根据管理员id查询管理员信息
    @Select("select * from admin_info where id = #{id} ")
    @Results({
        @Result(id = true, column = "id", property = "id"),
        @Result(column = "name", property = "name"),
        @Result(column = "id", property = "fs",
            many = @Many(select =
"com.mybatis.mapper.FunctionsMapper.findFunctionsByAid")) })
    public AdminInfo findAdminInfoById(int id);
}
```

在 AdminInfoMapper 接口中，声明了一个 findAdminInfoById 方法，使用@Select 注解映射一个 select 语句，根据管理员编号查询管理员对象。

如果想将 AdminInfo 关联的 Functions 对象也查出来，可以使用@Results 注解手工完成关联属性的映射。在@Results 注解中，添加了三个@Result 注解。前两个@Result 注解用于完成 AdminInfo 对象的基本属性 id 和 name 与数据表 admin_info 的 id 和 name 字段的映射，最后一个@Result 注解用于关联属性的映射。在最后一个@Result 注解中，property 属性用来指定关联属性，这里为 fs；many 属性用来指定数据表属于哪种关联关系，通过@Many 注解表明数据表 admin_info 和 functions 之间是一对多关联关系。在@Many 注解中，select 属性用于指定关联属性 fs 的值是通过执行 com.mybatis.mapper 包中 FunctionsMapper 接口里定义的 findFunctionsByAid 方法获得的。@Result 注解的 column 属性设置为 id，表示传入 findFunctionsByAid(int aid)方法的参数为从数据表 admin_info 查询出的 id 字段值。

(3) 添加接口文件的引用。

在 mybatis-config.xml 文件中，添加对接口 AdminInfoMapper 和 FunctionsMapper 的引用，如下所示：

```xml
<!-- 引用接口 -->
<mappers>
    ……
    <mapper class="com.mybatis.mapper.FunctionsMapper" />
    <mapper class="com.mybatis.mapper.AdminInfoMapper" />
</mappers>
```

(4) 测试多对多关联映射。

在测试类 MybatisTest 中，添加测试方法 testM2M，使用@Test 注解修饰，查看管理员及其功能权限。其代码如下：

```java
@Test
public void testM2M() {
    // 获得 AdminInfoMapper 接口的代理对象
    AdminInfoMapper aim = sqlSession.getMapper(AdminInfoMapper.class);
    // 直接调用接口的方法，查询 AdminInfo 对象及关联的 Functions 对象
    AdminInfo admin = aim.findAdminInfoById(1);
    System.out.println(admin);
}
```

执行测试方法 testM2M，控制台输出如下所示：

```
DEBUG [main] - ==>  Preparing: select * from admin_info where id = ?
DEBUG [main] - ==> Parameters: 1(Integer)
DEBUG [main] - ====>  Preparing: select * from functions where id in (select fid from powers where aid = ? )
DEBUG [main] - ====> Parameters: 1(Integer)
DEBUG [main] - <====      Total: 10
DEBUG [main] - <==      Total: 1
AdminInfo [id=1, name=admin, fs=[Functions [id=1, name=电子商城管理后台], Functions [id=2, name=商品管理], Functions [id=3, name=商品列表], Functions [id=4, name=商品类型列表], Functions [id=5, name=订单管理], Functions [id=6,
```

name=查询订单], Functions [id=7, name=创建订单], Functions [id=8, name=用户管理], Functions [id=9, name=用户列表], Functions [id=11, name=退出系统]]]

可以看到，查询 AdminInfo 对象时关联的 Functions 对象也查询出来了。

15.5 基于注解的动态 SQL

使用 MyBatis 的注解配置，除了可以实现单表的增删改查操作、多表的关联映射外，还可以实现动态 SQL 语句。MyBatis 提供了 @SelectProvider、@InsertProvider、@UpdateProvider 和 @DeleteProvider 等注解来构建动态 SQL 语句，然后再执行这些 SQL 语句。接下来将以数据表 user_info 的增删改查为例来讲解这四个注解。

15.5.1 @SelectProvider 注解

@SelectProvider 注解用于生成查询所用的 SQL 语句，与 @Select 注解不同，@SelectProvide 注解指定一个类及其方法，并通过调用类的这个方法来获得 SELECT 语句。使用 @SelectProvider 注解的好处在于，可以根据不同的需求产生不同的 SQL 语句，因此适用性更好。

以数据表 user_info 的查询操作为例，使用 @SelectProvider 注解动态查询数据的具体步骤如下：

(1) 将项目 mybatis1 复制并命名为 mybatis9，再导入到 Eclipse 开发环境中。
(2) 将项目 mybatis9 的 com.mybatis.pojo 包中的映射文件 UserInfoMapper.xml 删除。
(3) 在 com.mybatis.mapper 包中，新建一个接口 UserInfoMapper，并编写如下代码：

```
package com.mybatis.mapper;
import java.util.List;
import java.util.Map;
import org.apache.ibatis.annotations.SelectProvider;
import com.mybatis.pojo.UserInfo;
public interface UserInfoMapper {
    @SelectProvider(type = UserInfoDynaSqlProvider.class, method = "selectWithParam")
    List<UserInfo> findUserInfoByCond(Map<String, Object> param);
}
```

在 findUserInfoByCond 方法中，Map<String, Object> 类型的参数 param 用于封装查询条件。与 @Select 注解不同，@SelectProvider 注解没有直接提供映射的 SQL 语句，而是指定使用另一个 UserInfoDynaSqlProvider.java 类中定义的方法 selectWithParam，该方法提供需要执行的 SELECT 语句。

(4) 在 com.mybatis.mapper 包中，新建一个类 UserInfoDynaSqlProvider，并添加 selectWithParam 方法，如下所示：

```
package com.mybatis.mapper;
import java.util.Map;
import org.apache.ibatis.jdbc.SQL;
```

```java
public class UserInfoDynaSqlProvider {
    public String selectWithParam(Map<String, Object> param) {
        return new SQL() {
            {
                SELECT("*");
                FROM("user_info");
                if (param.get("id") != null) {
                    WHERE("id = #{id} ");
                }
                if (param.get("userName") != null) {
                    WHERE("userName = #{userName} ");
                }
                if (param.get("password") != null) {
                    WHERE("password = #{password} ");
                }
            }
        }.toString();
    }
}
```

在 selectWithParam(Map<String, Object> param)方法中，根据参数 param 中的内容构建动态 SELECT 语句。

(5) 添加接口文件的引用。

在 mybatis-config.xml 文件中，先将原先引用的映射文件 UserInfoMapper.xml 删除或注释掉，再添加对接口 UserInfoMapper 的引用，如下所示：

```xml
<!-- 引用接口文件 -->
<mappers>
<!-- <mapper resource="com/mybatis/mapper/UserInfoMapper.xml" /> -->
    <mapper class="com.mybatis.mapper.UserInfoMapper" />
</mappers>
```

(6) 添加测试方法。

在测试类 MybatisTest 中，添加一个测试方法 testfindUserInfoByCond，使用@Test 注解修饰，编写如下代码：

```java
// 测试基于注解的动态 SQL 语句之@SelectProvider 注解
@Test
public void testfindUserInfoByCond() {
    // 获得 UserInfoMapper 接口的代理对象
    UserInfoMapper uim = sqlSession.getMapper(UserInfoMapper.class);
    // 使用 Map 类型对象封装查询条件
    Map<String, Object> param = new HashMap<String, Object>();
    param.put("userName", "tom");
    param.put("password", "123456");
    // 直接调用接口的方法
    List<UserInfo> uis = uim.findtUserInfoByCond(param);
    for (UserInfo ui : uis) {
        // 打印输出结果
        System.out.println(ui.toString());
    }
}
```

执行测试方法 testfindUserInfoByCond，控制台输出如下：

```
DEBUG [main] - ==>  Preparing: SELECT * FROM user_info WHERE (userName = ? AND password = ? )
DEBUG [main] - ==> Parameters: tom(String), 123456(String)
DEBUG [main] - <==      Total: 1
UserInfo [id=1, userName=tom, password=123456]
```

可以看出，由于 Map 中设置了 userName 和 password 两个参数，因此执行的 SQL 语句中包含这两个条件。如果 Map 中只设置 userName 参数，则控制台输出的 SQL 语句中只包含 userName 这个条件，如下所示：

```
DEBUG [main] - ==>  Preparing: SELECT * FROM user_info WHERE (userName = ? )
```

当然，UserInfoDynaSqlProvider 类中的 selectWithParam 方法也可以传递 UserInfo 类型对象作为参数。也就是说，动态 SQL Provider 方法可无参，也可用 Java 对象或 Map 对象作为参数。

15.5.2 @InsertProvider 注解

@InsertProvider 注解用于生成插入数据所用的 SQL 语句，与@Insert 注解不同，@InsertProvider 注解指定一个类及其方法，并通过调用类的这个方法来获得 INSERT 语句。

以数据表 user_info 的插入操作为例，使用@InsertProvider 注解动态添加数据的具体步骤如下：

（1）在接口 UserInfoMapper 中，添加一个 insertUserInfo 方法，如下所示：

```
@InsertProvider(type = UserInfoDynaSqlProvider.class, method = "insertUserInfo")
@Options(useGeneratedKeys = true, keyProperty = "id")
int insertUserInfo(UserInfo ui);
```

数据表 user_info 有一个自增的主键 id，为了能在插入数据后自动获取该主键值，可以使用@Options 注解返回添加的主键值。@Options 注解的 keyProperty 属性用来设置主键对应的字段名，这里为 id；将 useGeneratedKeys 属性设置为 true。这样，在向数据表 user_info 插入数据时，自动将主键字段 id 的自增值赋值给对象 ui 的属性 id。

@InsertProvider 注解指定 UserInfoDynaSqlProvider 类中定义的方法 insertUserInfo，由该方法提供需要执行的 INSERT 语句。

（2）在 UserInfoDynaSqlProvider 类中，添加一个 insertUserInfo 方法，如下所示：

```
public String insertUserInfo(UserInfo ui) {
    return new SQL() {
        {
            INSERT_INTO("user_info");
            if (ui.getUserName() != null) {
                VALUES("userName", "#{userName}");
            }
            if (ui.getPassword() != null) {
```

```
                VALUES("password", "#{password}");
            }
        }
    }.toString();
}
```

在 insertUserInfo(UserInfo ui)方法中，根据参数 ui 中的内容构建动态 INSERT 语句。

（3）添加测试方法。

在测试类 MybatisTest 中，添加一个测试方法 testInsertUserInfo，使用@Test 注解修饰，编写如下代码。

```
// 测试基于注解的动态 SQL 语句之@InsertProvider 注解
@Test
public void testInsertUserInfo() {
    // 获得 UserInfoMapper 接口的代理对象
    UserInfoMapper uim = sqlSession.getMapper(UserInfoMapper.class);
    // 创建 UserInfo 对象并初始化
    UserInfo ui = new UserInfo();
    ui.setUserName("mybatis2");
    ui.setPassword("123456");
    // 直接调用接口的方法
    uim.insertUserInfo(ui);
    // 输出数据表 user_info 中新插入记录的 id
    System.out.println("插入的用户编号：" + ui.getId());
}
```

执行该测试方法，控制台输出如下：

```
DEBUG [main] - ==>  Preparing: INSERT INTO user_info (userName, password) VALUES (?, ?)
DEBUG [main] - ==> Parameters: mybatis2(String), 123456(String)
DEBUG [main] - <==    Updates: 1
插入的用户编号：9
```

读者可以通过只设置 userName 或 password 参数，来观察控制台中的 SQL 语句。

15.5.3 @UpdateProvider 注解

@UpdateProvider 注解用于生成更新所用的 SQL 语句，与@Update 注解不同，@UpdateProvider 注解指定一个类及其方法，并通过调用类的这个方法来获得 UPDATE 语句。

以数据表 user_info 的更新操作为例，使用@UpdateProvider 注解动态更新数据的具体步骤如下：

（1）在接口 UserInfoMapper 中，添加一个 updateUserInfo 方法，代码如下：

```
// 更新数据
@UpdateProvider(type = UserInfoDynaSqlProvider.class, method = "updateUserInfo")
int updateUserInfo(UserInfo ui);
```

@InsertProvider 注解指定了 UserInfoDynaSqlProvider 类中定义的方法 updateUserInfo，由该方法提供需要执行的 UPDATE 语句。

(2) 在 UserInfoDynaSqlProvider 类中，添加一个 updateUserInfo 方法，代码如下：

```java
public String updateUserInfo(UserInfo ui) {
    return new SQL() {
        {
            UPDATE("user_info");
            if (ui.getUserName() != null) {
                SET("userName = #{userName}");
            }
            if (ui.getPassword() != null) {
                SET("password = #{password}");
            }
            WHERE("id = #{id} ");
        }
    }.toString();
}
```

在 updateUserInfo(UserInfo ui)方法中，根据参数 ui 中的内容构建动态 UPDATE 语句。

(3) 添加测试方法。

在测试类 MybatisTest 中，添加一个测试方法 testUpdateUser，使用@Test 注解修饰，编写如下代码：

```java
// 测试基于注解的动态 SQL 语句之@UpdateProvider 注解
@Test
public void testUpdateUser() {
    // 获得 UserInfoMapper 接口的代理对象
    UserInfoMapper uim = sqlSession.getMapper(UserInfoMapper.class);
    // 使用 Map 封装查询条件
    Map<String, Object> param = new HashMap<String, Object>();
    param.put("id", "1");
    // 直接调用接口的方法，查询 id 为 1 的用户
    UserInfo ui = uim.findUserInfoByCond(param).get(0);
    // 修改该用户的密码
    ui.setPassword("666666");
    // 直接调用接口的方法，更新用户信息
    uim.updateUserInfo(ui);
}
```

在该测试方法中，使用 Map 对象封装了 id 为 1 的用户编号作为查询条件，先根据用户编号获取用户信息，再重新设置密码，最后更改用户信息。

执行该测试方法，控制台输出如下：

```
DEBUG [main] - ==>  Preparing: SELECT * FROM user_info WHERE (id = ? )
DEBUG [main] - ==> Parameters: 1(String)
DEBUG [main] - <==      Total: 1
DEBUG [main] - ==>  Preparing: UPDATE user_info SET userName = ?, password = ? WHERE (id = ? )
DEBUG [main] - ==> Parameters: tom(String), 666666(String), 1(Integer)
```

```
DEBUG [main] - <==      Updates: 1
```

此时,打开数据表 user_info,可以看到编号为 1 的用户的密码已被成功修改。

15.5.4 @DeleteProvider 注解

@DeleteProvider 注解用于生成删除所用的 SQL 语句,与@Delete 注解不同,@DeleteProvider 注解指定一个类及其方法,并通过调用类的这个方法来获得 DELETE 语句。

以数据表 user_info 的删除操作为例,使用@DeleteProvider 注解动态删除数据的具体步骤如下:

(1) 在接口 UserInfoMapper 中,添加一个 deleteUserInfo 方法,如下所示:

```
// 删除数据
@DeleteProvider(type = UserInfoDynaSqlProvider.class, method = 
"deleteUserInfo")
void deleteUserInfo(Map<String, Object> param);
```

@DeleteProvider 注解指定了 UserInfoDynaSqlProvider 类中定义的方法 deleteUserInfo,由该方法提供需要执行的 DELETE 语句。

(2) 在 UserInfoDynaSqlProvider 类中,添加一个 deleteUserInfo 方法,如下所示:

```
public String deleteUserInfo(Map<String, Object> param) {
    return new SQL() {
        {
            DELETE_FROM("user_info");
            if (param.get("id") != null) {
                WHERE("id = #{id} ");
            }
            if (param.get("userName") != null) {
                WHERE("userName = #{userName} ");
            }
            if (param.get("password") != null) {
                WHERE("password = #{password} ");
            }
        }
    }.toString();
}
```

在 deleteUserInfo(Map<String, Object> param)方法中,根据参数 param 中的内容构建动态 DELETE 语句。

(3) 添加测试方法。

在测试类 MybatisTest 中,添加一个测试方法 testDeleteUser,使用@Test 注解修饰,编写如下代码:

```
// 测试基于注解的动态 SQL 语句之@DeleteProvider 注解
@Test
public void testDeleteUser() {
    // 获得 UserInfoMapper 接口的代理对象
    UserInfoMapper uim = sqlSession.getMapper(UserInfoMapper.class);
```

```
        // 使用 Map 封装查询条件
        Map<String, Object> param = new HashMap<String, Object>();
        param.put("userName", "mybatis2");
        param.put("password", "123456");
        // 直接调用接口的方法，删除符合条件的用户
        uim.deleteUserInfo(param);
    }
```

在该测试方法中，首先使用 Map 封装了用户名和密码这两个查询条件，然后调用接口的方法，删除符合条件的用户。

执行该测试方法，控制台输出如下：

```
DEBUG [main] - ==>  Preparing: DELETE FROM user_info WHERE (userName = ? AND password = ? )
DEBUG [main] - ==> Parameters: mybatis2(String), 123456(String)
DEBUG [main] - <==    Updates: 1
```

此时，打开数据表 user_info，可以看到用户名为 mybatis2、密码为 123456 的用户已被成功删除。

15.6 小　　结

本章首先介绍了 MyBatis 框架的注解配置，然后结合具体示例，基于注解讲解了单表的增删改查、多表的关联映射和动态 SQL 语句等知识。在 MyBatis 框架中，熟练使用注解配置，可以简化程序的开发，从而提高开发效率。

第 16 章 MyBatis 缓存

为了有效地提高数据库查询的性能，MyBatis 提供了查询缓存。MyBatis 缓存分为一级缓存和二级缓存。

16.1 一级缓存

MyBatis 的一级缓存是 SqlSession 级别的缓存，当在同一个 SqlSession 中执行两次相同的 SQL 语句时，会将第一次执行查询的数据存入一级缓存中，第二次查询时会从缓存中获取数据，而不用再去数据库查询，从而提高了查询性能。但如果 SqlSession 执行 insert、delete 和 update 操作，并提交到数据库，或者 SqlSession 结束后，这个 SqlSession 中的一级缓存就不存在了。

下面通过示例测试 MyBatis 的一级缓存。

(1) 将项目 mybatis1 复制并命名为 mybatis10，再导入到 Eclipse 开发环境中。

(2) 在 com.mybatis.mapper 包中，新建一个接口 UserInfoMapper，并添加一个 findUserInfoById 方法，如下所示：

```java
package com.mybatis.mapper;
import com.mybatis.pojo.UserInfo;
public interface UserInfoMapper {
    // 根据id查询用户
    UserInfo findUserInfoById(int id);
}
```

(3) 添加测试方法。

在测试类 MybatisTest 中，添加一个测试方法 testFirstLevelCache，使用@Test 注解修饰，编写如下代码：

```java
// 测试一级缓存 1
@Test
public void testFirstLevelCache() {
    // 获得 UserInfoMapperd 代理对象
    UserInfoMapper uim = sqlSession.getMapper(UserInfoMapper.class);
    // 查询 id=1 的 UserInfo 对象
    UserInfo ui1 = uim.findUserInfoById(1);
    System.out.println(ui1.toString());
    // 再次查询 id=1 的 UserInfo 对象
    UserInfo ui2 = uim.findUserInfoById(1);
    System.out.println(ui2.toString());
}
```

执行该测试方法，控制台输出如下：

```
DEBUG [main] - ==>  Preparing: select * from user_info where id = ?
DEBUG [main] - ==> Parameters: 1(Integer)
DEBUG [main] - <==      Total: 1
UserInfo [id=1, userName=tom, password=666666]
UserInfo [id=1, userName=tom, password=666666]
```

可以看出，第一次查询 id 为 1 的 UserInfo 对象时发出了一条 SQL 语句，由于 MyBatis 默认开启了一级缓存，因此一级缓存 SqlSession 中缓存了 id 为 1 的 UserInfo 对象。第二次再查询 id 为 1 的 UserInfo 对象时，直接从一级缓存中获取数据，而不用查询数据库，所以第二次没有发出 select 语句。

在测试类 MybatisTest 中，添加一个测试方法 testFirstLevelCache_1，并使用@Test 注解修饰，编写如下代码：

```java
// 测试一级缓存 2
@Test
public void testFirstLevelCache_1() {
    // 获得 UserInfoMapperd 代理对象
    UserInfoMapper uim = sqlSession.getMapper(UserInfoMapper.class);
    // 查询 id=1 的 UserInfo 对象
    UserInfo ui1 = uim.findUserInfoById(1);
    System.out.println(ui1.toString());
    sqlSession.commit();
    // 关闭 sqlSession，即清空一级缓存
    sqlSession.close();
    // 开始一个新的 sqlSession
    sqlSession = sqlSessionFactory.openSession();
    // 再次获得 UserInfoMapperd 代理对象
    uim = sqlSession.getMapper(UserInfoMapper.class);
    // 再次查询 id=1 的 UserInfo 对象
    UserInfo ui2 = uim.findUserInfoById(1);
    System.out.println(ui2.toString());
}
```

在测试方法 testFirstLevelCache_1 中，第一次查询后关闭 SqlSession，然后开启一个新的 SqlSession，再次执行查询。

执行该测试方法，控制台输出如下：

```
DEBUG [main] - ==>  Preparing: select * from user_info where id = ?
DEBUG [main] - ==> Parameters: 1(Integer)
DEBUG [main] - <==      Total: 1
UserInfo [id=1, userName=tom, password=666666]
DEBUG [main] - Resetting autocommit to true on JDBC Connection
[com.mysql.jdbc.JDBC4Connection@5f0fd5a0]
DEBUG [main] - Closing JDBC Connection
[com.mysql.jdbc.JDBC4Connection@5f0fd5a0]
DEBUG [main] - Returned connection 1594873248 to pool.
DEBUG [main] - Opening JDBC Connection
DEBUG [main] - Checked out connection 1594873248 from pool.
DEBUG [main] - Setting autocommit to false on JDBC Connection
[com.mysql.jdbc.JDBC4Connection@5f0fd5a0]
DEBUG [main] - ==>  Preparing: select * from user_info where id = ?
DEBUG [main] - ==> Parameters: 1(Integer)
DEBUG [main] - <==      Total: 1
UserInfo [id=1, userName=tom, password=666666]
```

可以看出，关闭 SqlSession 后一级缓存会被清空，所以第二次查询时，一级缓存中查询不到数据，发出了第二条 select 语句查询数据库。

16.2 二级缓存

MyBatis 的二级缓存是 mapper 级别的缓存，多个 SqlSession 共用二级缓存，它们使用同一个 Mapper 的 SQL 语句操作数据库，获得的数据会存放在二级缓存中。

MyBatis 默认没有开启二级缓存，需要在 MyBatis 的配置文件 mybatis-config.xml 中开启二级缓存，配置如下：

```
<!-- 启用二级缓存 -->
<settings>
    <setting name="cacheEnabled" value="true" />
</settings>
```

需要注意的是，settings 元素要放在 properties 元素之后，typeAliases 元素之前，否则配置文件会报错。

此外，还需要在 UserInfoMapper.xml 映射文件中，使用<cache>元素开启当前 mapper 的 namespace 下的二级缓存，如下所示：

```
<cache eviction="LRU" flushInterval="30000" size="512" readOnly="true" />
```

这样，UserInfoMapper.xml 下的 SQL 语句执行结束后，会将结果存储到它的二级缓存中。<cache>元素配置在<mapper>元素内，<cache>元素的属性含义如下。

① flushInterval 属性：表示刷新间隔，可以被设置为任意正整数，单位是毫秒。默认

不设置，表示没有刷新间隔，仅仅调用语句时刷新缓存。

② size 属性：表示引用数目，可以被设置为任意正整数，默认值是 1024。readOnly 属性设置是否只读，true 表示只读，false 表示可读写。只读的缓存会给所有调用者返回缓存对象的相同实例，因此这些对象不能被修改，这提供了重要的性能优势。可读写的缓存会返回缓存对象的拷贝，这会慢一些，但是安全，因此默认是 false。

③ eviction 属性：表示收回策略，有 LRU、FIFO、SOFT、WEAK 等策略，默认为 LRU。LRU 表示最近最少使用策略，即移除最近最少使用的对象。FIFO 表示先进先出策略，按对象进入缓存的顺序来移除。SOFT 表示软引用策略，移除基于垃圾回收器状态和软引用规则的对象。WEAK 表示弱引用策略，更积极地移除基于垃圾收集器状态和弱引用规则的对象。

再次执行测试类 MybatisTest 中的测试方法 testFirstLevelCache_1，控制台输出如下：

```
DEBUG [main] - ==>  Preparing: select * from user_info where id = ?
DEBUG [main] - ==> Parameters: 1(Integer)
DEBUG [main] - <==      Total: 1
UserInfo [id=1, userName=tom, password=666666]
DEBUG [main] - Resetting autocommit to true on JDBC Connection
[com.mysql.jdbc.JDBC4Connection@e3c0e40]
DEBUG [main] - Closing JDBC Connection
[com.mysql.jdbc.JDBC4Connection@e3c0e40]
DEBUG [main] - Returned connection 238816832 to pool.
DEBUG [main] - Cache Hit Ratio [com.mybatis.mapper.UserInfoMapper]: 0.5
UserInfo [id=1, userName=tom, password=666666]
```

可以看出，第一次查询 id 为 1 的 UserInfo 对象时执行了一条 select 语句，然后关闭 SqlSession，一级缓存被清空。第二次查询 id 为 1 的 UserInfo 对象时，先查找一级缓存，没有找到 id 为 1 的 UserInfo 对象，再去查找二级缓存，找到了 id 为 1 的 UserInfo 对象，所以不会再次执行 select 语句。

在映射文件 UserInfoMapper.xml 中，如果给 id 为 findUserInfoById 的<select>元素添加 useCache="false"属性值，则表示禁用当前 select 语句的二级缓存，即每次查询都会发出 SQL 语句。useCache 属性默认值为 true，表示该 SQL 语句使用二级缓存。

再次执行测试方法 testFirstLevelCache_1，观察控制台输出。由于该 select 语句禁用了二级缓存，因此第二次查询时会再次发出 select 语句。

16.3 小 结

本章首先介绍了 MyBatis 框架的缓存概念，然后结合具体示例，详细讲解了 MyBatis 框架的一级缓存和二级缓存的用法。在 MyBatis 框架中，合理使用缓存，可以极大地减少资源的消耗，从而提高系统的性能。

第 17 章　Spring 整合 MyBatis

MyBatis 与 Hibernate 一样,也是非常优秀的持久层框架。在框架整合开发时,也会选择 MyBatis 替代 Hibernate。本章以登录功能为例,采用注解方式实现 Spring 与 MyBatis 框架的整合。从实质上来说,Spring 与 MyBatis 的整合也就是 Spring、Spring MVC 与 MyBatis 的整合,通常简称 SSM 框架整合。

17.1　环 境 搭 建

在 Eclipse 中,新建一个名为 ssm 的 Maven 项目,如图 17-1 所示。单击 Next 按钮,打开 New Maven Project 对话框,选中 Create a simple project 复选框,如图 17-2 所示。

图 17-1　创建一个 Maven 项目

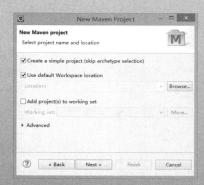

图 17-2　New Maven Project 对话框

单击 Next 按钮，进入项目配置界面，设置 Group Id 为 com.my，Artifact Id 为 ssm，Packaging 为 war，如图 17-3 所示。单击 Finish 按钮，完成项目的创建。项目最初目录结构如图 17-4 所示。在顶级目录上是工程的描述文件 pom.xml，它是 Maven 项目的核心配置文件。顶级目录还包括 src 和 target 两个子目录，target 目录是所有工程编译构建的输出目录，src 目录包含所有工程的源码文件、配置文件、资源文件等等。src 目录的子目录 main/java 用于存放 Java 源文件，子目录 main/resources 用于存放框架或其他工具的配置文件，子目录 test/java 用于存放 Java 测试的源文件，子目录 test/resources 用于存放测试的配置文件，子目录 main/webapp 是 Web 应用的目录，里面可以包含 WEB-INF、js 和 css 等内容。

在 Eclipse 包资源管理器窗口中，右击项目名 ssm，选择 Properties 命令，打开 Properties for ssm 对话框。在对话框的左侧区域，选择 Project Facets 菜单项，右侧会显示项目 ssm 的相关特性，如图 17-5 所示。创建项目 ssm 时，默认的 Dynamic Web Module 版本为 2.5，Java 版本为 1.5，可根据需要将它们设置为相应的版本。这里先将 Java 版本设置为 9；然后取消 Dynamic Web Module 复选框的选中状态，并单击一次 Apply 按钮；接着选中 Dynamic Web Module 复选框，将其版本选择为 3.1，此时会出现 Futher configuration available…这个链接。单击此链接，打开 Modify Faceted Project 对话框，如图 17-6 所示。

单击 Next 按钮，打开 Configure web module settings 界面，如图 17-7 所示。将 Content directory 设置为 src/main/webapp，选中 Generate web.xml deployment descriptor 复选框，最后单击 OK 按钮，此时会返回到图 17-5 所示的界面，单击 Apply And Close 按钮。

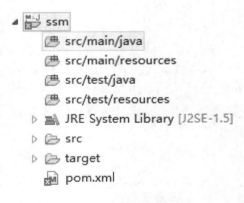

图 17-3 配置项目　　　　　　　　图 17-4 项目初始目录结构

第 17 章　Spring 整合 MyBatis

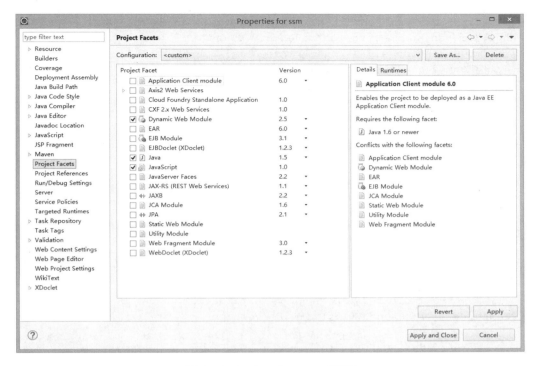

图 17-5　Project for ssm 对话框

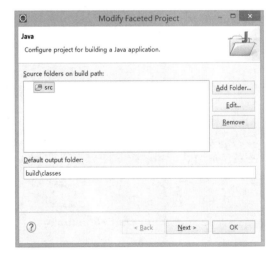

图 17-6　修改项目特性对话框

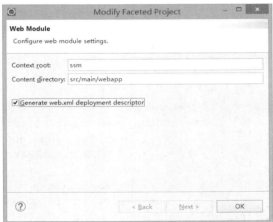

图 17-7　Configure web module settings 界面

至此，一个 Maven 项目就算真正创建好了。接下来，就需要给项目添加 SSM 框架所需的 jar 包了。打开 pom.xml 文件，在<project>元素中添加一个<dependencies>元素。在<dependencies>元素内，通过添加多个<dependency>子元素来引入多个 jar 包。最后，pom.xml 文件内容如下：

```
<project xmlns="http://maven.apache.org/POM/4.0.0"
xmlns:xsi="http://www.w3.org/2001/XMLSchema-instance"
```

```xml
        xsi:schemaLocation="http://maven.apache.org/POM/4.0.0
http://maven.apache.org/xsd/maven-4.0.0.xsd">
        <modelVersion>4.0.0</modelVersion>
        <groupId>com.my</groupId>
        <artifactId>ssm</artifactId>
        <version>0.0.1-SNAPSHOT</version>
        <packaging>war</packaging>
        <dependencies>
            <!-- Servlet API -->
            <dependency>
                <groupId>javax.servlet</groupId>
                <artifactId>servlet-api</artifactId>
                <version>3.0-alpha-1</version>
                <scope>provided</scope>
            </dependency>
<!-- https://mvnrepository.com/artifact/org.springframework/spring-web -->
            <dependency>
                <groupId>org.springframework</groupId>
                <artifactId>spring-web</artifactId>
                <version>5.0.3.RELEASE</version>
            </dependency>
            <!-- Spring SpringMVC -->
            <!-- https://mvnrepository.com/artifact/org.springframework/spring-webmvc -->
            <dependency>
                <groupId>org.springframework</groupId>
                <artifactId>spring-webmvc</artifactId>
                <version>5.0.3.RELEASE</version>
            </dependency>
            <!-- Spring JDBC -->
            <!-- https://mvnrepository.com/artifact/org.springframework/spring-jdbc -->
            <dependency>
                <groupId>org.springframework</groupId>
                <artifactId>spring-jdbc</artifactId>
                <version>5.0.3.RELEASE</version>
            </dependency>
            <!-- Spring Aspects -->
            <!-- https://mvnrepository.com/artifact/org.springframework/spring-aspects -->
            <dependency>
                <groupId>org.springframework</groupId>
                <artifactId>spring-aspects</artifactId>
                <version>5.0.3.RELEASE</version>
            </dependency>
            <!-- MyBatis -->
            <!-- https://mvnrepository.com/artifact/org.mybatis/mybatis -->
            <dependency>
                <groupId>org.mybatis</groupId>
                <artifactId>mybatis</artifactId>
                <version>3.4.5</version>
```

```xml
        </dependency>
        <!-- MyBatis Spring -->
    <!-- https://mvnrepository.com/artifact/org.mybatis/mybatis-spring -->
        <dependency>
            <groupId>org.mybatis</groupId>
            <artifactId>mybatis-spring</artifactId>
            <version>1.3.1</version>
        </dependency>
        <!-- C3P0 -->
        <!-- https://mvnrepository.com/artifact/c3p0/c3p0 -->
        <dependency>
            <groupId>c3p0</groupId>
            <artifactId>c3p0</artifactId>
            <version>0.9.1.2</version>
        </dependency>
        <!-- MySQL 驱动包 -->
    <!-- https://mvnrepository.com/artifact/mysql/mysql-connector-java -->
        <dependency>
            <groupId>mysql</groupId>
            <artifactId>mysql-connector-java</artifactId>
            <version>5.1.45</version>
        </dependency>
        <!-- JSTL -->
        <!-- https://mvnrepository.com/artifact/jstl/jstl -->
        <dependency>
            <groupId>jstl</groupId>
            <artifactId>jstl</artifactId>
            <version>1.2</version>
        </dependency>
        <!-- https://mvnrepository.com/artifact/com.googlecode.json-simple/json-simple -->
        <dependency>
            <groupId>com.googlecode.json-simple</groupId>
            <artifactId>json-simple</artifactId>
            <version>1.1</version>
        </dependency>
        <!-- https://mvnrepository.com/artifact/commons-fileupload/commons-fileupload -->
        <dependency>
            <groupId>commons-fileupload</groupId>
            <artifactId>commons-fileupload</artifactId>
            <version>1.3.3</version>
        </dependency>
    <!-- https://mvnrepository.com/artifact/com.google.code.gson/gson -->
        <dependency>
            <groupId>com.google.code.gson</groupId>
            <artifactId>gson</artifactId>
            <version>2.8.2</version>
        </dependency>
        <!-- https://mvnrepository.com/artifact/org.apache.commons/commons-lang3 -->
```

```xml
        <dependency>
            <groupId>org.apache.commons</groupId>
            <artifactId>commons-lang3</artifactId>
            <version>3.7</version>
        </dependency>
    </dependencies>
</project>
```

在 pom.xml 文件中，依次引入了 servlet-api、spring-web、spring-webmvc、spring-jdbc、spring-aspects、mybatis、mybatis-spring、c3p0、mysql-connector-java、jstl、json-simple、commons-fileupload、gson 和 commons-lang3 等资源包。在这些资源包中，spring 打头的是与 Spring 框架相关的资源包，servlet-api 用于编写 Servlet，mybatis 是 MyBatis 框架的核心包，mybatis-spring 是 Spring 与 MyBatis 框架整合时使用的包，c3p0 是一个库，它扩展了传统的 JDBC 数据库连接池。mysql-connector-java 是 MySQL 数据的驱动包，json-simple 是 Java 生成的 JSON 工具包，gson 是 Google 解析 JSON 的一个开源框架。读者可以根据实际项目的需要，添加其他必要的 jar 包。

保存 pom.xml 文件，如果此时电脑处于联网状态，这些资源包就会从指定网站下载到本地。可以在包资源管理器中观察项目 ssm，看看是否出现一个 Maven Dependencies 目录。展开这个目录，就可以看到下载好的 jar 包了。

17.2 编写 SSM 整合的相关配置文件

要想实现 Spring、Spring MVC 与 MyBatis 框架的整合，就需要编写 Web 应用程序主配置文件 web.xml、Spring 配置文件和 Spring MVC 配置文件。

1. 编写 web.xml 文件

web.xml 文件位于 src/main/webapp/WEB-INF 目录中，编写后的内容如下：

```xml
<?xml version="1.0" encoding="UTF-8"?>
<web-app xmlns:xsi="http://www.w3.org/2001/XMLSchema-instance"
    xmlns="http://xmlns.jcp.org/xml/ns/javaee"
    xsi:schemaLocation="http://xmlns.jcp.org/xml/ns/javaee
http://xmlns.jcp.org/xml/ns/javaee/web-app_3_1.xsd"
    id="WebApp_ID" version="3.1">
    <!--1、启动 Spring 的容器 -->
    <context-param>
        <param-name>contextConfigLocation</param-name>
        <param-value>classpath:applicationContext.xml</param-value>
    </context-param>
    <listener>
        <listener-class>org.springframework.web.context.ContextLoaderListener
</listener-class>
    </listener>
    <!--2、springmvc 的前端控制器，拦截所有请求 -->
    <servlet>
        <servlet-name>dispatcherServlet</servlet-name>
```

```xml
        <servlet-class>org.springframework.web.servlet.DispatcherServlet
</servlet-class>
        <load-on-startup>1</load-on-startup>
    </servlet>
    <servlet-mapping>
        <servlet-name>dispatcherServlet</servlet-name>
        <url-pattern>/</url-pattern>
    </servlet-mapping>
    <!-- 3、字符编码过滤器，一定要放在所有过滤器之前 -->
    <filter>
        <filter-name>CharacterEncodingFilter</filter-name>
        <filter-class>org.springframework.web.filter.CharacterEncodingFilter
</filter-class>
        <init-param>
            <param-name>encoding</param-name>
            <param-value>utf-8</param-value>
        </init-param>
        <init-param>
            <param-name>forceRequestEncoding</param-name>
            <param-value>true</param-value>
        </init-param>
        <init-param>
            <param-name>forceResponseEncoding</param-name>
            <param-value>true</param-value>
        </init-param>
    </filter>
    <filter-mapping>
        <filter-name>CharacterEncodingFilter</filter-name>
        <url-pattern>/*</url-pattern>
    </filter-mapping>
    <!-- 4、使用 Rest 风格的 URI，将页面普通的 post 请求转为指定的 delete 或者 put 请求 -->
    <filter>
        <filter-name>HiddenHttpMethodFilter</filter-name>
        <filter-class>org.springframework.web.filter.HiddenHttpMethodFilter
</filter-class>
    </filter>
    <filter-mapping>
        <filter-name>HiddenHttpMethodFilter</filter-name>
        <url-pattern>/*</url-pattern>
    </filter-mapping>
    <filter>
        <filter-name>HttpPutFormContentFilter</filter-name>
        <filter-class>org.springframework.web.filter.HttpPutFormContentFilter
</filter-class>
    </filter>
    <filter-mapping>
        <filter-name>HttpPutFormContentFilter</filter-name>
        <url-pattern>/*</url-pattern>
    </filter-mapping>
</web-app>
```

在 web.xml 文件中，要实现 Web 项目启动时让 Spring 容器也启动，可以配置一个监听器 ContextLoaderListener。当项目启动时，它会自动加载 Spring 的配置文件 applicationContext.xml，这个配置文件稍后会作介绍。为了能拦截所有请求，需要配置 Spring MVC 的前端控制器，这里使用了 DispatcherServlet，它能够加载 Spring MVC 的配置文件 dispatcherServlet-servlet.xml，这个配置文件稍后也会介绍。为了处理中文乱码，可以配置一个字符编码过滤器，它通过 CharacterEncodingFilter 类来实现。如果项目中要使用 Rest 风格的 URI，可配置一个 HiddenHttpMethodFilter，将页面普通的 post 请求转为指定的 delete 或者 put 请求。

2. 编写 Spring MVC 配置文件

在 src/main/webapp/WEB-INF 目录中，创建 Spring MVC 配置文件，文件名为 dispatcherServlet-servlet.xml，编写后的内容如下：

```xml
<?xml version="1.0" encoding="UTF-8"?>
<beans xmlns="http://www.springframework.org/schema/beans"
    xmlns:xsi="http://www.w3.org/2001/XMLSchema-instance"
    xmlns:context="http://www.springframework.org/schema/context"
    xmlns:mvc="http://www.springframework.org/schema/mvc"
    xsi:schemaLocation="http://www.springframework.org/schema/mvc
http://www.springframework.org/schema/mvc/spring-mvc-4.3.xsd
       http://www.springframework.org/schema/beans
http://www.springframework.org/schema/beans/spring-beans-4.3.xsd
       http://www.springframework.org/schema/context
http://www.springframework.org/schema/context/spring-context-4.3.xsd">
    <!-- 启动注解扫描功能 -->
    <context:component-scan base-package="com.my" use-default-filters="false">
        <!--只扫描控制器 -->
        <context:include-filter type="annotation" expression="org.springframework.stereotype.Controller"/>
    </context:component-scan>
    <!--配置视图解析器，方便页面返回 -->
    <bean class="org.springframework.web.servlet.view.InternalResourceViewResolver">
        <property name="prefix" value="/"/>
        <property name="suffix" value=".jsp"/>
    </bean>
    <!--两个标准配置 -->
    <!-- 将springmvc不能处理的请求交给tomcat -->
    <mvc:default-servlet-handler/>
    <!-- 能支持springmvc更高级的一些功能，JSR303校验，快捷的ajax...映射动态请求 -->
    <mvc:annotation-driven/>
</beans>
```

与 Spring 配置文件类似，在 Spring MVC 配置文件中，也需要启动注解扫描功能，让 Spring 容器自动扫描包含注解的类，然后将其注册到 Bean 容器中。为了避免重复扫描，Spring MVC 配置文件中只扫描控制器 Controller，而将 Dao 和 Service 的扫描交由 Spring 配置文件完成。在 Spring MVC 配置文件中，通过<context:component-scan>元素和<context:include-filter>

子元素配合，实现只对包含@Controller 注解的类进行扫描。在<context:component-scan>元素中，base-package 属性用于指定要扫描的包名，这里设置为 com.my，表示扫描 com.my 包及其子包中的类；use-default-filters 属性用于指示是否自动扫描带有@Component、@Repository、@Service 和@Controller 的类，默认为 true，即默认扫描。这里设置为 false，就是不自动扫描。<context:include-filter>子元素用来指定需要扫描的类，expression 属性设置为 Controller，表示扫描含有@Controller 注解的类，然后注册到 Bean 容器中。

在 Controller 控制器方法执行后，为了方便页面返回，需要配置视图解析器。这里使用 InternalResourceViewResolver 类，它包含 prefix 和 suffix 两个属性，分别用于指定返回的 URL 的前后缀。

在配置文件 web.xml 中，由于将 DispatcherServlet 请求映射配置为"/"，因此 Spring MVC 将捕获 Web 容器所有的请求，包括静态资源的请求。Spring MVC 会将这些静态资源当作一个普通请求进行处理，因此找不到对应处理器会导致错误。如何让 Spring 框架能够捕获所有 URL 的请求，同时又将静态资源的请求转由 Web 容器处理，Spring 团队提供了一些解决方案，其中之一就是使用<mvc:default-servlet-handler/>元素。在 Spring MVC 的配置文件中配置 <mvc:default-servlet-handler/> 后，会在 Spring MVC 上下文中定义一个 org.springframework.web.servlet.resource.DefaultServletHttpRequestHandler，它会像一个检查员，对进入 DispatcherServlet 的 URL 进行筛查，如果发现是静态资源的请求，就将该请求转由 Web 应用服务器默认的 Servlet 处理，如果不是静态资源的请求，才由 DispatcherServlet 继续处理。

为了能支持 Spring MVC 更高级的一些功能，例如 JSR303 校验、映射动态请求等，就需使用<mvc:annotation-driven/>，它会自动注册 RequestMappingHandlerMapping 和 RequestMappingHandlerAdapter 两个 Bean，这是 Spring MVC 为@Controller 分发请求所必需的，并且提供了数据绑定支持、@NumberFormatannotation 支持、@DateTimeFormat 支持、@Valid 支持、读写 XML 的支持(JAXB)和读写 JSON 的支持(默认 Jackson)等功能。

3. 编写 Spring 配置文件

在 src/main/resources 目录中，创建 Spring 配置文件，文件名为 applicationContext.xml，编写后的内容如下：

```
<?xml version="1.0" encoding="UTF-8"?>
<beans xmlns="http://www.springframework.org/schema/beans"
    xmlns:xsi="http://www.w3.org/2001/XMLSchema-instance"
    xmlns:context="http://www.springframework.org/schema/context"
    xmlns:aop="http://www.springframework.org/schema/aop"
    xmlns:tx="http://www.springframework.org/schema/tx"
    xsi:schemaLocation="http://www.springframework.org/schema/aop
http://www.springframework.org/schema/aop/spring-aop-4.3.xsd
        http://www.springframework.org/schema/beans
http://www.springframework.org/schema/beans/spring-beans-4.3.xsd
        http://www.springframework.org/schema/tx
http://www.springframework.org/schema/tx/spring-tx-4.3.xsd
        http://www.springframework.org/schema/context
http://www.springframework.org/schema/context/spring-context-4.3.xsd">
```

```xml
<context:component-scan base-package="com.my">
    <context:exclude-filter type="annotation"
        expression="org.springframework.stereotype.Controller" />
</context:component-scan>
<!-- Spring 的配置文件,这里主要配置和业务逻辑有关的 -->
<!--===================== 数据源,事务控制,xxx ==================-->
<context:property-placeholder location="classpath:dbconfig.properties" />
<bean id="dataSource"
    class="com.mchange.v2.c3p0.ComboPooledDataSource">
    <property name="jdbcUrl" value="${jdbc.jdbcUrl}"></property>
    <property name="driverClass" value="${jdbc.driverClass}"></property>
    <property name="user" value="${jdbc.user}"></property>
    <property name="password" value="${jdbc.password}"></property>
</bean>
<!-- 配置 SqlSessionFactoryBean -->
<bean id="sqlSessionFactory" class="org.mybatis.spring.SqlSessionFactoryBean">
    <property name="dataSource" ref="dataSource" />
</bean>
<!-- 配置 MapperScannerConfigurer,Dao 接口所在包名,Spring 会自动查找其下的类 -->
<bean class="org.mybatis.spring.mapper.MapperScannerConfigurer">
    <property name="basePackage" value="com.my.dao" />
    <property name="sqlSessionFactoryBeanName" value="sqlSessionFactory"></property>
</bean>
<!-- 配置 DataSourceTransactionManager(事务管理) -->
<bean id="transactionManager"
    class="org.springframework.jdbc.datasource.DataSourceTransactionManager">
    <property name="dataSource" ref="dataSource" />
</bean>
<!-- 启用基于注解的声明式事务管理配置 -->
<tx:annotation-driven transaction-manager="transactionManager" />
</beans>
```

在 Spring 配置文件 applicationContext.xml 中,首先需要启动对带有@Component、@Repository 和@Service 这些注解的类的自动扫描功能,但对包括@Controller 注解的控制器类则不进行扫描。为了实现这一目的,可以结合使用<context:component-scan>元素和<context:exclude-filter>子元素。

此外,还需要配置数据源 ComboPooledDataSource 的实例 dataSource,数据库的连接信息定义在 src/main/resources 目录下的属性文件 dbconfig.properties 中,然后再通过 dataSource 配置 SqlSessionFactoryBean 实例 sqlSessionFactory,接着再通过 sqlSessionFactory 进一步配置 MapperScannerConfigurer 实例。MapperScannerConfigurer 是 Spring 和 MyBatis 整合的 mybatis-spring.jar 包中提供的一个类,它将扫描 basePackage 指定的包下的所有接口类(包括子类),然后创建各自接口的动态代理类,并将它们动态定义为一个个 Spring Bean。之后使用 basePackage 所指定的包下的接口时,可以直接通过 Spring 注入相应的 Spring Bean,然后就可以直接使用了。因此,配置 MapperScannerConfigurer 是实现 Spring 与 MyBatis 整合的关键。

为了能让 Spring 框架进行事务管理，需要配置 DataSourceTransactionManager，它是 Spring 在 JDBC 中提供的一个事务管理组件。

使用事务管理的功能，与创建 Bean 一样，可以采用注解和 XML 配置两种方式。如果采用注解方式，则需要使用<tx:annotation-driven>元素来开启事务注解标记@Transactional。这样，当调用带有@Transactional 注解的方法时，会将事务管理功能切入进去。

至此，Spring、Spring MVC 与 MyBatis 整合的配置文件就全部编写好了。在接下来的小节中，将按照实体类创建、Dao 层开发、Service 层开发和表示层开发的流程具体讲解用户登录功能的实现过程。

17.3　创建实体类

在 src/main/java 目录下，新建一个 com.my.pojo 包。在包中新建一个实体类 UserInfo，编写如下代码：

```java
package com.my.pojo;
public class UserInfo {
    private int id;
    private String userName;
    private String password;
    // 此处省略上述属性的 get 和 set 方法
}
```

为了简单起见，UserInfo 类只添加了三个与数据表 user_info 的字段对应的属性。

17.4　数据访问层开发

在 src/main/java 目录中，新建一个 com.my.dao 包，用于存放数据访问层接口。在包中新建一个接口 UserInfoDao，在接口中声明方法，代码如下：

```java
package com.my.dao;
import org.apache.ibatis.annotations.Param;
import org.apache.ibatis.annotations.Select;
import com.my.pojo.UserInfo;
public interface UserInfoDao {
    // 根据用户名和密码查询
    @Select("select * from user_info where userName = #{userName} and password = #{password}")
    public UserInfo findUserInfoByCond(@Param("userName") String userName, @Param("password") String password);
}
```

在接口 UserInfoDao 中，声明了一个 findUserInfoByCond 方法，并通过@Select 注解映射了一个根据用户名和密码查询的 select 语句。

17.5 业务逻辑层开发

在 src/main/java 目录中，新建一个 com.my.service 包，用于存放业务逻辑层接口。在包中新建一个接口 UserInfoService，在接口中声明一个 login 方法，用于登录验证，代码如下：

```java
package com.my.service;
import com.my.pojo.UserInfo;
public interface UserInfoService {
    public UserInfo login(String userName, String password);
}
```

新建 UserInfoService 接口的实现类 UserInfoServiceImpl，存放在 com.my.service.impl 包中，以实现 login 方法。

```java
package com.my.service.impl;
import org.springframework.beans.factory.annotation.Autowired;
import org.springframework.stereotype.Service;
import com.my.dao.UserInfoDao;
import com.my.pojo.UserInfo;
import com.my.service.UserInfoService;
@Service("userInfoService")
public class UserInfoServiceImpl implements UserInfoService {
    @Autowired
    private UserInfoDao userInfoDao;
    @Override
    public UserInfo login(String userName, String password) {
        return userInfoDao.findUserInfoByCond(userName, password);
    }
}
```

在这个实现类中，首先使用@Service 注解标注 UserInfoServiceImpl 类，将这个类自动注册到 Spring 容器，这样就无需在 Spring 配置文件中定义 Bean 了；然后使用@Autowired 注解标注 UserInfoDao 类型的属性 userInfoDao，完成自动装配的工作。

17.6 控制器开发

在 src/main/java 目录中，创建包 com.my.controller，用于存放控制器类。在包中新建一个类 UserInfoController，编写如下代码：

```java
package com.my.controller;
import org.springframework.beans.factory.annotation.Autowired;
import org.springframework.stereotype.Controller;
import org.springframework.web.bind.annotation.RequestMapping;
import com.my.pojo.UserInfo;
import com.my.service.UserInfoService;
@Controller
```

```java
@RequestMapping("/userinfo")
public class UserInfoController {
    @Autowired
    private UserInfoService userInfoService;
    @RequestMapping("/login")
    public String login(UserInfo ui) {
        UserInfo tempUi = userInfoService.login(ui.getUserName(), ui.getPassword());
        if (tempUi != null && tempUi.getUserName() != null) {
            return "index";
        } else {
            return "redirect:/login.jsp";
        }
    }
}
```

在这个控制器类中，使用@Controller 和@RequestMapping 标注 UserInfoController 类，使用@Controller 注解标记一个类 Controller，使用@RequestMapping 注解定义 URL 请求和 Controller 方法之间的映射，这样的 Controller 就能被外部访问到了。

17.7 表示层开发

在 src/main/webapp 目录下，新建一个登录页 login.jsp，表单部分代码如下：

```html
<form action="userinfo/login" method="post">
    <table>
        <tr>
            <td>用户名：</td>
            <td><input type="text" name="userName" /></td>
        </tr>
        <tr>
            <td>密 码：</td>
            <td><input type="text" name="password" /></td>
        </tr>
        <tr>
            <td><input type="submit" value="登录" /></td>
            <td></td>
        </tr>
    </table>
</form>
```

再新建一个登录成功后跳转到的首页面 index.jsp，代码如下：

```html
<body>
    欢迎您，登录成功！
</body>
```

在包资源管理器中右击项目名，出现快捷菜单，依次选择 Run As、Run on Server 命令，打开 Run on Server 对话框。选择 Tomcat v9.0 Server 作为 Web 服务器，单击 Finish 按钮。

部署项目并启动 Web 服务器后，打开一个浏览器，在地址栏中输入 http://localhost:8080/ssm/login.jsp 地址，会显示一个登录页面，如图 17-8 所示。输入正确的用户名和密码，单击"登录"按钮，页面成功跳转到 index.jsp，如图 17-9 所示。

图 17-8　登录页

图 17-9　index.jsp 页

如果用户名或密码输入错误，则重定向到登录页。

17.8　小　　结

本章首先讲解了 Spring、Spring MVC 与 MyBatis 框架整合的环境搭建，以及相关配置文件的编写，然后针对数据表 user_info，以用户登录为例，遵循三层架构，按照实体类创建、数据访问层开发、业务逻辑层开发、控制器开发和表示层开发的流程，完整地描述了一个功能模块的实现过程。

第 18 章　前端 UI 框架

jQuery 是 JavaScript 的一个基础框架，考虑到框架的通用性和代码文件大小，jQuery 仅仅集成了 JavaScript 中最为核心和常用的功能。目前，在 jQuery 的基础上已开发出众多的插件，这些插件均以 jQuery 为核心编写而成。本章主要介绍 jQuery Easy UI、Bootstrap 和 Vue 三种当前流行的框架。

18.1　Easy UI 框架

Easy UI 是在 jQuery 的基础上开发的一个 UI 插件，目的在于让 Web 开发者快捷地构建出功能丰富且美观的用户界面。开发者无需编写复杂的 JavaScript，也无需对 CSS 样式有深入的了解。开发者只需有一些 HTML 和 jQuery 基础，就可以轻松地开发出较好的软件界面。Easy UI 控件的种类很多，由于篇幅所限，这里仅介绍本书项目案例篇中的项目用到的几种常用控件。可以从官方网站下载 jQuery EasyUI 插件，本书以 jquery-easyui-1.5.1 版本来介绍。下载 jquery-easyui-1.5.1.zip 文件，解压后的目录中主要包含 jquery.min.js、jquery.easyui.min.js 两个文件和 demo、locale、plugins 和 themes 四个目录。demo 目录下包含 jQuery EasyUI 官方提供的例子，locale 目录下包含语言本地化 JavaScript 文件，plugins 目录下包含 Easy UI 提供的各个功能的插件，themes 目录下包含样式和图片文件目录。

18.1.1 Layout 控件

使用 Easy UI 的 Layout 控件可以实现页面布局，布局是有五个区域(北区 north、南区 south、东区 east、西区 west 和中区 center)的容器。中间的区域面板是必需的，边缘的区域面板是可选的。每个边缘区域面板可通过拖曳边框调整尺寸，也可以通过点击折叠触发器来折叠面板。布局可以嵌套，因此用户可建立复杂的布局。

使用 Layout 控件实现一个简单布局的过程如下：

(1) 创建 Web 项目 easyui_demo，将 Easy UI 所需的文件事先存放到文件夹 EasyUI 中，再将该文件夹拷贝到项目的 WebRoot 目录下，EasyUI 文件夹的内容如图 18-1 所示。

(2) 新建页面 layout.jsp，在页面的<head></head>元素中引用相关的 css 和 js 文件，代码如下：

```
<head>
<link href="EasyUI/themes/default/easyui.css" rel="stylesheet"
    type="text/css" />
<link href="EasyUI/themes/icon.css" rel="stylesheet" type="text/css" />
<link href="EasyUI/demo.css" rel="stylesheet" type="text/css" />
<script src="EasyUI/jquery.min.js" type="text/javascript"></script>
<script src="EasyUI/jquery.easyui.min.js"
type="text/javascript"></script>
<script src="EasyUI/easyui-lang-zh_CN.js"
type="text/javascript"></script>
</head>
```

(3) 在页面 layout.jsp 的<body></body>元素中添加如下代码：

```
<body>
    <div class="easyui-layout" style="width:700px;height:350px;">
        <div data-options="region:'north'" style="height:50px">这是北区 north</div>
        <div data-options="region:'south',split:true"
style="height:50px;">这是南区 south</div>
        <div data-options="region:'east',split:true" title="East"
            style="width:100px;">这是东区 east</div>
        <div data-options="region:'west',split:true" title="West"
            style="width:100px;">这是西区 west</div>
        <div
            data-options="region:'center',title:'Main
Title',iconCls:'icon-ok'">
            这是中区 center</div>
    </div>
</body>
```

(4) 部署项目并启动 Tomcat，在浏览器中浏览页面 layout.jsp，效果如图 18-2 所示。

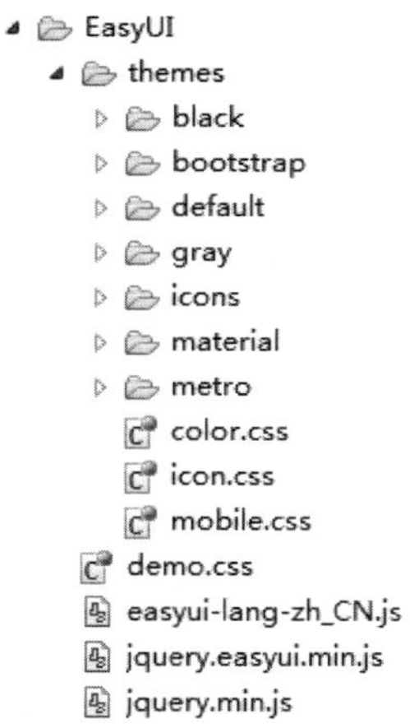

图 18-1　Easy UI 所需的文件 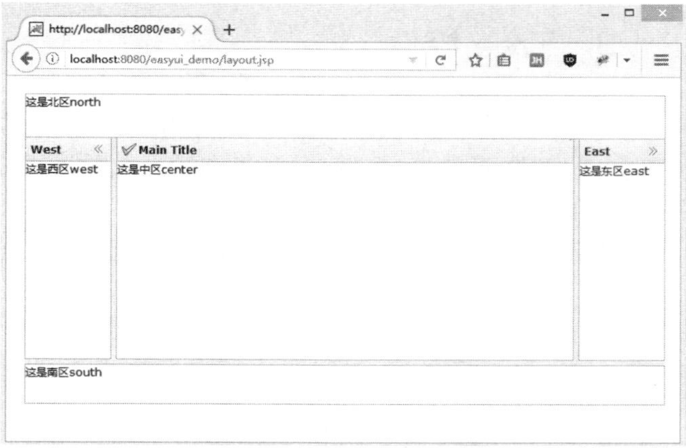　　　　图 18-2　Layout 控件效果

18.1.2　Tabs 控件

使用 Tabs 控件可以实现选项卡布局，一般用于中部选项卡。在项目 easyui_demo 中创建页面 tabs.jsp，在页面的<head></head>标签中引用相关的 css 和 js 文件。

在页面 tabs.jsp 的<body></body>标签中编写如下代码：

```
<body>
    <div class="easyui-tabs" style="width:700px;height:250px">
        <div title="选项卡 1" style="padding:10px">
            页面 1
        </div>
        <div title="选项卡 2" style="padding:10px">
            页面 1
        </div>
        <div title="选项卡 3"
data-options="iconCls:'icon-help',closable:true"
            style="padding:10px">页面 3</div>
    </div>
</body>
```

在浏览器中浏览页面 tabs.jsp，效果如图 18-3 所示。

图 18-3　Tabs 控件效果

18.1.3　Tree 控件

Tree 控件可以将数据分层以树形结构在 web 页面显示，Tree 控件在页面上以<ul></ul>标签标识。在项目 easyui_demo 中创建页面 tree.jsp，在页面的<head></head>标签中引用相关的 css 和 js 文件。

在页面 tree.jsp 的<body></body>标签中编写如下代码：

```
<body>
    <!-- 定义ul -->
    <ul id="tt"></ul>
    <script type="text/javascript">
        // 为Tree控件指定数据源
        $('#tt').tree({
            url : 'tree_data.json'
        });
    </script>
</body>
```

在项目的 WebRoot 目录下创建一个 JSON 格式的文件 tree_data.json，作为 Tree 控件的数据源，代码如下：

```
[
    {
        "id": 1,
        "text": "订餐系统管理后台",
        "fid": 0,
        "children": [
            {
                "id": 2,
                "text": "餐品管理",
                "fid": 0,
                "children": [
                    {
                        "id": 3,
                        "text": "餐品列表",
                        "fid": 0
                    },
                    {
                        "id": 4,
                        "text": "餐品类型列表",
                        "fid": 0
                    }
                ]
            }, {
                "id": 12,
                "text": "退出系统",
                "fid": 0
            }
        ]
```

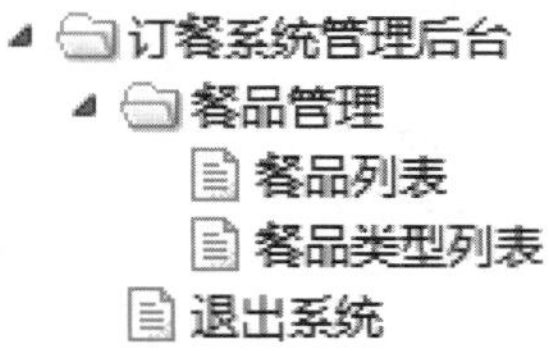

        }
]
```
在浏览器中浏览页面 tree.jsp，效果如图 18-4 所示。

```
▲ 🗀 订餐系统管理后台
   ▲ 🗀 餐品管理
         📄 餐品列表
         📄 餐品类型列表
      📄 退出系统
```

图 18-4　Tree 控件效果

18.1.4　DataGrid 控件

DataGrid 控件以表格格式显示数据，并为选择、排序、分组和编辑数据提供了丰富的支持。数据网格(DataGrid)的设计目的是为了减少开发时间，且不要求开发人员具备指定的知识。它是轻量级的，但是功能丰富。它的特性包括单元格合并、多列页眉、冻结列和页脚等。

在项目 easyui_demo 中创建页面 datagrid.jsp，在页面的<head></head>标签中引用相关的 css 和 js 文件。

在页面 datagrid.jsp 的<body></body>标签中编写如下代码：

```
<body>
    <table id="newsinfoDg" class="easyui-datagrid"></table>
    <script type="text/javascript">
        $(function() {
            $('#newsinfoDg').datagrid({
                fit : true,
                fitColumn : true,
                rownumbers : true,
                singleSelect : false,
                url : 'datagrid_data.txt',
                columns : [ [ {
                    title : '',
                    field : 'productid',
                    align : 'center',
                    checkbox : true
                }, {
                    field : 'unitcost',
                    title : 'unitcost',
                    width : 50
                }, {
                    field : 'status',
                    title : 'status',
                    width : 60
                }, {
                    field : 'listprice',
                    title : 'listprice',
```
```

```
                    width : 50
                } , {
                    field : 'attr1',
                    title : 'attr1',
                    width : 200
                } , {
                    field : 'itemid',
                    title : 'itemid',
                    width : 100
                } ] ]
            });
        });
    </script>
</body>
```

在项目的 WebRoot 目录下创建文件 datagrid_data.txt，作为 DataGrid 控件的数据源，代码如下：

```
[{"productid":"FI-SW-01","unitcost":10.00,"status":"P",
"listprice":36.50,"attr1":"Large","itemid":"EST-1"},
{"productid":"K9-DL-01","unitcost":12.00,"status":"P",
"listprice":18.50,"attr1":"Spotted Adult Female","itemid":"EST-10"},
// 由于篇幅，此处省略了其他数据
```

在浏览器中浏览页面 datagrid.jsp，效果如图 18-5 所示。

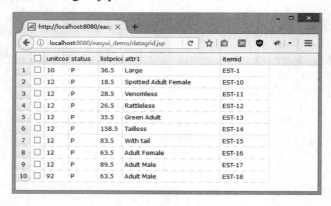

图 18-5 DataGrid 控件效果

18.2 Bootstrap 框架

Bootstrap 是美国 Twitter 公司的设计师 Mark Otto 和 Jacob Thornton 合作基于 HTML、CSS、JavaScript 开发的简洁、直观、强悍的前端开发框架，使得 Web 开发更加快捷。

18.2.1 Bootstrap 简介

Bootstrap 是一个用于快速开发 Web 应用程序和网站的前端框架，基于 HTML、CSS、

JavaScript。Bootstrap 中包含丰富的 Web 组件，根据这些组件，可以快速地搭建一个漂亮、功能完备的网站。Bootstrap 包括下拉菜单、按钮组、按钮下拉菜单、导航、导航条、路径导航、分页、排版、缩略图、警告对话框、进度条、媒体对象等组件。

18.2.2 环境安装

可以从 http://getbootstrap.com/ 上下载 Bootstrap 的最新版本，为了更好地了解和更方便地使用，本书中下载 Bootstrap 的预编译版本 bootstrap-3.3.7-dist.zip。解压缩这个压缩文件，里面包含 css 和 js 两个文件夹。其中，css 文件夹中主要包含*.css 文件，这里选择使用 bootstrap.min.css 文件，这是 Bootstrap 的基本样式；js 文件夹中包含*.js 文件，这里选择使用 bootstrap.min.js 文件，这是 Bootstrap 的 jQuery 插件源文件。

在 HBuilder 编辑器中，新建一个名为 BootstrapTest 的 Web 项目，将 bootstrap.min.css 文件添加到该项目的 css 目录中，将 bootstrap.min.js 文件添加到 js 目录中。

打开项目默认页 index.html，在页面中使用 Bootstrap 框架，代码如下：

```
<!DOCTYPE html>
<html>
    <head>
        <!-- 引入 Bootstrap -->
        <link href="css/bootstrap.min.css" rel="stylesheet">
        <!-- 引入 jQuery -->
        <script type="text/javascript" src="js/jquery-3.3.1.min.js"></script>
        <!-- 包括所有已编译的插件 -->
        <script type="text/javascript" src="js/bootstrap.min.js"></script>
        <meta charset="utf-8" />
        <title></title>
    </head>
    <body>
        <h1>Hello, world!</h1>
    </body>
</html>
```

在页面的 <head></head> 部分，依次引入 bootstrap.min.css、jquery-3.3.1.min.js 和 bootstrap.min.js。jquery-3.3.1.min.js 是 jQuery 库的基础文件，这个文件也需要添加到项目的 js 目录中。浏览页面 index.html，输出 Hello, world!。

18.2.3 Bootstrap 按钮

在项目 BootstrapTest 中，新建一个页面 button.html，用于实现 Bootstrap 的按钮效果，代码如下：

```
<!DOCTYPE html>
<html>
    <head>
        <!-- 引入 Bootstrap -->
```

```html
            <link href="css/bootstrap.min.css" rel="stylesheet">
            <!-- 引入 jQuery -->
            <script type="text/javascript"
    src="js/jquery-3.3.1.min.js"></script>
            <!-- 包括所有已编译的插件 -->
            <script type="text/javascript" src="js/bootstrap.min.js"></script>
            <meta charset="UTF-8">
            <title></title>
        </head>
        <body>
            <!-- 标准的按钮 -->
            <button type="button" class="btn btn-default">默认按钮</button>
            <!-- 提供额外的视觉效果，标识一组按钮中的原始动作 -->
            <button type="button" class="btn btn-primary">原始按钮</button>
            <br><br>
            <!-- 表示一个成功的或积极的动作 -->
            <button type="button" class="btn btn-success btn-sm">成功按钮
</button>
            <!-- 信息警告消息的上下文按钮 -->
            <button type="button" class="btn btn-info btn-lg">信息按钮</button>
            <br><br>
            <!-- 表示应谨慎采取的动作 -->
            <button type="button" class="btn btn-warning disabled">警告按钮
</button>
            <!-- 表示一个危险的或潜在的负面动作 -->
            <button type="button" class="btn btn-danger">危险按钮</button>
        </body>
    </html>
```

在页面 button.html 中，定义六个<button>按钮。任何带有 class .btn<button>的元素都会继承圆角灰色按钮的默认外观。此外，Bootstrap 提供了一些选项来定义按钮的样式，比如，btn-default 表示默认或标准按钮，btn-primary 表示原始按钮样式(未被操作)，btn-success 表示成功的动作，btn-info 表示需要弹出信息的按钮，btn-warning 表示需要谨慎操作的按钮，btn-danger 表示一个危险动作的按钮操作，btn-sm 表示制作一个小按钮，btn-lg 表示制作一个大按钮，disabled 表示禁用按钮。

浏览页面 button.html，效果如图 18-6 所示。

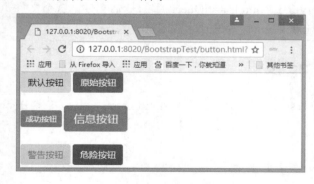

图 18-6　Bootstrap 按钮

18.2.4 Bootstrap 表格

在项目 BootstrapTest 中,新建一个页面 table.html,用于实现 Bootstrap 的表格效果,代码如下:

```html
<!DOCTYPE html>
<html>
    <head>
        <!-- 引入 Bootstrap -->
        <link href="css/bootstrap.min.css" rel="stylesheet">
        <!-- 引入 jQuery -->
        <script type="text/javascript" src="js/jquery-3.3.1.min.js"></script>
        <!-- 包括所有已编译的插件 -->
        <script type="text/javascript" src="js/bootstrap.min.js"></script>
        <meta charset="UTF-8">
        <title></title>
    </head>
    <body>
        <div class="table-responsive">
            <table class="table table-hover">
                <caption>响应式表格布局</caption>
                <thead>
                    <tr>
                        <th>产品</th>
                        <th>付款日期</th>
                        <th>状态</th>
                    </tr>
                </thead>
                <tbody>
                    <tr>
                        <td>产品 1</td>
                        <td>23/11/2013</td>
                        <td>待发货</td>
                    </tr>
                    <tr>
                        <td>产品 2</td>
                        <td>10/11/2013</td>
                        <td>发货中</td>
                    </tr>
                    <tr>
                        <td>产品 3</td>
                        <td>20/10/2013</td>
                        <td>待确认</td>
                    </tr>
                    <tr>
                        <td>产品 4</td>
                        <td>20/10/2013</td>
                        <td>已退货</td>
                    </tr>
```

```
            </tbody>
        </table>
    </div>
</body>
</html>
```

在页面 table.html 中，.table 类为任意<table>添加基本样式 (只有横向分隔线)。除了基本的表格标记和.table class，还有一些可以用来为标记定义样式的类。例如，添加.table-striped class，会在<tbody>内的行上出现条纹；添加.table-bordered class，会看到每个元素周围都有边框，且整个表格是圆角的；通过添加.table-hover class，当指针悬停在行上时会出现浅灰色背景；通过添加.table-condensed class，行内边距(padding)被切为两半，以便让表看起来更紧凑；通过把任意的.table 包在.table-responsive class 内，可以让表格水平滚动以适应小型设备(小于 768px)。当在大于 768px 宽的大型设备上查看时，将看不到任何差别。

运行页面 table.html，效果如图 18-7 所示。 

图 18-7　Bootstrap 表格

18.2.5　Bootstrap 网格系统

Bootstrap 提供了一套响应式、移动设备优先的流式网格系统，随着屏幕或视口尺寸的增加，系统会自动分为最多 12 列。

在项目 BootstrapTest 中，新建一个页面 grid.html，用于实现 Bootstrap 的网格结构，代码如下：

```
<!DOCTYPE html>
<html>
    <head>
        <!-- 引入 Bootstrap -->
        <link href="css/bootstrap.min.css" rel="stylesheet">
        <!-- 引入 jQuery -->
        <script type="text/javascript" src="js/jquery-3.3.1.min.js"></script>
        <!-- 包括所有已编译的插件 -->
        <script type="text/javascript" src="js/bootstrap.min.js"></script>
        <meta charset="UTF-8">
        <title></title>
    </head>
    <body>
        <div class="container">
            <div class="row">
                <div class="col-sm-3 col-md-6 col-lg-8" style="background-color: #dedef8;
```

```
            box-shadow: inset 1px -1px 1px #444,
            inset -1px 1px 1px #444;">
                    <p>Mr. Johnson had never been up in an aerophane before
and he had read a lot about air accidents, so one day when a friend offered
to take him for a ride in his own small phane, Mr. Johnson was very worried
about accepting. Finally, however, his friend persuaded him that it was very
safe, and Mr. Johnson boarded the plane.
                    </p>
                    <p>His friend started the engine and began to taxi onto
the runway of the airport. Mr. Johnson had heard that the most dangerous part
of a flight were the take-off and the landing, so he was extremely frightened
and closed his eyes.
                    </p>
                </div>
                <div class="col-sm-9 col-md-6 col-lg-4" style="background-color: #dedef8;
            box-shadow: inset 1px -1px 1px #444,
            inset -1px 1px 1px #444;">
                    <p>After a minute or two he opened them again, looked out
of the window of the plane, and said to his friend, "Look at those people
down there. They look as small as ants, don't they?"
                    </p>
                    <p> "Those are ants," answered his friend. "We're still
on the ground."
                    </p>
                </div>
            </div>
        </div>
    </body>
</html>
```

在页面 grid.html 中，提供了 3 种不同的列布局，分别适用于三种设备。在手机上，它将是左边 25%、右边 75%的布局；在平板电脑上，它将是 50%、50%的布局；在大型视口的设备上，它将是 33%、66%的布局。

浏览页面 grid.html，如果将页面调整到手机尺寸，效果如图 18-8 所示；如果将页面调整到平板电脑尺寸，效果如图 18-9 所示；如果将页面调整到大型视口的设备尺寸，效果如图 18-10 所示。

图 18-8 手机尺寸下网格效果　　　　图 18-9 平板电脑尺寸下网格效果

图 18-10　大型视口设备尺寸下网格效果

18.2.6　Bootstrap 下拉菜单

Bootstrap 提供的下拉菜单是可切换的，可以向任何组件(如导航栏、标签页、按钮等)添加下拉列表。如需使用下拉菜单，只需要在 class.dropdown 内加上下拉菜单即可。

在项目 BootstrapTest 中，新建一个页面 dropdowns.html，用于实现 Bootstrap 的下拉菜单，代码如下：

```html
<!DOCTYPE html>
<html>
    <head>
        <!-- 引入 Bootstrap -->
        <link href="css/bootstrap.min.css" rel="stylesheet">
        <!-- 引入 jQuery -->
        <script type="text/javascript"
src="js/jquery-3.3.1.min.js"></script>
        <!-- 包括所有已编译的插件 -->
        <script type="text/javascript" src="js/bootstrap.min.js"></script>
        <meta charset="UTF-8">
        <title></title>
        <style>
            .dropdown-menu li a {
                padding: 3px 20px;
                font-weight: normal;
                line-height: 1.42857;
                color: #000000;
            }
        </style>
    </head>
    <body>
        <div class="container" style="margin-top:20px">
            <div class="dropdown">
                <button class="btn dropdown-toggle" id="mydropdownmenu" data-toggle="dropdown">下拉菜单</button>
                <ul class="dropdown-menu">
                    <li>
                        <a href="#">社区</a>
                    </li>
                    <li>
                        <a href="#">服务</a>
```

```
            </li>
            <li>
                <a href="#">俱乐部</a>
            </li>
            <li class="divider"></li>
            <!--分割线 divider-->
            <li>
                <a href="#">交友</a>
            </li>
            <li class="dropdown-header">友情链接</li>
            <!--标题 dropdown-header-->
            <li>
                <a href="#">邮箱</a>
            </li>
            <li>
                <a rhref="#">苏宁</a>
            </li>
            <li>
                <a href="#">淘宝</a>
            </li>
            <li class="disabled">
                <a href="#">禁用</a>
            </li>
        </ul>
    </div>
  </div>
</body>
</html>
```

下拉菜单组件必须包含在 dropdown 类容器中,该容器包含下拉菜单的触发器(触发元素)和下拉菜单,下拉菜单必须包含在 dropdown-menu 容器中。

在页面 dropdowns.html 中,下拉菜单组件包含在 class="dropdown"的<div>标签中。在<div>标签中包含下拉菜单的触发器和下拉菜单。下拉菜单的触发器通过<button>标签来实现,在<button>标签中定义 data-toggle="dropdown"属性,用于激活下拉菜单的交互行为。下拉菜单通过<ul>标签来实现,在<ul>标签中定义 class="dropdown-menu"属性,用于设置下拉菜单的样式。

浏览页面 dropdowns.html,单击下拉菜单按钮,效果如图 18-11 所示。

图 18-11 Bootstrap 下拉菜单

18.2.7 Bootstrap 面板

Bootstrap 提供的面板组件 Panels 用于把 DOM 组件插入到一个盒子中,在项目 BootstrapTest 中,新建一个页面 panels.html,用于实现 Bootstrap 面板,代码如下:

```html
<!DOCTYPE html>
<html>
    <head>
        <meta charset="utf-8" />
        <title></title>
        <!-- 引入 Bootstrap -->
        <link href="css/bootstrap.min.css" rel="stylesheet">
        <!-- 引入 jQuery -->
        <script type="text/javascript" src="js/jquery-3.3.1.min.js"></script>
        <!-- 包括所有已编译的插件 -->
        <script type="text/javascript" src="js/bootstrap.min.js"></script>
    </head>
    <body>
        <div class="panel panel-default">
            <div class="panel-body">
                这是一个基本的面板
            </div>
        </div>
        <div class="panel panel-default">
            <div class="panel-heading">
                不带 title 的面板标题
            </div>
            <div class="panel-body">
                面板内容
            </div>
        </div>
        <div class="panel panel-default">
            <div class="panel-heading">
                <h3 class="panel-title">
                    带有 title 的面板标题
                </h3>
            </div>
            <div class="panel-body">
                面板内容
            </div>
        </div>
    </body>
</html>
```

创建一个基本的面板，只需要向<div>元素添加 class .panel 和 class .panel-default 即可。也可以通过以下两种方式来添加面板标题。

- 使用.panel-heading class 可以很简单地向面板添加标题容器。
- 使用带有.panel-title class 的<h1>～<h6>来添加预定义样式的标题。

浏览页面 panels.html，效果如图 18-12 所示。

图 18-12　Bootstrap 面板

18.2.8　Bootstrap 模态框

模态框(Modal)是覆盖在父窗体上的子窗体，用来显示来自一个单独的源的内容，可以在不离开父窗体的情况下有一些互动，子窗体可提供信息、交互等。

在项目 BootstrapTest 中，新建一个页面 modal.html，用于实现 Bootstrap 的模态框，代码如下：

```
<!DOCTYPE html>
<html>
    <head>
        <!-- 引入 Bootstrap -->
        <link href="css/bootstrap.min.css" rel="stylesheet">
        <!-- 引入 jQuery -->
        <script type="text/javascript" src="js/jquery-3.3.1.min.js"></script>
        <!-- 包括所有已编译的插件 -->
        <script type="text/javascript" src="js/bootstrap.min.js"></script>
        <meta charset="utf-8" />
        <title></title>
    </head>
    <body>
        <!-- 按钮触发模态框 -->
        <button class="btn btn-primary btn-lg" data-toggle="modal" data-target="#myModal">去登录...</button>
        <!-- 模态框 Modal -->
        <div class="modal fade" id="myModal" tabindex="-1" role="dialog" aria-labelledby="myModalLabel" aria-hidden="true">
            <div class="modal-dialog">
                <div class="modal-content">
                    <div class="modal-header">
                        <button type="button" class="close" data-dismiss="modal" aria-hidden="true">×</button>
                        <h4 class="modal-title" id="myModalLabel">登录窗口</h4>
                    </div>
                    <div class="modal-body">用户名</div><br>
                    <div class="modal-body">密码</div>
                    <div class="modal-footer">
                        <button type="button" class="btn btn-default" data-dismiss="modal">关闭</button>
```

```
                        <button type="button" class="btn btn-primary">
登录</button>
                    </div>
                </div>
                <!-- /.modal-content -->
            </div>
            <!-- /.modal-dialog -->
        </div>
        <!-- /.modal -->
        <script>
            $(function() {
                $('#myModal').modal('hide')
            });
        </script>
        <script>
            $(function() {
                $('#myModal').on('hide.bs.modal',
                    function() {
                        alert('你关闭了模态框登录窗口...');
                    })
            });
        </script>
    </body>
</html>
```

在页面 modal.html 中，定义了一个 id 为 myModal 的<div>标签。在这个<div>标签中定义 class="modal fade"属性，用来把<div>的内容识别为模态框，同时当模态框被切换时，它会引起内容淡入淡出；定义 aria-labelledby="myModalLabel"属性，用来引用模态框的标题；定义 aria-hidden="true"属性，用于保持模态窗口不可见，直到触发器被触发为止；定义 role="dialog"，用于指定模态框为对话框。

在模态框内，定义 class="modal-dialog"属性，用于窗口声明；定义 class="modal-content" 属性，用于内容声明；定义 class="modal-header"属性，用于为模态窗口的标题设置样式；定义 class="modal-body"属性，用于为模态窗口的主体设置样式；定义 class="modal-footer" 属性，用于为模态窗口的底部设置样式。

$(function(){})代码段内的代码用于隐藏模态框，也就是说页面加载时模态框是关闭的。为了打开模态框，定义了一个名为"去登录"的<button>标签。在这个<button>标签中定义 data-target="#myModal"属性，从而与模态框建立绑定关系。

浏览页面 modal.html，单击"去登录"按钮，就可以打开模态框了，效果如图 18-13 所示。

图 18-13　打开模态框

在这个模态框中，定义了"关闭"和"登录"两个按钮。在名为"关闭"的<button>标签中，定义 data-dismiss="modal"属性，用于关闭这个模态框。

在模态框中还可以处理一些事件，例如，当模态框关闭时会触发 hidden.bs.modal 事件，可以在回调函数中处理这个事件，可以弹出一个警告框，显示一些信息。

在打开的模态框中，单击"关闭"按钮。此时，会弹出一个警告框，显示"你关闭了模态框登录窗口..."提示，效果如图 18-14 所示。

图 18-14　处理模态框的关闭事件

18.2.9　Bootstrap 标签页

标签页(Tab)是 Bootstrap 的一个插件，标签页可用来显示多个面板，且面板之间可以切换。

在项目 BootstrapTest 中，新建一个页面 tab.html，用于实现 Bootstrap 的标签页，代码如下：

```
<!DOCTYPE html>
<html>
    <head>
        <!-- 引入 Bootstrap -->
        <link href="css/bootstrap.min.css" rel="stylesheet">
        <!-- 引入 jQuery -->
        <script type="text/javascript" src="js/jquery-3.3.1.min.js"></script>
        <!-- 包括所有已编译的插件 -->
        <script type="text/javascript" src="js/bootstrap.min.js"></script>
        <meta charset="utf-8" />
        <title></title>
    </head>
    <body>
        <ul id="myTab" class="nav nav-tabs">
            <li class="active">
                <a href="#page1" data-toggle="tab">
                    显示面板一
                </a>
            </li>
            <li>
                <a href="#page2" data-toggle="tab">显示面板二</a>
            </li>
        </ul>
        <div id="myTabContent" class="tab-content">
```

```
                    <div class="tab-pane fade in active" id="page1">
                        <p>标签页面板一</p>
                    </div>
                    <div class="tab-pane fade" id="page2">
                        <p>标签页面板二</p>
                    </div>
                </div>
            </body>
</html>
```

在页面 tab.html 中，定义了一个 id 为 myTabContent 的<div>标签。在这个<div>标签中定义 class="tab-content"属性，用于定义面板区容器。在这个<div>标签内，又定义了两个<div>子标签。在 id 为 page1 的<div>子标签中，定义了 class="tab-pane fade in active"属性，将这个<div>子标签创建成一个面板，并为标签页设置淡入淡出效果，且这个面板是激活的。在 id 为 page2 的<div>子标签中，定义了 class="tab-pane fade"属性，创建第二个面板，并设置淡入淡出效果。

为了能实现标签页中面板的切换，定义了一个 id 为 myTab 的<ul>标签，在这个<ul>标签中定义 class="nav nav-tabs"属性，用于创建一个标签式的导航菜单。在<ul>标签中定义了两个<a>标签，在<a>标签中定义 data-toggle="tab"属性，用于激活标签页插件；通过设置 href 属性值来指向对应面板的 id。

浏览页面 tab.html，效果如图 18-15 所示。此时，单击导航菜单"显示面板二"，可以切换到第二个面板，如图 18-16 所示。

图 18-15　标签页面板一

图 18-16　标签页面板二

18.3　Vue 框架

Vue.js 是一套构建用户界面的渐进式框架，只关注视图层，采用自底向上增量开发的设计，通过尽可能简单的 API 实现相应的数据绑定和组合的视图组件。

18.3.1　Vue 简介

与知名前端 Angular 一样，Vue.js 在设计上也使用了 MVVM(Model-View-View-Model) 模式。MVVM 模式本质上就是 MVC 的改进版，View 绑定到 ViewModel，然后执行一些命令，再向它请求一个动作，View 和 ViewModel 之间通过双向绑定建立联系。ViewModel

跟 Model 通讯，告诉它更新来响应 UI。这样便使得为应用构建 UI 非常容易。

18.3.2 第一个 Vue 应用

首先从 Vue.js 的官网上，直接下载 vue.min.js。然后在 HBuilder 编辑器中，新建一个名为 VueTest 的 Web 项目，将 vue.min.js 文件添加到该项目的 js 目录中。

打开项目默认页 index.html，在页面中使用<script>标签引入 vue.min.js 文件，再编写第一个 Vue 应用代码，如下所示：

```html
<!DOCTYPE html>
<html>
    <head>
        <meta charset="utf-8" />
        <title></title>
        <script type="text/javascript" src="js/vue.min.js"></script>
    </head>
    <body>
        <div id="app">
            {{ message }}
        </div>
        <script>
            var app = new Vue({
                el: '#app',
                data: {
                    message: 'Hello Vue!'
                }
            })
        </script>
    </body>
</html>
```

在页面 index.html 中的<script>部分，通过构造函数 Vue 创建了一个 Vue 的根实例，并启动 Vue 应用。Vue 的实例名为 app，实例内部可以包含多个选项。el 选项用于指定页面中用于挂载 Vue 实例的 DOM 元素，这里使用 id 为 app 的<div>元素来挂载 Vue 实例；data 选项用于定义数据对象，这个数据对象中有一个属性 message，它的值为 Hello Vue!，该对象被加入到一个 Vue 实例中。

在 id 为 app 的<div>元素中，使用双大括号{{ }}，这是最基本的文本插值方法，用于输出对象属性和函数返回值。这里，双大括号{{ }}里的内容会被替换为 Hello Vue!。如果数据对象中的 message 属性值发生改变，HTML 视图也会发生相应的变化。

浏览页面 index.html，页面显示"Hello Vue!"，效果如图 18-17 所示。

图 18-17　一个简单的 Vue 应用

18.3.3 生命周期

每个 Vue 实例在被创建时都要经过一系列的初始化过程,例如,需要设置数据监听、编译模板、将实例挂载到 DOM 并在数据变化时更新 DOM 等。同时在这个过程中也会运行一些叫做生命周期钩子的函数,用户可以在不同阶段添加自己的处理业务逻辑的代码。

Vue 生命周期钩子常用的有 created 和 mounted。created 函数在实例创建完成后调用,此阶段完成了 data 数据的初始化,但 el 还没有初始化;mounted 函数在 el 挂载到实例上之后调用,此阶段完成挂载,可以开始处理业务逻辑。

在项目 VueTest 中,新建一个页面 lifecycle.html,用于演示 Vue 生命周期,代码如下:

```html
<!DOCTYPE html>
<html>
    <head>
        <meta charset="utf-8" />
        <title></title>
        <script type="text/javascript" src="js/vue.min.js"></script>
    </head>
    <body>
        <div id="app">
            {{ message }}
        </div>
        <script>
            var app = new Vue({
                el: '#app',
                data: {
                    message: 'Hello Vue!'
                },
                created: function () {
                    console.group('created 创建完毕状态=======》');
                    // undefined
                    console.log("%c%s", "color:red", "el : " + this.$el);
                    // 已被初始化
                    console.log("%c%s", "color:red", "message: " + this.message);
                },
                mounted: function () {
                    console.group('mounted 挂载结束状态=======》');
                    // 已被初始化
                    console.log("%c%s", "color:red", "el : " + this.$el);
                    console.log(this.$el);
                    // 已被初始化
                    console.log("%c%s", "color:red", "message: " + this.message);
                }
            })
        </script>
    </body>
</html>
```

在 chrome 浏览器中浏览页面 lifecycle.html，按下 F12 键，打开开发者工具，在 Console 中查看函数的执行情况，如图 18-18 所示。

图 18-18　查看生命周期钩子

18.3.4　模板语法

Vue.js 使用了基于 HTML 的模板语法，可以采用简洁的模板语法声明式地将数据渲染进 DOM。通过结合响应系统，在应用状态改变时，能智能地计算重新渲染组件的最小代价并应用到 DOM 操作上。

1．插值

1) 文本

数据绑定最常见的形式就是使用{{...}}(双大括号)的文本插值，示例如下：

```
<div id="app">
  <p>{{ message }}</p>
</div>
```

2) 原始 HTML

双大括号会将数据解释为普通文本，而非 HTML 代码。为了输出真正的 HTML，需要使用 v-html 指令。

在项目 VueTest 中，新建一个页面 vhtml.html，使用 v-html 指令，代码如下：

```
<body>
    <div id="app">
        <div v-html="message"></div>
    </div>
    <script>
        var app = new Vue({
            el: '#app',
            data: {
                message: '<h1>Hello Vue!</h1>'
            }
        })
    </script>
</body>
```

浏览页面 vhtml.html，页面输出内容为 Hello Vue!。如果使用{{ message }}，则输出内容为<h1>Hello Vue!</h1>。

3) 属性

如果想动态更新 HTML 元素上的属性，比如 id、class 和 style 等，可以使用 v-bind 指令。

在项目 VueTest 中，新建一个页面 vbind.html，使用 v-bind 指令，代码如下：

```
<body>
    <div id="app">
        <div v-bind:style="styleObject">Hello Vue!</div>
    </div>
    <script>
        new Vue({
            el: '#app',
            data: {
                styleObject: {
                    color: 'red',
                    fontSize: '25px'
                }
            }
        });
    </script>
</body>
```

v-bind:style 的对象语法非常直观，它是一个 JavaScript 对象，很像 CSS。浏览页面 vbind.html，页面输出内容如图 18-19 所示。

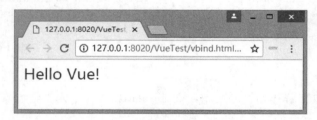

图 18-19　v-bind 指令示例效果

4) 表达式

Vue.js 提供了完全的 JavaScript 表达式支持，在项目 VueTest 中，新建一个页面 expression.html，在页面中使用表达式，代码如下：

```
<body>
    <div id="app">
        {{5+10}}<br> {{ flag ? 'YES' : 'NO' }}<br>
{{ message.split('').reverse().join('') }}
        <div v-bind:id="'list-' + id">Hello Vue! </div>
    </div>
    <script>
        var app = new Vue({
            el: '#app',
```

```
        data: {
            flag: true,
            message: 'Hello Vue!',
            id: 10
        }
    })
    </script>
</body>
```

这些表达式会在所属 Vue 实例的数据作用域下作为 JavaScript 脚本被解析，浏览页面 expression.html，页面输出内容如图 18-20 所示。

图 18-20　表达式使用效果

2. 指令

指令(Directives)是带有 v- 前缀的特殊特性，用于在表达式的值改变时，将某些行为应用到 DOM 上。

在项目 VueTest 中，新建一个页面 directives.html，在页面中使用指令，代码如下：

```
<body>
    <div id="app">
        <p v-if="show">这是一段文本</p>
    </div>
    <script>
        var app = new Vue({
            el: '#app',
            data: {
                show: true
            }
        })
    </script>
</body>
```

当数据 show 的值为 true 时，<p>标签会被插入，为 false 时则会被移除。运行页面 directives.html，页面输出内容为"这是一段文本"。如果将 show 修改为 false，则页面不显示这段文本。

一些指令能够接收一个参数，在指令名称之后以冒号隔开。例如 v-bind:href，这里 href 是参数，告知 v-bind 指令将该元素的 href 特性与表达式 url 的值绑定。

在项目 VueTest 中，新建一个页面 parameter.html，演示指令参数的使用，代码如下：

```
<body>
```

```
<div id="app">
    <pre><a v-bind:href="url">百度</a></pre>
</div>
<script>
    var app = new Vue({
        el: '#app',
        data: {
            url: 'http://www.baidu.com'
        }
    })
</script>
</body>
```

浏览页面 parameter.html，页面中显示一个百度链接，单击该连接，跳转到百度官网。

3．用户输入

在 input 输入框中，可以使用 v-model 指令实现双向数据绑定。在项目 VueTest 中，新建一个页面 vmodel.html，演示 v-model 指令的使用，代码如下：

```
<body>
    <div id="app">
        <p>{{ message }}</p>
        <input v-model="message">
    </div>
    <script>
        var app = new Vue({
            el: '#app',
            data: {
                message: 'Hello Vue!'
            }
        })
    </script>
</body>
```

v-model 指令可以自动让原生表单组件的值自动和用户输入的值绑定，在这个示例中输入框的值和数据 message 是绑定的，若输入框的值变化，则和它绑定的值也会发生变化。

浏览页面 vmodel.html，页面效果如图 18-21 所示。在输入框中修改内容后，和它绑定的值也会发生变化，页面效果如图 18-22 所示。

图 18-21　页面 vmodel.html 效果 1

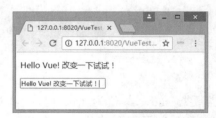

图 18-22　页面 vmodel.html 效果 2

4．过滤器

Vue.js 允许自定义过滤器，实现对输入数据的处理，并返回处理结果。过滤器在双花括号插值和 v-bind 表达式这两个地方使用。过滤器应该被添加在 JavaScript 表达式的尾部，由管道符号指示。

在项目 VueTest 中，新建一个页面 filter.html，演示 Vue 过滤器的使用，代码如下：

```
<body>
    <div id="app">
        {{ message|capitalize}}
    </div>
    <script>
        var app = new Vue({
            el: '#app',
            data: {
                message: 'welcome!'
            },
            filters: {
                capitalize: function(value) {
                    if(!value) return ''
                    value = value.toString()
                    return value.charAt(0).toUpperCase() + value.slice(1)
                }
            }
        })
    </script>
</body>
```

filters 选项用于在 Vue 实例内部注册一个过滤器，过滤器函数始终以表达式的值作为第一个参数。在这个示例中，capitalize 过滤器函数将会接收 message 的值作为第一个参数。

浏览页面 filter.html，数据对象 data 中的 message 属性值 welcome! 的首字母变为大写了，效果如图 18-23 所示。

图 18-23　Vue 过滤器示例效果

5．缩写

Vue.js 为 v-bind 和 v-on 这两个最常用的指令提供了特定简写，v-bind 指令缩写前后的对比如下：

```
<!-- 完整语法 -->
<a v-bind:href="url"></a>
<!-- 缩写 -->
<a :href="url"></a>
```

v-on 指令缩写前后的对比如下：

```
<!-- 完整语法 -->
<a v-on:click="doSomething"></a>
<!-- 缩写 -->
<a @click="doSomething"></a>
```

18.3.5 计算属性

模板内的表达式非常便利，常用于简单运算。但在模板中放入大量的逻辑会让模板难以维护。为了解决这个问题，可以使用计算属性。

在项目 VueTest 中，新建一个页面 computed_properties.html，使用计算属性显示变量 message 的反转字符串，代码如下：

```
<body>
    <div id="app">
        {{ reversedMessage }}
    </div>
    <script>
        var app = new Vue({
            el: '#app',
            data: {
                message: 'Hello Vue!'
            },
            computed: {
                // 计算属性的 getter
                reversedMessage: function() {
                    // this 指向当前 Vue 的实例
                    return this.message.split('').reverse().join('')
                }
            }
        })
    </script>
</body>
```

在 Vue 的 computed 选项中，定义了一个计算属性 reversedMessage，它提供的函数将作为属性 reversedMessage 的 getter 方法，返回变量 message 反转后的字符串，然后通过 {{ reversedMessage }} 显示出来。

浏览页面 computed_properties.html，页面显示 message 变量反转后的字符串，如图 18-24 所示。

图 18-24 Vue 计算属性示例效果

18.3.6 条件渲染

与 JavaScript 的条件语句 if、else、else if 类似，Vue.js 提供的条件渲染指令 v-if、v-else、v-else-if 可以根据表达式的值将 DOM 中的元素或组件渲染或销毁。

在项目 VueTest 中，新建一个页面 vif.html，使用 Vue.js 的条件渲染，代码如下：

```
<body>
    <div id="app">
        <p v-if="score>=90">优秀</p>
        <p v-else-if="score>=80">良好</p>
        <p v-else-if="score>=70">中等</p>
        <p v-else-if="score>=60">及格</p>
        <p v-else>不及格</p>
    </div>
    <script>
        var app = new Vue({
            el: '#app',
            data: {
                score: 85
            }
        })
    </script>
</body>
```

v-if、v-else、v-else-if 作为指令,必须将它们添加到一个标签上,这里在<p>标签上使用这些指令。v-else-if 要紧跟在 v-if 之后,v-else 要紧跟在 v-if 或 v-else-if 之后。哪个<p>标签上的条件成立,就渲染哪个标签,而其他不满足条件的<p>标签就会被销毁。

浏览页面 vif.html,页面输出如图 18-25 所示。

图 18-25　Vue 条件渲染示例效果

18.3.7　列表渲染

如果想要循环显示一个数组或一个对象属性时,可以使用 v-for 指令。在项目 VueTest 中,新建一个页面 vfor.html,使用 v-for 指令对数组类型数据进行渲染,代码如下:

```
<body>
    <div id="app">
        <ul>
            <li v-for="student in students">{{ student.name }}</li>
        </ul>
    </div>
    <script>
        var app = new Vue({
            el: '#app',
            data: {
                students: [
                    { name: 'zhangsan'},
                    { name: 'lisi'},
```

```
                    { name: 'wangwu'},
                    { name: 'zhaoliu'},
                ]
            }
        })
    </script>
</body>
```

在 data 选项中，定义了一个数组类型的数据 students，用 v-for 指令将<li>标签循环渲染。在 v-for 指令的表达式中，students 是源数据数组，student 是数组元素迭代的别名。

浏览页面 vfor.html，页面输出如图 18-26 所示。

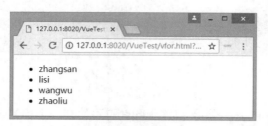

图 18-26　数组渲染

除了数组外，对象的属性也可以使用 v-for 指令进行渲染。在项目 VueTest 中，新建一个页面 vfor2.html，使用 v-for 指令渲染对象的属性，代码如下：

```
<body>
    <div id="app">
        <ul>
            <li v-for="(value,key,index) in stu">
{{ index }}-{{ key }}:{{ value }}</li>
        </ul>
    </div>
    <script>
        var app = new Vue({
            el: '#app',
            data: {
                stu: {
                    name: 'zhangsan',
                    age: 21,
                    gender: '男'
                }
            }
        })
    </script>
</body>
```

在 Vue 的 data 选项中，定义了一个对象 stu，它有三个属性。遍历对象属性时，有两个可选参数，index 是索引，key 是键名。

浏览页面 vfor2.html，页面输出如图 18-27 所示。

图 18-27　对象的属性渲染

18.3.8　方法和事件

在 Vue 实例中，可以在 methods 选项内定义一个方法。那么，如何通过一个按钮调用这个方法呢？可以使用 v-on 指令监听这个按钮的点击事件，并在触发时调用这个方法。

在项目 VueTest 中，新建一个页面 method_event.html，演示方法和事件的使用，代码如下：

```
<body>
    <div id="app">
        购买数量：{{ quantity }}
        <button @click="add()"> + </button>
        <button v-if="quantity>1" @click="subtract()"> - </button>
    </div>
    <script>
        var app = new Vue({
            el: '#app',
            data: {
                quantity: 1
            },
            methods: {
                add: function(quantity) {
                    quantity = quantity || 1;
                    // this 指向当前 Vue 实例 app
                    this.quantity += 1;
                },
                subtract: function(quantity) {
                    quantity = quantity || 1;
                    // this 指向当前 Vue 实例 app
                    this.quantity -= 1;
                }
            }
        })
    </script>
</body>
```

在 Vue 实例的 methods 选项内，定义了 add 和 subtract 两个方法，add 方法将 quantity 属性值加 1，subtract 方法将 quantity 属性值减 1。在 methods 中定义的方法可供@click 调用，在+和-两个<button>标签中，就是通过@click 调用 methods 内的方法的。

浏览页面 method_event.html，点击+或-按钮，可对购买数量计数，如图 18-28 所示。

图 18-28　方法和事件示例效果

18.3.9　Vue 组件

组件(Component)是 Vue.js 最强大的功能之一，组件可以扩展 HTML 元素，封装可重用的代码，通过可复用的小组件可以构建大型应用。

组件是可复用的 Vue 实例，所以它们与 new Vue 接收相同的选项，例如 data、computed、watch、methods 以及生命周期钩子等，只是没有 el 选项，el 是根实例特有的选项。

组件需要注册后才能使用，注册有全局注册和局部注册两种方式。全局注册后，所有的 Vue 实例都可以使用该组件。

在项目 VueTest 中，新建一个页面 component1.html，演示如何注册全局组件，代码如下：

```
<body>
    <div id="app">
        <my-counter></my-counter>
        <my-counter></my-counter>
    </div>
    <script>
        // 注册全局组件
        Vue.component('my-counter', {
            data: function() {
                return {
                    count: 0
                }
            },
            template: '<button @click="count++"> 点击了{{ count }}次</button>'
        });
        // 创建根实例
        var app = new Vue({
            el: '#app'
        })
    </script>
</body>
```

在页面 component1.html 中，注册了一个全局组件 my-counter，这个组件显示的内容是在 template 选项中定义的，这里定义了一个<button>模板。单击按钮时，会对这个组件内 data 函数返回的数据 count 执行递增操作。

全局组件 my-counter 注册后，在父实例中可以用<my-counter></my-counter>这样的形式来使用组件，这里组件 my-counter 使用了 2 次。

浏览页面 component1.html，出现两个按钮。单击这两个按钮，它们之间互不影响，可实现各自计数功能，如图 18-29 所示。

图 18-29 注册全局组件

使用组件可以把模板中的内容进行复用，然而组件之间还需要进行通信。通常父组件中会包含子组件，父组件要将数据传递给子组件，子组件会根据接收的数据来渲染不同的内容或执行不同的操作。要让子组件使用父组件的数据，需要通过子组件的 props 选项，也就是说子组件要显式地用 props 选项声明它期待获得的数据。使用 props 传递数据包括静态和动态两种形式。

在项目 VueTest 中，新建一个页面 component2.html，使用静态 props 实现将父组件数据传递给子组件，代码如下：

```
<body>
    <div id="app">
        <child message="Hello Vue!"></child>
    </div>
    <script>
        // 注册全局组件
        Vue.component('child', {
            // 声明 props,期望从父组件获取数据 message
            props: ['message'],
            template: '<span>{{ message }}</span>'
        })
        // 创建根实例
        var app = new Vue({
            el: '#app'
        })
    </script>
</body>
```

在子组件 child 的 template 模板中，{{ message }}显示的内容 message 就是父组件传递来的 Hello Vue!。浏览页面 component2.html，显示 Hello Vue!。

类似于用 v-bind 绑定 HTML 特性到一个表达式，也可以用 v-bind 动态绑定 props 的值到父组件的数据中。每当父组件的数据变化时，该变化也会传导给子组件。

在项目 VueTest 中，新建一个页面 component3.html，使用动态 props 实现将父组件数据传递给子组件，代码如下：

```
<body>
    <div id="app">
```

```
            <input v-model="parentMessage">
            <br>
            <child v-bind:message="parentMessage"></child>
        </div>
        <script>
            // 注册全局组件
            Vue.component('child', {
                // 声明 props,期望从父组件获取数据 message
                props: ['message'],
                template: '<span>{{ message }}</span>'
            })
            // 创建根实例
            var app = new Vue({
                el: '#app',
                data: {
                    parentMessage: ''
                }
            })
        </script>
    </body>
```

需要注意，props 是单向绑定的，当父组件的属性变化时，将传导给子组件，但是不会反过来。浏览页面 component3.html，页面效果如图 18-30 所示。

图 18-30　动态 props

在输入框中，输入 Hello, Vue，这个父组件的数据将实时传递给子组件 child。子组件将接收这个数据，并显示在自己的模板中，输入框下方显示的是子组件的内容，数据来自父组件。

18.3.10　Vue 脚手架

vue-cli 是 Vue 的脚手架工具，它大大降低了 webpack 的使用难度，支持热更新，有 webpack-dev-server 的支持，相当于启动了一个请求服务器，搭建了一个测试环境。

使用 vue-cli 前，要确保已经安装了最新版的 Node.js 和 NPM。由于篇幅所限，此处不再介绍，读者可以参照本节配套视频学习安装过程。

下面介绍如何使用 vue-cil 构建一个项目。

首先需要创建自己的工作空间，并在命令端口切换至刚刚创建好的工作空间，这里以 D 盘根目录为工作空间。

然后安装 vue-cil，可以直接在 cmd 命令端口输入如下命令。

```
npm install -g vue-cli
```

这个命令是全局安装 vue-cli，只需运行一次就可以了，以后就不用安装了。

安装完成后，在命令端口输入命令，创建一个基于 webpack 模板的新项目，如下所示：

```
vue init webpack vuedemo
```

vuedemo 是项目名称，输入命令后，会进入安装阶段，需要用户输入或确认一些信息，如图 18-31 所示。

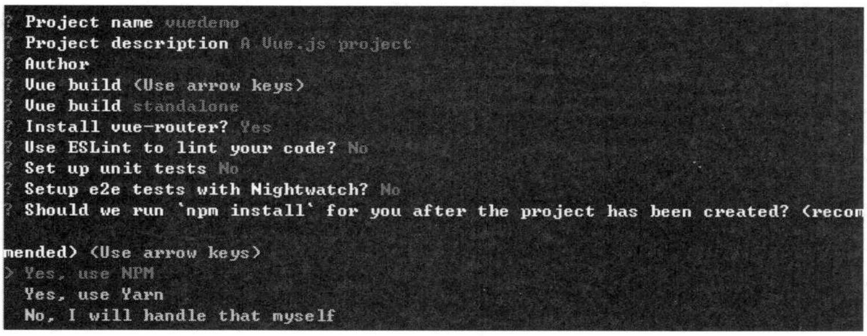

图 18-31　项目安装阶段输入或确认信息

项目创建后，需要运行 npm install 安装依赖模块，这里先选择 No,I will handle that myself 选项，表示自己稍后安装。

在命令端口中，切换到项目路径，如下所示：

```
cd vuedemo
```

再使用命令给项目添加依赖模块，如下所示：

```
npm install
```

安装完成后，在命令端口输入命令，就可以启动这个项目了，如下所示：

```
npm run dev
```

打开浏览器，输入地址 http://localhost:8080，页面效果如图 18-32 所示。

图 18-32　使用 vue-cli 创建的项目

18.3.11 Vue 路由

使用 Vue.js 路由可以根据不同的 URL 访问不同的内容，从而实现单页面富应用(SPA)。使用 Vue.js 路由前需要安装 vue-router。在 "Vue 脚手架" 一节创建的项目 vuedemo 中，已经添加了 vue-router。如果项目中没有安装，可以在命令端口输入如下命令：

```
npm install --save vue-router
```

在 SPA 应用中，每个页面对应一个.vue 文件。使用 Hbuilder 打开项目 vuedemo，在 src 目录下创建 views 子目录，然后在 views 里面创建 cart.vue 和 list.vue 两个 Vue 文件。

然后，在 src/router 目录下，创建 router.js 文件，内容如下：

```
// 引入组件
import cart from "../views/cart.vue";
import list from "../views/list.vue";
const routers = [
    {
        path: '/list',
        meta: {
            title: '商品列表'
        },
        component: list
    }, {
        path: '/cart',
        meta: {
            title: '购物车'
        },
        component: cart
    }
];
export default routers;
```

在 router.js 文件中，首先引入 cart.vue 和 list.vue 两个组件，然后创建一个数组 routers 来指定路由匹配列表，每个路由映射一个组件。

在 main.js 文件中，首先引入 vue-router 和 src/router 目录下的 router.js 文件，再加载 vue-router 插件，如下所示：

```
import Vue from 'vue'
import App from './App'
import router from './router'
// 引入 vue-router
import VueRouter from 'vue-router';
// 引入 src/router 目录下的 router.js 文件
import Routers from './router/router';
// 加载 vue-router 插件
Vue.use(VueRouter)
```

然后完成路由配置，并定义路由组件，如下所示：

```
// 路由配置
```

```
const RouterConfig = {
    // 使用 HTML5 的 History 路由模式
    mode: 'history',
    routes: Routers
};
// 定义路由组件
const router = new VueRouter(RouterConfig);
```

在 RouterConfig 里，设置 mode 为 history，使用 HTML5 的 History 路由模式，通过"/"设置路径。如果不配置，则使用"#"设置路径。设置 routes 为 Routers，Routers 就是之前引入的 src/router 目录下的 router.js 文件。以 RouterConfig 为参数，通过实例化 VueRouter 来创建一个路由组件 router。

最后在 Vue 的根实例中引用这个路由组件，如下所示：

```
new Vue({
    el: '#app',
    router: router,  // 引用路由实例 router
    render: h => {
        return h(App)
    }
});
```

至此，路由就配置好了。接下来就可以设置跳转了，可以使用 vue-router 提供的 <router-link>，它会被渲染为一个<a>标签。

在 App.vue 组件的模板中，添加两个<router-link>，用作购物车页和商品列表页的链接，如下所示：

```
<template>
    <div id="app">
        <!--<img src="./assets/logo.png">-->
        <router-view/>
        <router-link to="/cart">购物车</router-link>
        <router-link to="/list">商品列表</router-link>
    </div>
</template>
```

在命令端口启动服务，打开浏览器，输入地址 http://localhost:8080/vuedemo，页面效果如图 18-33 所示。在页面中，单击购物车链接，打开购物车页，如图 18-34 所示。单击商品列表链接，则打开商品列表页。

图 18-33　App.vue 组件内容

图 18-34　购物车页

18.3.12　Vuex 状态管理

Vuex 是一个专为 Vue.js 应用程序开发的状态管理模式，它采用集中式存储管理应用的所有组件的状态，并以相应的规则保证状态以一种可预测的方式发生变化。

一个组件通常包括数据和视图，数据改变时，视图也会随之更新。在视图中，可以绑定一些事件。例如，可以给一个<button>按钮绑定 click 事件，然后通过 methods 选项中定义的方法处理这个事件，从而改变数据、更新视图。但是一个组件内的数据和方法只能在这个组件里被使用，其他组件无法使用。为了能让其他组件共享这些数据和方法，可以通过 Vuex 来统一管理组件状态。

首先需要安装 Vuex，在命令端口中，切换到项目 vuedemo 的路径，这里为 d:\vuedemo。在命令端口中输入命令，通过 NPM 安装 Vuex，如下所示：

```
npm intall --save vuex
```

然后在 main.js 文件中，引入并加载 Vuex 插件，如下所示：

```
// 引入 Vuex
import Vuex from 'vuex';
// 加载 Vuex 插件
Vue.use(Vuex);
```

接着在 main.js 文件中配置 Vuex，如下所示：

```
// Vuex 配置
const store = new Vuex.Store({
  state: {
    count: 0
  },
  mutations: {
    increment (state) {
      state.count++
    }
  }
})
```

最后，还需要在根实例中引用这个 Vuex 实例 store，如下所示：

```
new Vue({
    el: '#app',
    router: router,
    store:store,     // 引用 Vuex 实例 store
    render: h => {
        return h(App)
    }
});
```

仓库 store 包含应用数据(状态)和操作过程，数据保存在 Vuex 的 state 选项中，这里定义了一个数据 count，初始值为 0。

任何一个组件都可以使用这个数据 count，在 App.vue 组件的模板<template>中使用数据 count，代码如下：

```
<template>
    <div id="app">
        ……
        <!-- 使用 Vuex 中的数据 -->
        {{ $store.state.count }}
    </div>
</template>
```

此时，在命令端口启动服务，打开浏览器，输入地址 http://localhost:8080/vuedemo，页面显示计数 0，如图 18-35 所示。

如果想修改 state 中的数据 count，可以在 Vuex 的 mutations 选项中定义方法，这里定义了 increment 方法，实现数据 count 的自增。

在组件中，可以通过 this.$store.commit 方法调用 Vuex 的 mutations 选项中定义的方法。在 App.vue 组件中，首先在模板<template>中添加一个<button>按钮，如下所示：

```
<button @click="add">+</button>
```

然后在<script></script>部分，添加一个 methods 选项，在选项中添加一个 add 方法，用来处理按钮的 click 事件。在这个事件处理函数中，调用 Vuex 的 mutations 选项中定义的 increment 方法，如下所示：

```
<script>
    export default {
        name: 'App',
        methods: {
            add() {
                this.$store.commit('increment');
            }
        }
    }
</script>
```

此时，页面多了一个按钮，单击这个按钮，计数会增加，如图 18-36 所示。

图 18-35　访问 state 中的数据

图 18-36　访问 mutations 中的方法

假设在 Vuex 的 state 选项中有一个数组类型的数据 priceList，里面存放了不同商品的价格。在组件中，如果不想直接使用这个数据，而是希望对价格过滤后再使用，则可以使用 Vuex 的 getters 选项。

在 main.js 文件的 Vuex 中，首先在 state 选项中定义一个数组类型的数据 priceList，值为[624, 134, 453, 567, 245]；然后添加一个 getters 选项，以价格小于 500 为条件，计算 priceList 过滤后的数据，并存储在数据 filterPrice 中，如下所示：

```
// Vuex 配置
const store = new Vuex.Store({
    state: {
        count: 0,
        // 价格列表
        priceList: [624, 134, 453, 567, 245]
    },
    mutations: {
        increment(state) {
            state.count++
        }
    },
    getters: {
        filterPrice: state => {
            return state.priceList.filter(item => item < 500);
        }
    }
})
```

在 App.vue 组件的模板中，使用过滤后的数据 filterPrice，如下所示：

```
<!-- 获取过滤后的价格 -->
{{ $store.getters.filterPrice }}
```

此时，页面显示了价格小于 500 的数字，如图 18-37 所示。

图 18-37　访问 getters 中的方法

之前讲解的 mutations 是同步改变 state 状态的，如果想异步改变，可以使用 Vuex 的 actions 选项。

在 main.js 文件的 Vuex 中，添加一个 actions 选项，如下所示：

```
// Vuex 配置
const store = new Vuex.Store({
    ……
    mutations: {
        increment(state) {
            state.count++
        }
    },
    ……
    actions: {
```

```
        // 异步调用，3 秒后再提交 mutations 中的 increment 方法
        asycIncrement(context, callback) {
            setTimeout(() => {
                context.commit('increment');
                callback();
            }, 3000);
        }
    }
})
```

在 actions 选项中，定义一个 asycIncrement 方法，通过普通的回调来实现异步调用。actions 与 mutations 很像，不同的是 actions 里提交的是 mutations。由于实际开发中，通过异步调用向后台服务器请求数据是一个耗时的操作。为了模拟这个效果，这里设置了 3 秒后再提交 mutations 中的 increment 方法。

在 App.vue 组件的<script></script>部分，修改 methods 选项中的 add 方法，如下所示：

```
<script>
    export default {
        name: 'App',
        methods: {
            add() {
                /*this.$store.commit('increment');*/
                this.$store.dispatch('asycIncrement', () => {
                    console.log(this.$store.state.count);
                })
            }
        }
    }
</script>
```

在 App.vue 组件中，通过 this.$store.dispatch 来触发 Vuex 的 actions 选项中定义的 asycIncrement 方法。

此时，浏览页面，单击+按钮，可以看到等待 3 秒后，计数才会增加。

18.4 小　　结

本章介绍了当前流行的 jQuery EasyUI、Bootstrap 和 Vue 三个前端 UI 框架，希望读者能够初步了解这三个框架。本书最后将基于 SSM 整合，结合这三个前端 UI 框架，详细讲解三个典型的项目案例的开发过程。

第 19 章 电商平台后台管理系统

第 17 章以用户登录功能为例,详细讲解了 Spring、Spring MVC 与 MyBatis 框架(SSM)整合的流程。本章将使用 SSM 整合,并结合前端 Easy UI 框架实现电商平台后台管理系统。

19.1 需求与系统分析

电商平台后台管理系统用于管理员登录系统后,对商品信息、商品类型、订单信息和客户信息进行管理。在这个系统中,管理员用例图如图 19-1 所示。

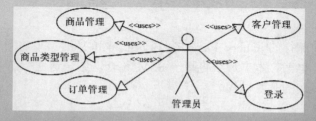

图 19-1 管理员用例图

根据需求分析,管理员拥有如下功能权限。
(1) 商品管理,包括添加商品、商品下架、修改商品、查询商品。
(2) 商品类型管理,包括添加商品类型、修改商品类型。
(3) 订单管理,包括创建订单、查询订单、删除订单、查看订单明细。

(4) 客户管理，查询客户、禁用或启用客户。

根据上述分析，可以得到系统的模块结构，如图 19-2 所示。

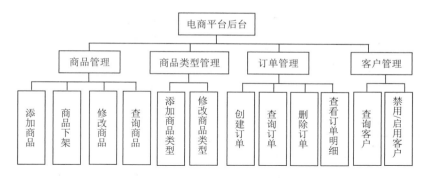

图 19-2　系统的模块结构

19.2　数据库设计

根据系统需求，创建名称为 eshop 的数据库，创建 8 张数据表，如下所示。
(1) 客户信息表 user_info，用于记录前台客户基本信息。
(2) 管理员信息表 admin_info，用于记录管理员基本信息。
(3) 商品类型表 type，用于记录各种商品类型。
(4) 商品信息表 product_info，用于记录商品信息。
(5) 订单信息表 order_info，用于记录订单主要信息。
(6) 订单明细表 order_detail，用于记录订单详细信息。
(7) 系统功能表 functions，用于记录系统功能信息。
(8) 权限表 powers，用于记录管理员权限。

客户信息表 user_info 的字段说明如表 19-1 所示。

表 19-1　客户信息表 user_info

字段名	类型	主外键	说明
id	int(4)	PK	客户 id 标识，主键，自增
userName	varchar(16)		登录名
password	varchar(16)		登录密码
realName	varchar(8)		真实姓名
sex	varchar(4)		性别
address	varchar(255)		联系地址
email	varchar(50)		电子邮件
regDate	date		注册时间
status	int(4)		客户状态

管理员信息表 admin_info 的字段说明如表 19-2 所示。

表 19-2　管理员信息表 admin_info

字段名	类型	主外键	说明
id	int(4)	PK	管理员 id 标识，主键，自增
name	varchar(16)		管理员姓名
pwd	varchar(50)		管理员密码

商品类型表 type 的字段说明如表 19-3 所示。

表 19-3　商品类型表 type

字段名	类型	主外键	说明
id	int(4)	PK	类型 id 标识，主键，自增
name	varchar(20)		类型名称

商品信息表 product_info 的字段说明如表 19-4 所示。

表 19-4　商品信息表 product_info

字段名	类型	主外键	说明
id	int(4)	PK	商品 id 标识，主键，自增
code	varchar(16)		商品编号
name	varchar(255)		商品名称
tid	int(4)	FK	商品类别 id
brand	varchar(20)		品牌
pic	varchar(255)		商品图片
num	int(4)		商品数量
price	decimal(10,0)		商品价格
intro	longtext		商品描述
status	int(4)		商品状态

订单信息表 order_info 的字段说明如表 19-5 所示。

表 19-5　订单信息表 order_info

字段名	类型	主外键	说明
id	int(4)	PK	订单 id 标识，主键，自增
uid	int(4)	FK	客户 id
status	varchar(16)		订单状态
ordertime	datetime		订单下单时间
orderprice	decimal(8,2)		订单金额

订单明细表 order_detail 的字段说明如表 19-6 所示。

表 19-6　订单明细表 order_detail

字段名	类型	主外键	说明
id	int(4)	PK	订单明细 id，主键，自增
oid	int(4)	FK	订单 id
pid	int(4)	FK	商品 id
num	int(4)		购买数量

系统功能表 functions 的字段说明如表 19-7 所示。

表 19-7　系统功能表 functions

字段名	类型	主外键	说明
id	int(4)	PK	系统功能 id，主键，自增
name	varchar(20)		功能菜单名称
parentid	int(4)		父结点 id
url	varchar(50)		功能页面
isleaf	bit(1)		是否叶子结点
nodeorder	int(4)		结点顺序

权限表 powers 的字段说明如表 19-8 所示。

表 19-8　权限表 powers

字段名	类型	主外键	说明
aid	int(4)	PK	管理员 id，主键
fid	int(4)	PK	系统功能 id，主键

创建数据表时，还需要设置数据表之间的关联关系，如图 19-3 所示。

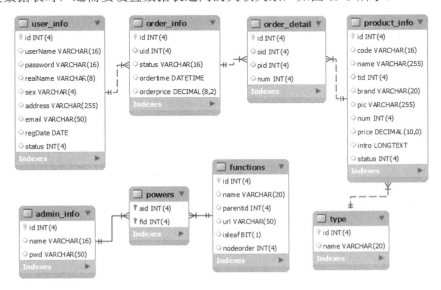

图 19-3　系统数据表之间关系图

19.3 环境搭建与配置文件

可以参照第 17 章 Spring 整合 MyBatis，完成电商平台后台管理系统 ecpbm 的框架搭建及相关配置文件的编写，项目最终的目录结构如图 19-4 所示。

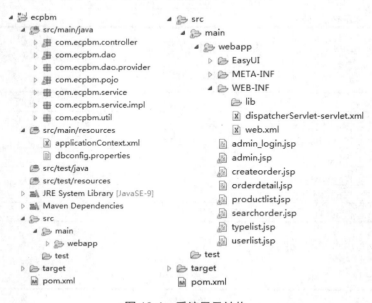

图 19-4　系统目录结构

com.ecpbm.controller 包用于存放控制器类，com.ecpbm.service 包用于存放业务逻辑层接口，com.ecpbm.service.impl 包用于存放业务逻辑层接口的实现类，com.ecpbm.dao 包用于存放数据访问层接口，com.ecpbm.dao.provider 包用于存放 DynaSqlProvider 类，com.ecpbm.pojo 包用于存放实体类。dbconfig.properties 为存储数据库连接信息的属性文件，applicationContext.xml 为 Spring 框架的配置文件，dispatcherServlet-servlet.xml 为 Spring MVC 框架的配置文件，admin_login.jsp 为管理员登录页，admin.jsp 为后台管理首页面，productlist.jsp 为商品列表页，createorder.jsp 为创建订单页，searchorder.jsp 为订单查询页，orderdetail.jsp 为订单明细页，userlist.jsp 为客户列表页，typelist.jsp 为商品类型列表页，Easyui 目录下的文件或子目录下的文件为使用 Easy UI 控件所需的 js、css 等文件。

19.4 创建实体类

在 com.ecpbm.pojo 包中，依次创建实体类 UserInfo、AdminInfo、Functions、Powers、ProductInfo、Type、OrderInfo 和 OrderDetail。

实体类 UserInfo 用于封装客户信息，代码如下：

```
package com.ecpbm.pojo;
public class UserInfo {
    private int id;
```

```
    private String userName;
    private String password;
    private String realName;
    private String sex;
    private String address;
    private String email;
    private String regDate;
    private int status;
    // 省略上述属性的 getter 和 setter 方法
}
```

实体类 AdminInfo 用于封装管理员信息，代码如下：

```
package com.ecpbm.pojo;
import java.util.List;
public class AdminInfo {
    private int id;
    private String name;
    private String pwd;
    // 关联的属性
    private List<Functions> fs;
    // 省略上述属性的 getter 和 setter 方法
}
```

实体类 Functions 用于封装系统功能信息，代码如下：

```
package com.ecpbm.pojo;
import java.util.HashSet;
import java.util.Set;
public class Functions implements Comparable<Functions> {
    private int id;
    private String name;
    private int parentid;
    private boolean isleaf;
    // 关联的属性
    private Set ais = new HashSet();
    // 省略上述属性的 getter 和 setter 方法
    @Override
    public int compareTo(Functions arg0) {
        return ((Integer) this.getId()).compareTo((Integer)
(arg0.getId()));
    }
}
```

在实体类 Functions 中，重写了 compareTo(Functions arg0)方法。该方法用于在排序时将两个 Functions 对象的 id 进行比较，根据比较的结果是小于、等于或者大于而返回一个负数、零或者正数。

实体类 Powers 用于封装权限信息，代码如下：

```
package com.ecpbm.pojo;
public class Powers {
    private AdminInfo ai;
```

```java
    private Functions f;
    // 省略上述属性的getter和setter方法
    @Override
    public String toString() {
        return "Powers [ai=" + ai + ", f=" + f + "]";
    }
}
```

实体类 ProductInfo 用于封装商品信息，代码如下：

```java
package com.ecpbm.pojo;
public class ProductInfo {
    // 商品基本信息(部分)
    private int id; // 商品编号
    private String code; // 商品编码
    private String name; // 商品名称
    // 关联属性
    private Type type;  // 商品类型
    private String brand; // 商品品牌
    private String pic; // 商品小图
    private int num; // 商品数量
    private double price; // 商品价格
    private String intro; // 商品介绍
    private int status; // 商品状态
    private double priceFrom;
    private double priceTo;
// 省略上述属性的getter和setter方法
}
```

实体类 Type 用于封装商品类型信息，代码如下：

```java
package com.ecpbm.pojo;
public class Type {
    private int id; // 产品类型编号
    private String name; // 产品类型名称
    // 省略上述属性的getter和setter方法
}
```

实体类 OrderInfo 用于封装订单信息，代码如下：

```java
package com.ecpbm.pojo;
public class OrderInfo {
    private Integer id;
    private int uid;
    private UserInfo ui;
    private String status;
    private String ordertime;
    private double orderprice;
    private String orderTimeFrom;
    private String orderTimeTo;
    // 省略上述属性的getter和setter方法
}
```

实体类 OrderDetail 用于封装订单明细信息，代码如下：

```java
package com.ecpbm.pojo;
public class OrderDetail {
    private int id;
    private int oid;
    private OrderInfo oi;
    private int pid;
    private ProductInfo pi;
    private int num;
    private double price;
    private double totalprice;
    // 省略上述属性的 getter 和 setter 方法
}
```

此外，还创建了三个辅助类：Pager、SearchProductInfo 和 TreeNode。Pager 类用于封装分页信息，代码如下：

```java
package com.ecpbm.pojo;
public class Pager {
    private int curPage;// 待显示页
    private int perPageRows;// 每页显示的记录数
    private int rowCount;  // 记录总数
    private int pageCount;  // 总页数
    // 省略 curPage、perPageRows、rowCount 属性的 getter 和 setter 方法
    // 根据 rowCount 和 perPageRows 计算总页数
    public int getPageCount() {
        return (rowCount + perPageRows - 1) / perPageRows;
    }
    // 分页显示时，获取当前页的第一条记录的索引
    public int getFirstLimitParam() {
        return (this.curPage - 1) * this.perPageRows;
    }
}
```

SearchProductInfo 类用于封装商品查询条件，代码如下：

```java
package com.ecpbm.pojo;
public class SearchProductInfo {
    // 产品基本信息(部分)
    private int id; // 产品编号
    private String code; // 产品编码
    private String name; // 产品名称
    private String brand; // 产品品牌
    private double priceFrom;
    private double priceTo;
    private int tid;
    // 省略上述属性的 getter 和 setter 方法
}
```

系统使用 Easy UI 提供的 Tree 控件来显示菜单，TreeNode 类用于封装树形控件的结点信息，代码如下：

```java
package com.ecpbm.pojo;
import java.util.List;
public class TreeNode {
    private int id;    // 结点 id
    private String text;    // 结点名称
    private int fid;    // 父结点 id
    private List<TreeNode> children;    // 包含的子结点
    // 省略上述属性的 getter 和 setter 方法
}
```

19.5 创建几个 Dao 接口及动态提供类

在 com.ecpbm.dao 包中，创建数据访问层接口 UserInfoDao、AdminInfoDao、ProductInfoDao、TypeDao、OrderInfoDao、FunctionDao。在这些 Dao 接口中，使用 MyBatis 注解完成数据表的操作。在接口 UserInfoDao 中，编写如下代码：

```java
package com.ecpbm.dao;
import java.util.List;
import java.util.Map;
import org.apache.ibatis.annotations.Param;
import org.apache.ibatis.annotations.Select;
import org.apache.ibatis.annotations.SelectProvider;
import org.apache.ibatis.annotations.Update;
import com.ecpbm.dao.provider.UserInfoDynaSqlProvider;
import com.ecpbm.pojo.UserInfo;
public interface UserInfoDao {
    // 获取系统合法客户，即数据表 user_info 中 status 字段为 1 的客户列表
    @Select("select * from user_info where status=1")
    public List<UserInfo> getValidUser();
    // 根据客户 id 号获取客户对象
    @Select("select * from user_info where id=#{id}")
    public UserInfo getUserInfoById(int id);
    // 分页获取客户信息
    @SelectProvider(type = UserInfoDynaSqlProvider.class, method = "selectWithParam")
    List<UserInfo> selectByPage(Map<String, Object> params);
    // 根据条件查询客户总数
    @SelectProvider(type = UserInfoDynaSqlProvider.class, method = "count")
    Integer count(Map<String, Object> params);
    // 更新客户状态
    @Update("update user_info set status=#{flag} where id in (${ids})")
    void updateState(@Param("ids") String ids, @Param("flag") int flag);
}
```

在接口 UserInfoDao 中，getValidUser 方法通过@Select 注解映射一条 SELECT 语句，获取系统合法客户。getUserInfoById 方法通过@Select 注解映射一条 SELECT 语句，根据客户 id 号获取客户对象。selectByPage 方法通过@SelectProvider 注解调用 UserInfoDynaSqlProvider 类中的 selectWithParam 方法获得一条 SELECT 语句，分页获取客

户信息，Map<String, Object>类型的参数 params 用于封装查询条件。接口 UserInfoDao 中的 count 方法通过@SelectProvider 注解调用 UserInfoDynaSqlProvider 类中的 count 方法获得一条 SELECT 语句，根据条件查询客户总数，Map<String, Object>类型的参数 params 用于封装查询条件。updateState 方法通过@Update 注解映射一条 UPDATE 语句，更新客户状态，并使用@Param 注解将前端页面传递来的参数 ids 和 flag 的值分别赋值给 String 类型的变量 ids 和 int 类型的变量 flag。

在 com.ecpbm.dao.provider 包中，创建动态 SQL 提供类 UserInfoDynaSqlProvider，并添加 selectWithParam 和 count 两个方法，代码如下：

```java
package com.ecpbm.dao.provider;
import java.util.Map;
import org.apache.ibatis.jdbc.SQL;
import com.ecpbm.pojo.UserInfo;
public class UserInfoDynaSqlProvider {
    // 分页动态查询
    public String selectWithParam(Map<String, Object> params) {
        String sql = new SQL() {
            {
                SELECT("*");
                FROM("user_info");
                if (params.get("userInfo") != null) {
                    UserInfo userInfo = (UserInfo) params.get("userInfo");
                    if (userInfo.getUserName() != null
&& !userInfo.getUserName().equals("")) {
                        WHERE(" userName LIKE CONCAT ('%',#{userInfo.userName},'%') ");
                    }
                }
            }
        }.toString();
        if (params.get("pager") != null) {
            sql += " limit #{pager.firstLimitParam} , #{pager.perPageRows} ";
        }
        return sql;
    }
    // 根据条件动态查询总记录数
    public String count(Map<String, Object> params) {
        return new SQL() {
            {
                SELECT("count(*)");
                FROM("user_info");
                if (params.get("userInfo") != null) {
                    UserInfo userInfo = (UserInfo) params.get("userInfo");
                    if (userInfo.getUserName() != null
&& !userInfo.getUserName().equals("")) {
                        WHERE(" userName LIKE CONCAT ('%',#{userInfo.userName},'%') ");
                    }
```

```
            }
        }
    }.toString();
}
```

在接口 AdminInfoDao 中，编写如下代码：

```
package com.ecpbm.dao;
import org.apache.ibatis.annotations.Many;
import org.apache.ibatis.annotations.Result;
import org.apache.ibatis.annotations.Results;
import org.apache.ibatis.annotations.Select;
import org.apache.ibatis.mapping.FetchType;
import com.ecpbm.pojo.AdminInfo;
public interface AdminInfoDao {
    // 根据登录名和密码查询管理员
    @Select("select * from admin_info where name = #{name} and pwd = #{pwd}")
    public AdminInfo selectByNameAndPwd(AdminInfo ai);
    // 根据管理员id获取管理员对象及关联的功能集合
    @Select("select * from admin_info where id = #{id}")
    @Results({ @Result(id = true, column = "id", property = "id"),
@Result(column = "name", property = "name"),
        @Result(column = "pwd", property = "pwd"),
        @Result(column = "id", property = "fs", many = @Many(select =
"com.ecpbm.dao.FunctionDao.selectByAdminId", fetchType =
FetchType.EAGER)) })
    AdminInfo selectById(Integer id);
}
```

在接口 AdminInfoDao 中，selectByNameAndPwd 方法通过 @Select 注解映射一条 SELECT 语句，根据登录名和密码查询管理员，AdminInfo 类型的参数 ai 用于封装查询条件。selectById 方法通过 @Select 注解映射一条 SELECT 语句，根据管理员 id 获取管理员对象及关联的功能集合，并通过 @Results 注解映射查询结果。

在接口 ProductInfoDao 中，编写如下代码：

```
package com.ecpbm.dao;
import java.util.List;
import java.util.Map;
import org.apache.ibatis.annotations.*;
import org.apache.ibatis.mapping.FetchType;
import com.ecpbm.dao.provider.ProductInfoDynaSqlProvider;
import com.ecpbm.pojo.ProductInfo;
public interface ProductInfoDao {
    // 分页获取商品
    @Results({ @Result(id = true, column = "id", property = "id"),
@Result(column = "code", property = "code"),
        @Result(column = "name", property = "name"),
@Result(column = "brand", property = "brand"),
        @Result(column = "pic", property = "pic"),
 @Result(column = "num", property = "num"),
```

```java
            @Result(column = "price", property = "price"),
    @Result(column = "intro", property = "intro"),
            @Result(column = "status", property = "status"),
            @Result(column = "tid", property = "type", one = @One(select =
"com.ecpbm.dao.TypeDao.selectById", fetchType = FetchType.EAGER)) })
    @SelectProvider(type = ProductInfoDynaSqlProvider.class,
method = "selectWithParam")
    List<ProductInfo> selectByPage(Map<String, Object> params);
    // 根据条件查询商品总数
    @SelectProvider(type = ProductInfoDynaSqlProvider.class, method =
"count")
    Integer count(Map<String, Object> params);
    // 添加商品
    @Insert("insert into product_info(code,name,tid,brand,pic,
num,price,intro,status) "
            + "values(#{code},#{name},#{type.id},#{brand},#{pic},
#{num},#{price},#{intro},#{status})")
    @Options(useGeneratedKeys = true, keyProperty = "id")
    void save(ProductInfo pi);
    // 修改商品
    @Update("update product_info set
code=#{code},name=#{name},tid=#{type.id}," +
"brand=#{brand},pic=#{pic},num=#{num},price=#{price},intro=#{intro}," +
"status=#{status} where id=#{id}")
    void edit(ProductInfo pi);
    // 更新商品状态
    @Update("update product_info set status=#{flag} where id in (${ids})")
    void updateState(@Param("ids") String ids, @Param("flag") int flag);
    // 获取在售商品列表
    @Select("select * from product_info where status=1")
    List<ProductInfo> getOnSaleProduct();
    // 根据商品 id 获取商品对象
    @Select("select * from product_info where id=#{id}")
    ProductInfo getProductInfoById(int id);
}
```

在接口 ProductInfoDao 中，selectByPage 方法通过@SelectProvider 注解调用 ProductInfoDynaSqlProvider 类中的 selectWithParam 方法获得一条 SELECT 语句，分页获取商品信息，Map<String, Object>类型的参数 params 用于封装查询条件，并通过@Results 注解映射查询结果。count 方法通过@SelectProvider 注解调用 ProductInfoDynaSqlProvider 类中的 count 方法获得一条 SELECT 语句，根据条件查询商品总数，Map<String, Object>类型的参数 params 用于封装查询条件。save 方法通过@Insert 注解映射一条 INSERT 语句，添加商品。edit 方法通过@Update 注解映射一条 UPDATE 语句，修改商品。updateState 方法通过@Update 注解映射一条 UPDATE 语句，更新商品状态。getOnSaleProduct 方法通过@Select 注解映射一条 SELECT 语句，获取在售商品列表。getProductInfoById 方法通过@Select 注解映射一条 SELECT 语句，根据商品 id 获取商品对象。

在 com.ecpbm.dao.provider 包中，创建动态 SQL 提供类 ProductInfoDynaSqlProvider，并添加 selectWithParam 和 count 两个方法，代码如下：

```java
package com.ecpbm.dao.provider;
import java.util.Map;
import org.apache.ibatis.jdbc.SQL;
import com.ecpbm.pojo.ProductInfo;
public class ProductInfoDynaSqlProvider {
    // 分页动态查询
    public String selectWithParam(Map<String, Object> params) {
        String sql = new SQL() {
            {
                SELECT("*");
                FROM("product_info");
                if (params.get("productInfo") != null) {
                    ProductInfo productInfo = (ProductInfo) params.get("productInfo");
                    if (productInfo.getCode() != null && !"".equals(productInfo.getCode())) {
                        WHERE(" code = #{productInfo.code} ");
                    }
                    if (productInfo.getName() != null && !productInfo.getName().equals("")) {
                        WHERE(" name LIKE CONCAT ('%',#{productInfo.name},'%') ");
                    }
                    if (productInfo.getBrand() != null && !productInfo.getBrand().equals("")) {
                        WHERE(" brand LIKE CONCAT ('%',#{productInfo.brand},'%') ");
                    }
                    if (productInfo.getType() != null && productInfo.getType().getId() > 0) {
                        WHERE(" tid = #{productInfo.type.id} ");
                    }
                    if (productInfo.getPriceFrom() > 0) {
                        WHERE(" price > #{productInfo.priceFrom} ");
                    }
                    if (productInfo.getPriceTo() > 0) {
                        WHERE(" price <= #{productInfo.priceTo} ");
                    }
                }
            }
        }.toString();
        if (params.get("pager") != null) {
            sql += " limit #{pager.firstLimitParam} , #{pager.perPageRows} ";
        }
        return sql;
    }
    // 根据条件动态查询商品总记录数
    public String count(Map<String, Object> params) {
        return new SQL() {
            {
```

```java
                SELECT("count(*)");
                FROM("product_info");
                if (params.get("productInfo") != null) {
                    ProductInfo productInfo = (ProductInfo) params.get("productInfo");
                    if (productInfo.getCode() != null && !"".equals(productInfo.getCode())) {
                        WHERE(" code = #{productInfo.code} ");
                    }
                    if (productInfo.getName() != null && !productInfo.getName().equals("")) {
                        WHERE(" name LIKE CONCAT ('%',#{productInfo.name},'%') ");
                    }
                    if (productInfo.getBrand() != null && !productInfo.getBrand().equals("")) {
                        WHERE(" brand LIKE CONCAT ('%',#{productInfo.brand},'%') ");
                    }
                    if (productInfo.getType() != null && productInfo.getType().getId() > 0) {
                        WHERE(" tid = #{productInfo.type.id} ");
                    }
                    if (productInfo.getPriceFrom() > 0) {
                        WHERE(" price > #{productInfo.priceFrom} ");
                    }
                    if (productInfo.getPriceTo() > 0) {
                        WHERE(" price <= #{productInfo.priceTo} ");
                    }
                }
            }.toString();
    }
}
```

在接口 TypeDao 中，编写如下代码：

```java
package com.ecpbm.dao;
import java.util.List;
import org.apache.ibatis.annotations.Insert;
import org.apache.ibatis.annotations.Options;
import org.apache.ibatis.annotations.Select;
import org.apache.ibatis.annotations.Update;
import com.ecpbm.pojo.Type;
public interface TypeDao {
    // 查询所有商品类型
    @Select("select * from type")
    public List<Type> selectAll();
    // 根据类型编号查询商品类型
    @Select("select * from type where id = #{id}")
    Type selectById(int id);
```

```java
        // 添加商品类型
        @Insert("insert into type(name) values(#{name})")
        @Options(useGeneratedKeys = true, keyProperty = "id")
        public int add(Type type);
        // 更新商品类型
        @Update("update type set name = #{name} where id = #{id}")
        public int update(Type type);
}
```

在接口 TypeDao 中，selectAll 方法通过@Select 注解映射一条 SELECT 语句，查询所有商品类型。selectById 方法通过@Select 注解映射一条 SELECT 语句，根据类型编号查询商品类型对象。add 方法通过@Insert 注解映射一条 INSERT 语句，添加商品类型。update 方法通过@Update 注解映射一条 UPDATE 语句，更新商品类型。

在接口 OrderInfoDao 中，编写如下代码：

```java
package com.ecpbm.dao;
import java.util.List;
import java.util.Map;
import org.apache.ibatis.annotations.*;
import org.apache.ibatis.mapping.FetchType;
import com.ecpbm.dao.provider.OrderInfoDynaSqlProvider;
import com.ecpbm.pojo.OrderDetail;
import com.ecpbm.pojo.OrderInfo;
public interface OrderInfoDao {
    // 分页获取订单信息
    @Results({
            @Result(column = "uid", property = "ui", one = @One(select = "com.ecpbm.dao.UserInfoDao.getUserInfoById", fetchType = FetchType.EAGER)) })
    @SelectProvider(type = OrderInfoDynaSqlProvider.class, method = "selectWithParam")
    List<OrderInfo> selectByPage(Map<String, Object> params);
    // 根据条件查询订单总数
    @SelectProvider(type = OrderInfoDynaSqlProvider.class, method = "count")
    Integer count(Map<String, Object> params);
    // 保存订单主表信息
    @Insert("insert into order_info(uid,status,ordertime,orderprice) "+
    "values(#{uid},#{status},#{ordertime},#{orderprice})")
    @Options(useGeneratedKeys = true, keyProperty = "id")
    int saveOrderInfo(OrderInfo oi);
    // 保存订单明细
    @Insert("insert into order_detail(oid,pid,num) values(#{oid},#{pid},#{num})")
    @Options(useGeneratedKeys = true, keyProperty = "id")
    int saveOrderDetail(OrderDetail od);
    // 根据订单编号获取订单对象
    @Results({
```

```java
        @Result(column = "uid", property = "ui", one = @One(select =
"com.ecpbm.dao.UserInfoDao.getUserInfoById", fetchType =
FetchType.EAGER)) })
        @Select("select * from order_info where id = #{id}")
        public OrderInfo getOrderInfoById(int id);
        // 根据订单编号获取订单明细
        @Results({
                @Result(column = "pid", property = "pi", one = @One(select =
"com.ecpbm.dao.ProductInfoDao.getProductInfoById", fetchType =
FetchType.EAGER)) })
        @Select("select * from order_detail where oid = #{oid}")
        public List<OrderDetail> getOrderDetailByOid(int oid);
        // 根据订单编号删除订单主表记录
        @Delete("delete from order_info where id=#{id}")
        public int deleteOrderInfo(int id);
        // 根据订单编号删除订单明细记录
        @Delete("delete from order_detail where oid=#{id}")
        public int deleteOrderDetail(int id);
}
```

在接口 OrderInfoDao 中，selectByPage 方法通过 @SelectProvider 注解调用 OrderInfoDynaSqlProvider 类中的 selectWithParam 方法获得一条 SELECT 语句，分页获取订单信息，Map<String, Object>类型的参数 params 用于封装查询条件，并通过@Results 注解映射查询结果。count 方法通过@SelectProvider 注解调用 OrderInfoDynaSqlProvider 类中的 count 方法获得一条 SELECT 语句，根据条件查询订单总数，Map<String, Object>类型的参数 params 用于封装查询条件。saveOrderInfo 方法通过@Insert 注解映射一条 INSERT 语句，保存订单主表信息。saveOrderDetail 方法通过@Insert 注解映射一条 INSERT 语句，保存订单明细。getOrderInfoById 方法通过@Select 注解映射一条 SELECT 语句，根据订单编号获取订单对象，并通过@Results 注解映射查询结果。getOrderDetailByOid 方法通过@Select 注解映射一条 SELECT 语句，根据订单编号获取订单明细，并通过@Results 注解映射查询结果。deleteOrderInfo 方法通过@Delete 注解映射一条 DELETE 语句，根据订单编号删除订单主表记录。deleteOrderDetail 方法通过@Delete 注解映射一条 DELETE 语句，根据订单编号删除订单明细记录。

在 com.ecpbm.dao.provider 包中，创建动态 SQL 提供类 OrderInfoDynaSqlProvider，并添加 selectWithParam 和 count 两个方法，代码如下：

```java
package com.ecpbm.dao.provider;
import java.util.Map;
import org.apache.ibatis.jdbc.SQL;
import com.ecpbm.pojo.OrderInfo;
public class OrderInfoDynaSqlProvider {
    // 分页动态查询
    public String selectWithParam(Map<String, Object> params) {
        String sql = new SQL() {
            {
                SELECT("*");
                FROM("order_info");
```

```java
                    if (params.get("orderInfo") != null) {
                        OrderInfo orderInfo = (OrderInfo)
params.get("orderInfo");
                        if (orderInfo.getId() != null && orderInfo.getId() > 0) {
                            WHERE(" id = #{orderInfo.id} ");
                        } else {
                            if (orderInfo.getStatus() != null && !"请选择".equals(orderInfo.getStatus())) {
                                WHERE(" status = #{orderInfo.status} ");
                            }
                            if (orderInfo.getOrderTimeFrom() != null
&& !"".equals(orderInfo.getOrderTimeFrom())) {
                                WHERE(" ordertime >= #{orderInfo.orderTimeFrom} ");
                            }
                            if (orderInfo.getOrderTimeTo() != null
&& !"".equals(orderInfo.getOrderTimeTo())) {
                                WHERE(" ordertime < #{orderInfo.orderTimeTo} ");
                            }
                            if (orderInfo.getUid() > 0) {
                                WHERE(" uid = #{orderInfo.uid} ");
                            }
                        }
                    }
        }.toString();
        if (params.get("pager") != null) {
            sql += " limit #{pager.firstLimitParam} , #{pager.perPageRows} ";
        }
        return sql;
    }
    // 根据条件动态查询订单总记录数
    public String count(Map<String, Object> params) {
        return new SQL() {
            {
                SELECT("count(*)");
                FROM("order_info");
                if (params.get("orderInfo") != null) {
                    OrderInfo orderInfo = (OrderInfo)
params.get("orderInfo");
                    if (orderInfo.getId() != null && orderInfo.getId() > 0) {
                        WHERE(" id = #{orderInfo.id} ");
                    } else {
                        if (orderInfo.getStatus() != null && !"请选择".equals(orderInfo.getStatus())) {
                            WHERE(" status = #{orderInfo.status} ");
                        }
                        if (orderInfo.getOrderTimeFrom() != null
&& !"".equals(orderInfo.getOrderTimeFrom())) {
                            WHERE(" ordertime >= #{orderInfo.orderTimeFrom} ");
```

```
                        }
                        if (orderInfo.getOrderTimeTo() != null
&& !"".equals(orderInfo.getOrderTimeTo())) {
                            WHERE(" ordertime < #{orderInfo.orderTimeTo} ");
                        }
                        if (orderInfo.getUid() > 0) {
                            WHERE(" uid = #{orderInfo.uid} ");
                        }
                    }
                }
            }
        }.toString();
    }
}
```

在接口 FunctionDao 中，编写如下代码：

```
package com.ecpbm.dao;
import java.util.List;
import org.apache.ibatis.annotations.Select;
import com.ecpbm.pojo.Functions;
public interface FunctionDao {
// 根据管理员 id 获取功能权限
    @Select("select * from functions where id in (select fid from powers where aid = #{aid} )")
    public List<Functions> selectByAdminId(Integer aid);
}
```

在接口 FunctionDao 中，selectByAdminId 方法通过 @Select 注解映射一个 SELECT 语句，根据管理员 id 获取功能权限。

19.6　创建 Service 接口及实现类

在 com.ecpbm.service 包中，创建业务逻辑层接口 UserInfoService、AdminInfoService、ProductInfoService、TypeService 和 OrderInfoService。

在接口 UserInfoService 中，编写如下代码：

```
package com.ecpbm.service;
import java.util.List;
import java.util.Map;
import com.ecpbm.pojo.Pager;
import com.ecpbm.pojo.UserInfo;
public interface UserInfoService {
    // 获取合法客户
    public List<UserInfo> getValidUser();
    // 根据客户编号查询客户
    public UserInfo getUserInfoById(int id);
    // 分页显示客户
    List<UserInfo> findUserInfo(UserInfo userInfo, Pager pager);
```

```java
    // 客户计数
    Integer count(Map<String, Object> params);
    // 修改指定编号的用户状态
    void modifyStatus(String ids, int flag);
}
```

在接口 AdminInfoService 中,编写如下代码:

```java
package com.ecpbm.service;
import com.ecpbm.pojo.AdminInfo;
public interface AdminInfoService {
    // 登录验证
    public AdminInfo login(AdminInfo ai);
    // 根据管理员编号,获取管理员对象及关联的功能权限
    public AdminInfo getAdminInfoAndFunctions(Integer id);
}
```

在接口 ProductInfoService 中,编写如下代码:

```java
package com.ecpbm.service;
import java.util.List;
import java.util.Map;
import com.ecpbm.pojo.Pager;
import com.ecpbm.pojo.ProductInfo;
public interface ProductInfoService {
    // 分页显示商品
    List<ProductInfo> findProductInfo(ProductInfo productInfo, Pager pager);
    // 商品计数
    Integer count(Map<String, Object> params);
    // 添加商品
    public void addProductInfo(ProductInfo pi);
    // 修改商品
    public void modifyProductInfo(ProductInfo pi);
    // 更新商品状态
    void modifyStatus(String ids, int flag);
    // 获取在售商品列表
    List<ProductInfo> getOnSaleProduct();
    // 根据商品 id 获取商品对象
    ProductInfo getProductInfoById(int id);
}
```

在接口 TypeService 中,编写如下代码:

```java
package com.ecpbm.service;
import java.util.List;
import com.ecpbm.pojo.Type;
public interface TypeService {
    // 获取所有商品类型
    public List<Type> getAll();
    // 添加商品类型
    public int addType(Type type);
    // 更新商品类型
    public void updateType(Type type);
}
```

在接口 OrderInfoService 中，编写如下代码：

```java
package com.ecpbm.service;
import java.util.List;
import java.util.Map;
import com.ecpbm.pojo.*;
public interface OrderInfoService {
    // 分页显示订单
    List<OrderInfo> findOrderInfo(OrderInfo orderInfo, Pager pager);
    // 订单计数
    Integer count(Map<String, Object> params);
    // 添加订单主表
    public int addOrderInfo(OrderInfo oi);
    // 添加订单明细
    public int addOrderDetail(OrderDetail od);
    // 根据订单编号获取订单信息
    public OrderInfo getOrderInfoById(int id);
    // 根据订单编号获取订单明细信息
    public List<OrderDetail> getOrderDetailByOid(int oid);
    // 删除订单
    public int deleteOrder(int id);
}
```

接下来，在 com.ecpbm.service.impl 包中，创建上述接口的实现类 UserInfoServiceImpl、AdminInfoServiceImpl、ProductInfoServiceImpl、TypeServiceImpl 和 OrderInfoServiceImpl，实现接口中的方法。

在实现类 UserInfoServiceImpl 中，编写如下代码：

```java
package com.ecpbm.service.impl;
……
@Service("userInfoService")
@Transactional(propagation = Propagation.REQUIRED, isolation =
Isolation.DEFAULT)
public class UserInfoServiceImpl implements UserInfoService {
    @Autowired
    UserInfoDao userInfoDao;
    @Override
    public List<UserInfo> getValidUser() {
        return userInfoDao.getValidUser();
    }
    @Override
    public UserInfo getUserInfoById(int id) {
        return userInfoDao.getUserInfoById(id);
    }
    @Override
    public List<UserInfo> findUserInfo(UserInfo userInfo, Pager pager) {
        Map<String, Object> params = new HashMap<String, Object>();
        params.put("userInfo", userInfo);
        int recordCount = userInfoDao.count(params);
        pager.setRowCount(recordCount);
        if (recordCount > 0) {
```

```java
            params.put("pager", pager);
        }
        return userInfoDao.selectByPage(params);
    }
    @Override
    public Integer count(Map<String, Object> params) {
        return userInfoDao.count(params);
    }
    @Override
    public void modifyStatus(String ids, int flag) {
        userInfoDao.updateState(ids, flag);
    }
}
```

在实现类 UserInfoServiceImpl 中，将@Transactional 注解标注在 UserInfoServiceImpl 类前，将类中的所有方法都进行事务处理。@Transactional 注解的 Propagation 属性用于指定事务传播行为，Propagation.REQUIRED 表示如果有事务，那么加入事务，没有事务则新建一个事务；Propagation.NOT_SUPPORTED 表示容器不为方法开启事务；Propagation.REQUIRES_NEW 表示不管是否存在事务，都会创建一个新的事务，将原来的事务挂起，当新的事务执行完毕后，再继续执行原来的事务；Propagation.MANDATORY 表示必须在一个已有的事务中执行，否则抛出异常；Propagation.NEVER 表示必须在一个没有的事务中执行，否则抛出异常；Propagation.SUPPORTS 表示如果其他 Bean 调用这个方法，在其他 Bean 中声明事务，那就用事务，如果其他 Bean 没有声明事务，那就不用事务；默认时，Propagation 属性为 Propagation.REQUIRED。isolation 属性用于指定事务隔离级别，Isolation.READ_UNCOMMITTED 表示读取未提交数据，由于会出现脏读且不可重复读，基本不使用；Isolation.READ_COMMITTED 表示读取已提交数据，但会出现不可重复读和幻读；Isolation.REPEATABLE_READ 表示可重复读，但会出现幻读；Isolation.SERIALIZABLE 表示串行化；Isolation.DEFAULT 表示使用数据库默认的隔离级别，一般情况使用这种配置。

在实现类 AdminInfoServiceImpl 中，编写如下代码：

```java
package com.ecpbm.service.impl;
……
@Service("adminInfoService")
@Transactional(propagation = Propagation.REQUIRED, isolation = Isolation.DEFAULT)
public class AdminInfoServiceImpl implements AdminInfoService {
    @Autowired
    private AdminInfoDao adminInfoDao;
    @Override
    public AdminInfo login(AdminInfo ai) {
        return adminInfoDao.selectByNameAndPwd(ai);
    }
    @Override
    public AdminInfo getAdminInfoAndFunctions(Integer id) {
        return adminInfoDao.selectById(id);
    }
}
```

实现类 AdminInfoServiceImpl 中的方法并没有实现其他功能，而是直接调用 AdminInfoDao 接口提供的方法。

在实现类 ProductInfoServiceImpl 中，编写如下代码：

```java
package com.ecpbm.service.impl;
……
@Service("productInfoService")
@Transactional(propagation = Propagation.REQUIRED, isolation = Isolation.DEFAULT)
public class ProductInfoServiceImpl implements ProductInfoService {
    @Autowired
    ProductInfoDao productInfoDao;
    @Override
    public List<ProductInfo> findProductInfo(ProductInfo productInfo, Pager pager) {
        // 创建对象 params
        Map<String, Object> params = new HashMap<String, Object>();
        // 将封装有查询条件的 productInfo 对象放入 params
        params.put("productInfo", productInfo);
        // 根据条件计算商品总数
        int recordCount = productInfoDao.count(params);
        // 给 pager 对象设置 rowCount 属性值（记录总数）
        pager.setRowCount(recordCount);
        if (recordCount > 0) {
            // 将 page 对象放入 params
            params.put("pager", pager);
        }
        // 分页获取商品信息
        return productInfoDao.selectByPage(params);
    }
    @Override
    public Integer count(Map<String, Object> params) {
        return productInfoDao.count(params);
    }
    @Override
    public void addProductInfo(ProductInfo pi) {
        productInfoDao.save(pi);
    }
    @Override
    public void modifyProductInfo(ProductInfo pi) {
        productInfoDao.edit(pi);
    }
    @Override
    public void modifyStatus(String ids, int flag) {
        productInfoDao.updateState(ids, flag);
    }
    @Override
    public List<ProductInfo> getOnSaleProduct() {
        return productInfoDao.getOnSaleProduct();
    }
    @Override
```

```java
    public ProductInfo getProductInfoById(int id) {
        return productInfoDao.getProductInfoById(id);
    }
}
```

在实体类 ProductInfoServiceImpl 中，findProductInfo 方法有两个参数，一个是 ProductInfo 类型的参数 productInfo，用于封装前端页面传递来的查询条件；另一个是 Pager 类型的参数 pager，用于封装从前端页面传递来的页码和每页记录数。在 findProductInfo 方法中，创建了一个 Map<String, Object>类型的对象 params，用来存放两个对象，一个是 productInfo 对象，另一个是 pager 对象。对于 pager 对象来说，放入 params 前，还需设置 pager 对象的 rowCount 属性值，这个值是通过调用 ProductInfoDao 接口的 count 方法获得的。设置好 params 对象后，再调用 ProductInfoDao 接口的 selectByPage 方法获得当前页的商品列表。实体类 ProductInfoServiceImpl 中其他方法的实现比较简单，直接调用 ProductInfoDao 接口提供的方法。

在实现类 TypeServiceImpl 中，编写如下代码：

```java
package com.ecpbm.service.impl;
……
@Service("typeService")
@Transactional(propagation = Propagation.REQUIRED, isolation = Isolation.DEFAULT)
public class TypeServiceImpl implements TypeService {
    @Autowired
    private TypeDao typeDao;
    @Override
    public List<Type> getAll() {
        return typeDao.selectAll();
    }
    @Override
    public int addType(Type type) {
        return typeDao.add(type);
    }
    @Override
    public void updateType(Type type) {
        typeDao.update(type);
    }
}
```

实体类 TypeServiceImpl 中的方法都是直接调用 TypeDao 接口提供的方法。

在实现类 OrderInfoServiceImpl 中，编写如下代码：

```java
package com.ecpbm.service.impl;
……
@Service("orderInfoService")
@Transactional(propagation = Propagation.REQUIRED, isolation = Isolation.DEFAULT)
public class OrderInfoServiceImpl implements OrderInfoService {
    @Autowired
    OrderInfoDao orderInfoDao;
    @Override
```

```java
public List<OrderInfo> findOrderInfo(OrderInfo orderInfo, Pager pager) {
    Map<String, Object> params = new HashMap<String, Object>();
    params.put("orderInfo", orderInfo);
    int recordCount = orderInfoDao.count(params);
    pager.setRowCount(recordCount);
    if (recordCount > 0) {
        params.put("pager", pager);
    }
    return orderInfoDao.selectByPage(params);
}
@Override
public Integer count(Map<String, Object> params) {
    return orderInfoDao.count(params);
}
@Override
public int addOrderInfo(OrderInfo oi) {
    return orderInfoDao.saveOrderInfo(oi);
}
@Override
public int addOrderDetail(OrderDetail od) {
    return orderInfoDao.saveOrderDetail(od);
}
@Override
public OrderInfo getOrderInfoById(int id) {
    return orderInfoDao.getOrderInfoById(id);
}
@Override
public List<OrderDetail> getOrderDetailByOid(int oid) {
    return orderInfoDao.getOrderDetailByOid(oid);
}
@Override
public int deleteOrder(int id) {
    int result = 1;
    try {
        orderInfoDao.deleteOrderDetail(id);
        orderInfoDao.deleteOrderInfo(id);
    } catch (Exception e) {
        result = 0;
    }
    return result;
}
}
```

OrderInfoServiceImpl 类中的 findOrderInfo 方法与 ProductInfoServiceImpl 类中 findProductInfo 方法的实现思路类似。

19.7 后台登录与管理首页面

系统后台登录页为 admin_login.jsp，页面效果如图 19-5 所示。

图 19-5 后台登录页

在 admin_login.jsp 页面中，使用了 Easy UI 框架进行布局，代码如下：

```jsp
<%@ page language="java" import="java.util.*" pageEncoding="UTF-8"%>
<html>
<head>
<title>电子商务平台——后台登录页</title>
<!-- 引入 EasyUI 的相关 css 和 js 文件 -->
<link href="EasyUI/themes/default/easyui.css" rel="stylesheet"
    type="text/css" />
<link href="EasyUI/themes/icon.css" rel="stylesheet" type="text/css" />
<link href="EasyUI/demo.css" rel="stylesheet" type="text/css" />
<script src="EasyUI/jquery.min.js" type="text/javascript"></script>
<script src="EasyUI/jquery.easyui.min.js"
type="text/javascript"></script>
<script src="EasyUI/easyui-lang-zh_CN.js"
type="text/javascript"></script>
</head>
<body>
    <script type="text/javascript">
        function clearForm() {
            $('#adminLoginForm').form('clear');
        }
    function checkAdminLogin() {
            $("#adminLoginForm").form("submit", {
                // 向控制器类 AdminInfoController 中的 login 方法发送请求
                url : 'admininfo/login',
                success : function(result) {
                    var result = eval('(' + result + ')');
                    if (result.success == 'true') {
                        window.location.href = 'admin.jsp';
                        $("#adminLoginDlg").dialog("close");
                    } else {
                        $.messager.show({
                            title : "提示信息",
                            msg : result.message
```

```html
                    });
                }
            }
        });
    }
</script>
<div id="adminLoginDlg" class="easyui-dialog"
    style="left: 550px; top: 200px;width: 300;height: 200"
    data-options="title:'后台登录',buttons:'#bb',modal:true">
    <form id="adminLoginForm" method="post">
        <table style="margin:20px;font-size: 13;">
            <tr>
                <th >用户名</th>
                <td><input class="easyui-textbox" type="text" id="name" name="name" data-options="required:true" value="admin"></input></td>
            </tr>
            <tr>
                <th>密码</th>
                <td><input class="easyui-textbox" type="text" id="pwd" name="pwd" data-options="required:true" value="123456"></input></td>
            </tr>
        </table>
    </form>
</div>
<div id="bb">
    <a href="javascript:void(0)" class="easyui-linkbutton"
        onclick="checkAdminLogin()">登录</a> <a
    href="javascript:void(0)" class="easyui-linkbutton"
    onclick="clearForm();">重置</a>
</div>
</body>
</html>
```

为了使用 Easy UI 框架，在页面开始部分的<head></head>标签中，需要引入 Easy UI 的相关 css 和 js 文件。在一个 id 为 adminLoginDlg 的<div>标签中，将 class 属性设置为 easyui-dialog，从而使用 Easy UI 对话框控件 Dialog 将这个<div>标签创建为一个对话框。在这个对话框中，包含 id 为 adminLoginForm 的登录表单，表单中使用 Easy UI 文本框控件 TextBox 创建用户名和密码两个文本域。在 id 为 bb 的<div>标签中，定义两个<a>标签，将它们的 class 属性都设置为 easyui-linkbutton，从而使用 Easy UI 的链接按钮控件 LinkButton 将这两个<a>标签创建为登录和重置两个按钮。

在后台登录表单中输入用户名和密码，单击登录按钮，执行 JavaScript 函数 checkAdminLogin。在函数 checkAdminLogin 中，通过 jQuery 向后台服务器发送请求，请求的地址为 admin/login，这个 url 请求将映射到控制器类 AdminInfoController 中的 login 方法；success 参数的类型为 Function，表示请求成功后的回调函数，服务器返回的数据将传给该函数的参数 result，然后判断 result 参数中 success 名称所对应的值是否等于 true，如果等于 true，表示登录成功，则打开后台管理首页面 admin.jsp，并关闭后台登录表单对话框；否则通过消息框给出错误提示。

在控制类 AdminInfoController 中，login 方法代码如下：

```java
package com.ecpbm.controller;
……
@SessionAttributes(value = { "admin" })
@Controller
@RequestMapping("/admininfo")
public class AdminInfoController {
    @Autowired
    private AdminInfoService adminInfoService;
    @RequestMapping(value = "/login", produces = "text/html;charset=UTF-8")
    @ResponseBody
    public String login(AdminInfo ai, ModelMap model) {
        // 后台登录验证
        AdminInfo admininfo = adminInfoService.login(ai);
        if (admininfo != null && admininfo.getName() != null) {
            // 验证通过后，再判断是否已为该管理员分配功能权限
            if (adminInfoService.getAdminInfoAndFunctions(admininfo.getId()).getFs().size() > 0) {
                // 验证通过且已分配功能权限，则将 admininfo 对象存入 model 中
                model.put("admin", admininfo);
                // 以 JSON 格式向页面发送成功信息
                return "{\"success\":\"true\",\"message\":\"登录成功\"}";
            } else {
                return "{\"success\":\"false\",\"message\":\"您没有权限，请联系超级管理员设置权限！\"}";
            }
        } else
            return "{\"success\":\"false\",\"message\":\"登录失败\"}";
    }
}
```

在 AdminInfoController 类中，通过@Controller 注解指示该类是一个控制器，通过@RequestMapping 注解将用户对 admin/login 这个 url 的请求映射到 login 方法。

login 方法包含两个参数，一个是 AdminInfo 类型的参数 ai，用于封装从前端登录表单传递来的用户名和密码；另一个是 ModelMap 类型的参数 model，用于存放登录成功后的管理员对象信息。

在 login 方法中，首先调用业务接口 AdminInfoService 中的 login 方法进行登录验证；验证通过后，再判断是否已为该管理员分配功能权限，如果没有，则以 JSON 格式返回"您没有权限，请联系超级管理员设置权限！"的提示信息。只有验证通过且分配了权限的管理员登录，才返回"登录成功"的提示，并将该管理员对象以 admin 为名称存入 ModelMap 类型的参数 model 中，并通过在 AdminInfoController 类名上修饰的@SessionAttributes 注解将其存入 Session 范围。登录成功后，跳转到后台管理首页面 admin.jsp，如图 19-6 所示。

图 19-6　后台管理首页面

admin.jsp 页面的主要代码如下：

```jsp
<%@ page language="java" import="java.util.*" contentType="text/html;
charset=utf-8" pageEncoding="utf-8" %>
<%
    if (session.getAttribute("admin") == null)
        response.sendRedirect("/ecpbm/admin_login.jsp");
%>
<html>
<head>
<title>后台管理首页面</title>
<link href="EasyUI/themes/default/easyui.css" rel="stylesheet"
    type="text/css" />
<link href="EasyUI/themes/icon.css" rel="stylesheet" type="text/css" />
<link href="EasyUI/demo.css" rel="stylesheet" type="text/css" />
<script src="EasyUI/jquery.min.js" type="text/javascript"></script>
<script src="EasyUI/jquery.easyui.min.js"
type="text/javascript"></script>
<script src="EasyUI/easyui-lang-zh_CN.js"
type="text/javascript"></script>
</head>
<body class="easyui-layout">
    <div data-options="region:'north',border:false"
        style="height: 60px; background: #B3DFDA; padding: 10px">
        <div align="left">
            <div style="font-family: Microsoft YaHei; font-size: 16px;">
电商平台后台管理系统</div>
        </div>
        <div align="right">
            欢迎您，<font color="Red">${sessionScope.admin.name}</font>
        </div>
    </div>
    <div data-options="region:'west',split:true,title:'功能菜单'"
        style="width: 180px">
        <ul id="tt"></ul>
```

```html
        </div>
        <div data-options="region:'south',border:false"
            style="height: 50px; background: #A9FACD; padding: 10px; text-align: center">powered
            by miaoyong</div>
        <div data-options="region:'center'">
            <div id="tabs" data-options="fit:true" class="easyui-tabs"
                style="width: 500px; height: 250px;"></div>
        </div>
        <script type="text/javascript">
            // 为tree指定数据
            $('#tt').tree({
                url : 'admininfo/getTree?adminid=${sessionScope.admin.id}'
            });
            $('#tt').tree({
                onClick : function(node) {
                    if ("商品列表" == node.text) {
                        if ($('#tabs').tabs('exists', '商品列表')) {
                            $('#tabs').tabs('select', '商品列表');
                        } else {
                            $('#tabs').tabs('add', {
                                title : node.text,
                                href : 'productlist.jsp',
                                closable : true
                            });
                        }
                    } else if ("商品类型列表" == node.text) {
                        if ($('#tabs').tabs('exists', '商品类型列表')) {
                            $('#tabs').tabs('select', '商品类型列表');
                        } else {
                            $('#tabs').tabs('add', {
                                title : node.text,
                                href : 'typelist.jsp',
                                closable : true
                            });
                        }
                    } else if ("查询订单" == node.text) {
                        if ($('#tabs').tabs('exists', '查询订单')) {
                            $('#tabs').tabs('select', '查询订单');
                        } else {
                            $('#tabs').tabs('add', {
                                title : node.text,
                                href : 'searchorder.jsp',
                                closable : true
                            });
                        }
                    } else if ("创建订单" == node.text) {
                        if ($('#tabs').tabs('exists', '创建订单')) {
                            $('#tabs').tabs('select', '创建订单');
                        } else {
                            $('#tabs').tabs('add', {
```

```
                        title : node.text,
                        href : 'createorder.jsp',
                        closable : true
                    });
                }
            } else if ("客户列表" == node.text) {
                if ($('#tabs').tabs('exists', '客户列表')) {
                    $('#tabs').tabs('select', '客户列表');
                } else {
                    $('#tabs').tabs('add', {
                        title : node.text,
                        href : 'userlist.jsp',
                        closable : true
                    });
                }
            } else if ("管理员列表" == node.text) {
                if ($('#tabs').tabs('exists', '管理员列表')) {
                    $('#tabs').tabs('select', '管理员列表');
                } else {
                    $('#tabs').tabs('add', {
                        title : node.text,
                        href : 'adminlist.jsp',
                        closable : true
                    });
                }
            } else if ("退出系统" == node.text) {
                $.ajax({
                    url : 'admininfo/logout',
                    success : function(data) {
                        window.location.href = "admin_login.jsp";
                    }
                })
            }
        }
    });
</script>
</body>
</html>
```

在 admin.jsp 页面中，为了使用 Easy UI 控件，同样需要在页面开始部分的<head></head>标签中引入 Easy UI 相关的 css 和 js 文件。在<body>标签中，将 class 属性设置为 easyui-layout，从而使用 Easy UI 的 Layout 控件生成页面的布局。

在 Layout 控件的左侧，定义了一个 id 为 tt 的<ul>标签，Easy UI 的 Tree 控件可以定义在<ul>标签中。通过 JavaScript 将这个<ul>标签定义为一个 Easy UI 的 Tree 控件，并为 Tree 控件指定数据源，用来显示系统功能菜单。Tree 控件的数据源是通过 url 属性指定的，这里为 admininfo/getTree?adminid=${sessionScope.admin.id}，这个请求将映射到控制器类 AdminInfoController 中的 getTree 方法，其代码如下：

```
@RequestMapping("getTree")
```

```java
@ResponseBody
public List<TreeNode> getTree(@RequestParam(value = "adminid") String adminid) {
    // 根据管理员编号,获取 AdminInfo 对象
    AdminInfo admininfo =
adminInfoService.getAdminInfoAndFunctions(Integer.parseInt(adminid));
    List<TreeNode> nodes = new ArrayList<TreeNode>();
    // 获取关联的 Functions 对象集合
    List<Functions> functionsList = admininfo.getFs();
    // 对 List<Functions>类型的 Functions 对象集合排序
    Collections.sort(functionsList);
    // 将排序后的 Functions 对象集合转换到 List<TreeNode>类型的列表 nodes
    for (Functions functions : functionsList) {
        TreeNode treeNode = new TreeNode();
        treeNode.setId(functions.getId());
        treeNode.setFid(functions.getParentid());
        treeNode.setText(functions.getName());
        nodes.add(treeNode);
    }
    // 调用自定义的工具类 JsonFactory 的 buildtree 方法,为 nodes 列表中的各个
    // TreeNode 元素中的 children 属性赋值(该结点包含的子结点)
    List<TreeNode> treeNodes = JsonFactory.buildtree(nodes, 0);
    return treeNodes;
}
```

通过在 getTree 方法上标注@RequestMapping 注解,将用户对 admininfo/getTree 这个 url 的请求映射到 getTree 方法。

getTree 方法有一个参数 adminid,用于封装管理员编号。在 getTree 方法中,首先调用业务接口 AdminInfoService 中的 getAdminInfoAndFunctions 方法,根据管理员编号获取 AdminInfo 对象;然后获取 AdminInfo 对象关联的 Functions 对象集合,并对 Functions 对象集合进行排序,从而保证绑定 Easy UI 的 Tree 控件时顺序一致;接着将排序后的 Functions 对象集合转换到 List<TreeNode>类型的列表 nodes 中,TreeNode 是用于描述菜单树的每个结点的实体类;最后调用自定义工具类 JsonFactory 中的 buildtree 方法,为列表 nodes 中的各个 TreeNode 元素中的 children 赋值,即设置各个结点所包含的子结点。

JsonFactory 类位于 com.ecpbm.util 包中,buildtree 方法如下:

```java
package com.ecpbm.util;
import java.util.ArrayList;
import java.util.List;
import com.ecpbm.pojo.TreeNode;
public class JsonFactory {
    public static List<TreeNode> buildtree(List<TreeNode> nodes, int id) {
        List<TreeNode> treeNodes = new ArrayList<TreeNode>();
        for (TreeNode treeNode : nodes) {
            TreeNode node = new TreeNode();
            node.setId(treeNode.getId());
            node.setText(treeNode.getText());
            if (id == treeNode.getFid()) {
                // 递给调用 buildtree 方法给 TreeNode 中的 children 属性赋值
```

```
                node.setChildren(buildtree(nodes, node.getId()));
                treeNodes.add(node);
            }
        }
        return treeNodes;
    }
}
```

为了给列表 nodes 中的 TreeNode 元素中的 children 属性赋值，递归调用了 buildtree 方法。

执行 getTree 方法后，通过@ResponseBody 注解将 List<TreeNode>类型的对象 treeNodes 转换成 JSON 格式数据返回前端页面，从而为 Tree 控件提供数据源。

在 Layout 控件的中部，定义了一个 id 为 tabs 的<div>标签，将其 class 属性设置为 easyui-tabs，从而将<div>标签创建为一个 Easy UI 的 Tabs 控件。Tabs 控件中可以添加多个标签页，每个标签页可用来显示一个页面。

在 admin.jsp 页面中，还为 Tree 控件添加了 onClick 事件。单击商品列表结点时，如果 Tabs 控件中没有商品列表标签页，则向 Tabs 控件中添加一个标签页，用于显示商品列表页 productlist.jsp，否则选中该标签页。单击商品类型列表、查询订单、创建订单、客户列表这些结点时，效果与商品列表结点类似。

单击退出系统结点时，通过 Ajax 方式向后台服务器发送一个请求，请求的地址为 admininfo/logout，这个 url 请求将映射到控制器类 AdminInfoController 中的 logout 方法，其代码如下：

```
@RequestMapping(value = "/logout", method = RequestMethod.GET)
@ResponseBody
public String logout(SessionStatus status) {
    // @SessionAttributes 清除
    status.setComplete();
    return "{\"success\":\"true\",\"message\":\"注销成功\"}";
}
```

通过在 logout 方法上标注@RequestMapping 注解，将用户对 admininfo/logout 这个 url 的请求映射到 logout 方法。logout 方法有一个 SessionStatus 类型的参数 status，调用 setComplete 后，将保存在 Session 中的管理员对象 admin 清除。退出系统后，页面跳转到后台登录页 admin_login.jsp。

19.8 商品管理

商品管理包括商品列表显示、查询商品、添加商品、商品下架和修改商品等功能。

19.8.1 商品列表显示

在 admin.jsp 页面中，单击 Tree 控件上的商品列表结点，打开商品列表页 productlist.jsp，如图 19-7 所示。

图 19-7　商品列表页 productlist.jsp

在页面 productlist.jsp 的<body></body>部分，首先定义了一个 id 为 dg_productinfo 的<table>标签，将其 class 属性设置为 easyui-datagrid，从而将<table>标签创建为一个 Easy UI 的 Datagrid 控件，用来显示商品记录，代码如下：

```
<table id="dg_productinfo" class="easyui-datagrid"></table>
```

然后定义了一个 id 为 tb_productinfo 的<div>标签，充当 Datagrid 控件的工具栏。在<div>标签中，定义了三个<a>标签，将每个<a>标签的 class 属性都设置为 easyui-linkbutton，从而将这三个<a>标签创建为添加、修改和删除三个 Easy UI 的链接按钮控件，代码如下：

```
<!-- 创建 Datagrid控件的工具栏 -->
<div id="tb_productinfo" style="padding: 2px 5px;">
    <a href="javascript:void(0)" class="easyui-linkbutton"
        iconCls="icon-add" plain="true" onclick="addProduct();">添加</a> <a
        href="javascript:void(0)" class="easyui-linkbutton"
        iconCls="icon-edit" plain="true" onclick="editProduct();">修改</a> <a
        href="javascript:void(0)" class="easyui-linkbutton"
        iconCls="icon-remove" onclick="removeProduct();" plain="true">删除
    </a>
</div>
```

接着定义一个 id 为 searchtb_productinfo 的<div>标签，充当 Datagrid 控件的搜索栏。在这个<div>标签中，定义了一个 id 为 searchForm_productinfo 的<form>标签。在<form>标签中，使用 Easy UI 的 TextBox 控件定义了商品编号、商品名称和商品品牌三个文本框控件，使用 Easy UI 的 ComboBox 控件定义了一个商品类型组合框控件，使用 Easy UI 的 Numberbox 控件定义了两个价格数字框控件，代码如下：

```
<!-- 创建查询工具栏 -->
<div id="searchtb_productinfo" style="padding: 2px 5px;">
    <form id="searchForm_productinfo" method="post">
        <div style="padding: 3px">
            商品编号  <input class="easyui-textbox"
                name="productinfo_search_code"
id="productinfo_search_code"   style="width: 110px" />
```

```html
        </div>
        <div style="padding: 3px">
            商品名称  <input class="easyui-textbox"
                name="productinfo_search_name"
id="productinfo_search_name"  style="width: 110px" />  商品类型  <input style="width: 110px;" id="productinfo_search_tid"
                class="easyui-combobox" name="productinfo_search_tid"
       data-options="valueField:'id',textField:'name',url:'type/getType/1'"
                value="0">   商品品牌  <input
                class="easyui-textbox" name="productinfo_search_brand"
                id="productinfo_search_brand" style="width: 110px" />
              价格: <input class="easyui-numberbox"
                name="productinfo_search_priceFrom"
                id="productinfo_search_priceFrom" style="width: 80px;" /> ~
<input class="easyui-numberbox" name="productinfo_search_priceTo"
                id="productinfo_search_priceTo" style="width: 80px;" />
              <a href="javascript:void(0)"
class="easyui-linkbutton" iconCls="icon-search" plain="true"
onclick="searchProduct();">查找</a>
        </div>
    </form>
</div>
```

在搜索栏中,由于商品类型组合框显示的数据来自于数据表 type,因此在该组合框的 data-options 属性中,将 url 属性设置为 type/getType/1,为组合框控件指定数据源,type/getType/1 这个请求将映射到一个控制器类中的一个方法。接下来,在 com.ecpbm.controller 包中创建一个类 TypeController,并编写 getType 方法,代码如下:

```java
package com.ecpbm.controller;
……
@Controller
@RequestMapping("/type")
public class TypeController {
    @Autowired
    private TypeService typeService;
    @RequestMapping("/getType/{flag}")
    @ResponseBody
    public List<Type> getType(@PathVariable("flag") Integer flag) {
        List<Type> typeList = typeService.getAll();
        if (flag == 1) {
            Type t = new Type();
            t.setId(0);
            t.setName("请选择…");
            typeList.add(0, t);
        }
        return typeList;
    }
    ……
}
```

在 TypeController 类中,通过@Controller 注解指示该类是一个控制器,通过

@RequestMapping 注解将用户对 type/getType/1 这个 url 的请求映射到 getType 方法。

getType 方法有一个 Integer 类型的参数 flag，通过@PathVariable 注解将请求 url 中的占位符参数绑定到控制器处理方法的参数中，这里将 url 中的 0 绑定到参数 flag。

在 getType 方法中，调用 TypeService 接口的 getAll 方法，获取所有商品类型列表。如果 flag 为 1，还会在列表中添加一个"请选择..."选项。最后通过@ResponseBody 注解，将结果进行 JSON 格式转换，并返回给前端页面，作为商品类型组合框控件的数据源。有了数据源，需要在组合框控件的 data-options 属性中，通过 valueField 属性指定从数据源中获取哪个属性作为组合框的返回值，这里为 id；通过 textField 属性指定从数据源中获取哪个属性作为组合框的显示文本，这里为 name。

最后通过 JavaScript 对 id 为 dg_productinfo 的<table>标签进行初始化，代码如下：

```javascript
<script type="text/javascript">
    $(function() {
        $('#dg_productinfo').datagrid({
            singleSelect : false, //设置datagrid为单选
            url : 'productinfo/list', //为datagrid设置数据源
            pagination : true, //启用分页
            pageSize : 10, //设置初始每页记录数(页大小)
            pageList : [ 10, 15, 20 ], //设置可供选择的页大小
            rownumbers : true, //显示行号
            fit : true, //设置自适应
            toolbar : '#tb_productinfo', //为datagrid添加工具栏
            header : '#searchtb_productinfo', //为datagrid添加搜索栏
            columns : [ [ { //编辑datagrid的列
                title : '序号',
                field : 'id',
                align : 'center',
                checkbox : true
            }, {
                field : 'name',
                title : '商品名称',
                width : 200
            }, {
                field : 'type',
                title : '商品类型',
                formatter : function(value, row, index) {
                    if (row.type) {
                        return row.type.name;
                    } else {
                        return value;
                    }
                },
                width : 60
            }, {
                field : 'status',
                title : '商品状态',
                formatter : function(value, row, index) {
                    if (row.status == 1) {
```

```
                    return "在售";
                } else {
                    return "下架";
                }
            },
            width : 60
        }, {
            field : 'code',
            title : '商品编码',
            width : 100
        }, {
            field : 'brand',
            title : '品牌',
            width : 120
        }, {
            field : 'price',
            title : '价格',
            width : 50
        }, {
            field : 'num',
            title : '库存',
            width : 50
        }, {
            field : 'intro',
            title : '商品描述',
            width : 450
        } ] ]
    });
});
……
</script>
```

　　Datagrid 控件的初始化是通过相关属性设置来完成的，singleSelect 属性用于设置是否单选，这里为 false，表示允许多选；pagination 属性用于设置是否分页，这里为 true，表示允许分页；pageSize 属性用于设置每页初始记录数，这里为 10；pageList 用于设置可供选择的每页记录数，这里为[10, 15, 20]，表示可以选择每页显示 10、15 或 20 条记录；rownumbers 属性用于设置是否显示行号，这里为 true，表示显示行号；fit 属性用于设置是否自适应显示数据，这里为 true，允许自适应显示；toolbar 属性用于设置工具栏，这里为#tb_productinfo，表示要将 id 为 tb_productinfo 的<div>标签作为 Datagrid 控件的工具栏；header 属性用于设置表头，这里为#searchtb_productinfo，表示要将 id 为 searchtb_productinfo 的<div>标签作为 Datagrid 控件的标题头；columns 属性用于设置 Datagrid 控件显示的列；url 属性用于指定 Datagrid 控件的数据源，这里为 productinfo/list，这个 url 请求会映射到一个控制器类中的一个方法。接下来，在 com.ecpbm.controller 包中，创建一个名为 ProductInfoController 的控制器类，并编写一个 list 方法，代码如下：

```
package com.ecpbm.controller;
……
@Controller
```

```java
@RequestMapping("/productinfo")
public class ProductInfoController {
    @Autowired
    ProductInfoService productInfoService;
    // 后台商品列表分页显示
    @RequestMapping(value = "/list")
    @ResponseBody
    public Map<String, Object> list(Integer page, Integer rows, ProductInfo productInfo) {
        // 初始化分页类对象 pager
        Pager pager = new Pager();
        pager.setCurPage(page);
        pager.setPerPageRows(rows);
        // 创建 params 对象，封装查询条件
        Map<String, Object> params = new HashMap<String, Object>();
        params.put("productInfo", productInfo);
        // 获取满足条件的商品总数
        int totalCount = productInfoService.count(params);
        // 获取满足条件的商品列表
        List<ProductInfo> productinfos =
    productInfoService.findProductInfo(productInfo, pager);
        // 创建 result 对象，保存查询结果数据
        Map<String, Object> result = new HashMap<String, Object>(2);
        result.put("total", totalCount);
        result.put("rows", productinfos);
        // 将结果以 JSON 格式发送到前端控制器
        return result;
    }
    ……
}
```

在 ProductInfoController 类中，通过@Controller 注解指示该类是一个控制器，通过@RequestMapping 注解将用户对 productinfo/list 这个 url 的请求映射到 list 方法。

list 方法有三个参数，一个是 ProductInfo 类型的参数 productInfo，用于封装表单传递来的查询条件；另外两个参数是 page 和 rows，用于接收从 Datagrid 控件传递来的页码和每页显示的记录数。

在 list 方法中，首先初始化一个分页类对象 pager，给其设置 curPage 和 perPageRows 两个属性值；然后创建 Map<String, Object>类型的对象 params，用于封装查询条件；接着依次调用 ProductInfoService 接口的 count 方法获取满足条件的商品总数，调用 findProductInfo 方法获取满足条件的商品列表；再创建 Map<String, Object>类型的对象 result，保存查询结果数据；最后将返回结果转为 JSON 格式，以字符串形式发送到前端页面 productlist.jsp，为 Datagrid 控件提供数据源。

19.8.2 查询商品

在商品列表页 productlist.jsp 的搜索栏中，输入商品编号、商品名称、商品类型、商品品牌或价格范围，单击"查找"按钮，将执行 JavaScript 函数 searchProduct，代码如下：

```javascript
// 查询商品
function searchProduct() {
    var productinfo_search_code = $('#productinfo_search_code')
            .textbox("getValue");
    var productinfo_search_name = $('#productinfo_search_name')
            .textbox("getValue");
    var productinfo_search_tid = $('#productinfo_search_tid').combobox(
            "getValue");
    var productinfo_search_brand = $('#productinfo_search_brand')
            .textbox("getValue");
    var productinfo_search_priceFrom;
    if ($("#productinfo_search_priceFrom").val() != null
            && $("#productinfo_search_priceFrom").val() != "") {
        productinfo_search_priceFrom = $(
                '#productinfo_search_priceFrom').textbox("getValue");
    } else {
        productinfo_search_priceFrom = "0";
    }
    var productinfo_search_priceTo;
    if ($("#productinfo_search_priceTo").val() != null
            && $("#productinfo_search_priceTo").val() != "") {
        productinfo_search_priceTo = $('#productinfo_search_priceTo')
                .textbox("getValue");
    } else {
        productinfo_search_priceTo = "0";
    }
    $("#dg_productinfo").datagrid('load', {
        "code" : productinfo_search_code,
        "name" : productinfo_search_name,
        "type.id" : productinfo_search_tid,
        "brand" : productinfo_search_brand,
        "priceFrom" : productinfo_search_priceFrom,
        "priceTo" : productinfo_search_priceTo
    });
}
```

在函数 searchProduct 中，首先获取用户输入的查询条件，然后执行 Datagrid 控件的 load 方法，再次将请求发送到 productinfo/list，即再次执行控制器类 ProductInfoController 中的 list 方法，并将搜索栏中输入的查询条件传递过去，list 方法根据传递来的查询参数重新获取商品列表以更新 Datagrid 控件的数据源。

以查询商品类型为冰箱、价格在 2000 到 4000 的商品为例。执行查询后，结果如图 19-8 所示。

图 19-8　商品查询结果

19.8.3 添加商品

在商品列表页 productlist.jsp 的工具栏中，单击"添加"按钮，执行 JavaScript 函数 addProduct，打开新增商品对话框，代码如下：

```
// 打开新增商品对话框
function addProduct() {
    $('#dlg_productinfo').dialog('open').dialog('setTitle', '新增商品');
    $('#dlg_productinfo').form('clear');
    urls = 'productinfo/addProduct';
}
```

新增商品对话框的布局代码如下：

```
<div id="dlg_productinfo" class="easyui-dialog" title="添加商品"
    closed="true" style="width: 500px;">
    <div style="padding: 10px 60px 20px 60px">
        <form id="ff_productinfo" method="POST" action=""
            enctype="multipart/form-data">
            <table cellpadding="5">
                <tr>
                    <td>商品状态:</td>
                    <td><select id="status" class="easyui-combobox"
 name="status"   style="width: 150px;">
                            <option value="1">在售</option>
                            <option value="0">下架</option>
                        </select></td>
                </tr>
                <tr>
                    <td>商品类型:</td>
                    <td><input style="width: 150px;" id="type.id"
                        class="easyui-combobox" name="type.id"
                        data-options="valueField:'id',textField:'name',
url:'type/getType/0'"></input>
                    </td>
                </tr>
                <tr>
                    <td>商品名称:</td>
                    <td><input class="easyui-textbox" type="text"
id="name" name="name" data-options="required:true"></input></td>
                </tr>
                <tr>
                    <td>商品编码:</td>
                    <td><input class="easyui-textbox" type="text"
id="code" name="code" data-options="required:true"></input></td>
                </tr>
                <tr>
                    <td>商品品牌:</td>
                    <td><input class="easyui-textbox" type="text"
id="brand"   name="brand" data-options="required:true"></input></td>
                </tr>
                <tr>
```

```html
            <td>商品数量:</td>
            <td><input class="easyui-textbox" type="text"
id="num"    name="num" data-options="required:true"></input></td>
        </tr>
        <tr>
            <td>商品价格:</td>
            <td><input class="easyui-textbox" type="text"
id="price"  name="price" data-options="required:true"></input></td>
        </tr>
        <tr>
            <td>商品描述:</td>
            <td><input class="easyui-textbox" name="intro"
id="intro"  data-options="multiline:true" style="height:
60px"></input></td>
        </tr>
        <tr>
            <td>商品图片:</td>
            <td><input class="easyui-filebox" id="file" name="file"
style="width: 200px" value="选择图片"></input></td>
        </tr>
    </table>
</form>
<div style="text-align: center; padding: 5px">
    <a href="javascript:void(0)" class="easyui-linkbutton"
        onclick="saveProduct();">保存</a> <a
href="javascript:void(0)" class="easyui-linkbutton"
onclick="clearForm();">清空</a>
</div>
    </div>
</div>
```

从布局上来说，新增商品对话框使用了一个 id 为 dlg_productinfo 的<div>标签，通过将其 easyui-dialog 属性设置为 easyui-dialog，从而将这个<div>标签创建为 Easy UI 的 Dialog 控件。在<div>标签中，定义了一个 id 为 ff_productinfo 的<form>标签。在<form>标签中，依次定义了商品状态和商品类型两个组合框控件，商品名称、商品编码、商品品牌、商品数量、商品价格和商品描述六个文本框控件，商品图片文件上传控件。最后使用 Easy UI 的 Link Button 控件定义了保存和清空两个链接按钮，新增商品对话框的效果如图 19-9 所示。

图 19-9　新增商品对话框

在新增商品对话框中，由于商品类型组合框显示的数据来自于数据表 type，因此在其 data-options 属性中，将 url 属性设置为 type/getType/0，为组合框控件指定数据源。

在新增商品对话框，输入商品信息，单击"保存"按钮，执行 JavaScript 函数 saveProduct，代码如下：

```javascript
// 保存商品信息
function saveProduct() {
    $("#ff_productinfo").form("submit", {
        url : urls, //使用参数
        success : function(result) {
            var result = eval('(' + result + ')');
            if (result.success == 'true') {
                $("#dg_productinfo").datagrid("reload");
                $("#dlg_productinfo").dialog("close");
            }
            $.messager.show({
                title : "提示信息",
                msg : result.message
            });
        }
    });
}
```

在打开新增商品对话框时，已经将 urls 设置为 productinfo/addProduct。因此，在函数 saveProduct 中，通过 JavaScript 将表单提交到 url 属性指定的 urls。productinfo/addProduct 这个 url 请求将映射到控制器类 ProductInfoController 中的 addProduct 方法；success 参数的类型为 Function，表示请求成功后的回调函数，服务器返回的数据将传给该函数的参数 result，然后判断 result 参数中 success 名称所对应的值是否等于 true。如果等于 true，表示添加成功，则调用 Easy UI 的 Datagrid 控件的 reload 方法，重新获取商品列表信息，再关闭新增商品对话框。

在控制器类 ProductInfoController 中，addProduct 方法的代码如下：

```java
// 添加商品
@RequestMapping(value = "/addProduct", produces = "text/html;charset=UTF-8")
@ResponseBody
public String addProduct(ProductInfo pi, @RequestParam(value = "file", required = false) MultipartFile file, HttpServletRequest request, ModelMap model) {
    // 服务器端upload文件夹物理路径
    String path = request.getSession().getServletContext().getRealPath("product_images");
    // 获取文件名
    String fileName = file.getOriginalFilename();
    // 实例化一个File对象，表示目标文件(含物理路径)
    File targetFile = new File(path, fileName);
    if (!targetFile.exists()) {
        targetFile.mkdirs();
    }
```

```
        try {
            // 将上传文件写到服务器上指定的文件
            file.transferTo(targetFile);
            pi.setPic(fileName);
            productInfoService.addProductInfo(pi);
            return "{\"success\":\"true\",\"message\":\"商品添加成功\"}";
        } catch (Exception e) {
            return "{\"success\":\"false\",\"message\":\"商品添加失败\"}";
        }
    }
```

在 addProduct 方法上标注@RequestMapping 注解，将 productinfo/addProduct 这个 url 请求映射到 addProduct 方法。

addProduct 方法有四个参数，第一个是 ProductInfo 类型的参数 pi，用于封装新增商品对话框中输入的商品信息。第二个参数是 MultipartFile 类型的参数 file，通过@RequestParam 注解将表单中文件上传控件中选择的图片文件绑定到参数 file 中。第三个参数是 HttpServletRequest 类型的对象 request，用于封装客户端浏览器发出的请求。第四个参数是 ModelMap 类型的参数 model，用于将数据传递到前端页面。

在 addProduct 方法中，首先要将选择的图片文件上传到服务器上指定的文件，然后还需要调用 ProductInfoService 接口中的 addProductInfo 方法，将新增商品添加到数据表 product_info 中。如果添加成功，则向前端页面发送成功信息，否则发送失败的信息。

19.8.4 商品下架

在商品列表页 productlist.jsp 的 Datagrid 控件中，选中一个或多条记录前的复选框，再单击工具栏中的删除按钮，执行 JavaScript 函数 removeProduct，代码如下：

```
// 删除商品(商品下架)
function removeProduct() {
    var rows = $("#dg_productinfo").datagrid('getSelections');
    if (rows.length > 0) {
        $.messager.confirm('Confirm', '确认要删除么?', function(r) {
            if (r) {
                var ids = "";
                for (var i = 0; i < rows.length; i++) {
                    ids += rows[i].id + ",";
                }
                $.post('productinfo/deleteProduct', {
                    id : ids,
                    flag : 0
                }, function(result) {
                    if (result.success == 'true') {
                        $("#dg_productinfo").datagrid('reload');
                        $.messager.show({
                            title : '提示信息',
                            msg : result.message
                        });
                    } else {
```

```
                    $.messager.show({
                        title : '提示信息',
                        msg : result.message
                    });
                }
            }, 'json');
        }
    });
} else {
    $.messager.alert('提示', '请选择要删除的行', 'info');
}
}
```

在函数 removeProduct 中，首先获取 Datagrid 控件上选中的商品记录，将选中的商品编号保存到变量 ids 中，并以逗号分隔，然后通过$.post 发送一个 productinfo/deleteProduct 请求，这个 url 请求将映射到控制器类 ProductInfoController 中的 deleteProduct 方法，并且会向这个方法传递 id 和 flag 两个参数。如果这个方法的返回值中包含的 success 属性值为 true，则表示下架成功，此时会调用 Easy UI 的 Datagrid 控件的 reload 方法，重新获取商品列表信息，否则商品下架失败。

在控制器类 ProductInfoController 中，deleteProduct 方法的代码如下：

```
// 商品下架(删除商品)
@RequestMapping(value = "/deleteProduct", produces = 
"text/html;charset=UTF-8")
@ResponseBody
public String deleteProduct(@RequestParam(value = "id") String id,
@RequestParam(value = "flag") String flag) {
    String str = "";
    try {
        productInfoService.modifyStatus(id.substring(0, id.length() - 1),
Integer.parseInt(flag));
        str = "{\"success\":\"true\",\"message\":\"删除成功\"}";
    } catch (Exception e) {
        str = "{\"success\":\"false\",\"message\":\"删除失败\"}";
    }
    return str;
}
```

deleteProduct 方法有两个参数，第一个参数是 String 类型的参数 id，通过@RequestParam 注解将前端页面传递来的参数 id 绑定到 deleteProduct 方法的参数 id 中。第二个参数是 String 类型的参数 flag，通过@RequestParam 注解将前端页面传递来的参数 flag 绑定到 deleteProduct 方法的参数 flag 中。

在 deleteProduct 方法中，调用了业务接口 ProductInfoService 中的 modifyStatus 方法，根据选中商品的 id 号，将其状态设置为 0。如果执行成功，则向前端页面发送下架成功信息，否则发送下架失败信息。

19.8.5 修改商品

在商品列表页 productlist.jsp 的 Datagrid 控件中，选中某一条记录前的复选框，再单击工具栏中的修改按钮，将执行 JavaScript 函数 editProduct，代码如下：

```javascript
// 打开修改商品对话框(与新增商品对话框共用)
function editProduct() {
    var rows = $("#dg_productinfo").datagrid('getSelections');
    if (rows.length > 0) {
        var row = $("#dg_productinfo").datagrid("getSelected");
        if (row) {
            $("#dlg_productinfo").dialog("open").dialog('setTitle',
                '修改商品信息');
            $("#ff_productinfo").form("load", {
                "type.id" : row.type.id,
                "name" : row.name,
                "code" : row.code,
                "brand" : row.brand,
                "num" : row.num,
                "price" : row.price,
                "intro" : row.intro,
                "status" : row.status,
            });
            urls = "productinfo/updateProduct?id=" + row.id;
        }
    } else {
        $.messager.alert('提示', '请选择要修改的行', 'info');
    }
}
```

在函数 editProduct 中，首先获取 Datagrid 控件中选中的行，然后打开修改商品对话框(与添加商品使用同一个对话框和表单)，并将要修改的商品信息绑定到对话框中的表单文本域，其效果如图 19-10 所示。

图 19-10　修改商品信息对话框

修改完商品信息后，单击"保存"按钮，发送一个 productinfo/updateProduct 请求，这个 url 请求将映射到控制器类 ProductInfoService 中的 updateProduct 方法，其代码如下：

```java
// 修改商品
@RequestMapping(value = "/updateProduct", produces = "text/html;charset=UTF-8")
@ResponseBody
public String updateProduct(ProductInfo pi, @RequestParam(value = "file", required = false) MultipartFile file, HttpServletRequest request, ModelMap model) {
    // 服务器端upload文件夹物理路径
    String path = request.getSession().getServletContext().getRealPath("product_images");
    // 获取文件名
    String fileName = file.getOriginalFilename();
    // 实例化一个File对象,表示目标文件(含物理路径)
    File targetFile = new File(path, fileName);
    if (!targetFile.exists()) {
        targetFile.mkdirs();
    }
    try {
        // 将上传文件写到服务器上指定的文件
        file.transferTo(targetFile);
        pi.setPic(fileName);
        productInfoService.modifyProductInfo(pi);
        return "{\"success\":\"true\",\"message\":\"商品修改成功\"}";
    } catch (Exception e) {
        return "{\"success\":\"false\",\"message\":\"商品修改失败\"}";
    }
}
```

在控制器类ProductInfoService中，updateProduct方法的参数与addProduct方法的参数相同，只是在ProductInfo类型的参数pi中，还封装了通过url传递来的商品id号。

在updateProduct方法中，调用了业务接口ProductInfoService中的modifyProductInfo方法，将对象pi的属性值更新到数据表product_info中。

至此，商品管理的功能就讲解完了。商品类型管理与商品管理实现过程类似，由于篇幅所限，在此不再赘述。

19.9 订单管理

订单管理包括创建订单、查询订单、删除订单和查看订单明细等功能。

19.9.1 创建订单

在admin.jsp页面中，单击Tree控件上的创建订单结点，打开创建订单页createorder.jsp，如图19-11所示。

图 19-11 创建订单页

在页面 createorder.jsp 的<body></body>部分，首先定义一个 id 为 odbox 的 table，创建 Easy UI 的 Datagrid 控件，用来录入订单明细信息，代码如下：

```
<table id="odbox"></table>
```

然后创建一个 id 为 ordertb 的<div>标签，作为 Datagrid 控件的工具栏。在<div>标签中，定义了三个<a>标签，将每个<a>标签的 class 属性都设置为 easyui-linkbutton，从而将这三个<a>标签创建为添加订单明细、保存订单和删除订单明细三个 Easy UI 的链接按钮控件，代码如下：

```
<div id="ordertb" style="padding: 2px 5px;">
    <a href="javascript:void(0)" class="easyui-linkbutton"
        iconCls="icon-add" plain="true" onclick="addOrderDetail();">添加订单明细</a>
    <a href="javascript:void(0)" class="easyui-linkbutton"
        iconCls="icon-save" plain="true" onclick="saveorder();">保存订单
    </a>
    <a href="javascript:void(0)" class="easyui-linkbutton"
        iconCls="icon-remove" plain="true" onclick="removeOrderDetail();">删除订单明细</a>
</div>
```

接着定义一个 id 为 divOrderInfo 的<div>标签，用于创建订单信息录入布局，代码如下：

```
<div id="divOrderInfo">
    <div style="padding: 3px">
        客户名称 <input style="width: 115px;" id="create_uid"
            class="easyui-combobox" name="create_uid" value="0"
        data-options="valueField:'id',textField:'userName',
url:'userinfo/getValidUser'">   
        订单金额 <input type="text" name="create_orderprice"
            id="create_orderprice" class="easyui-textbox"
readonly="readonly" style="width: 115px" />   
    </div>
    <div style="padding: 3px">
```

```
            订单日期 <input type="text" name="create_ordertime"
                id="create_ordertime" class="easyui-datebox" style="width:
115px" value="<%=new Date().toLocaleString()%>" />   
            订单状态 <select id="create_status" class="easyui-combobox"
name="create_status" style="width: 115px;">
                <option value="未付款" selected>未付款</option>
                <option value="已付款">已付款</option>
                <option value="待发货">待发货</option>
                <option value="已发货">已发货</option>
                <option value="已完成">已完成</option>
            </select>
    </div>
</div>
```

在这个<div>标签中，客户名称使用了 Easy UI 的 ComboBox 控件，其绑定的数据源为 userinfo/getValidUser，这个 url 请求将映射到控制器类中的一个方法。接下来，在 com.ecpbm.controller 包中，创建一个 UserInfoController 类，并在类中编写一个 getValidUser 方法，代码如下：

```
package com.ecpbm.controller;
……
@Controller
@RequestMapping("/userinfo")
public class UserInfoController {
    @Autowired
    UserInfoService userInfoService;
    @RequestMapping("/getValidUser")
    @ResponseBody
    public List<UserInfo> getValidUser() {
        List<UserInfo> uiList = userInfoService.getValidUser();
        UserInfo ui = new UserInfo();
        ui.setId(0);
        ui.setUserName("请选择…");
        uiList.add(0, ui);
        return uiList;
    }
}
```

在 UserInfoController 类中，通过@Controller 注解指示该类是一个控制器，通过@RequestMapping 注解将用户对 userinfo/getValidUser 这个 url 的请求映射到 getValidUser 方法。

在 getValidUser 方法中，首先调用业务接口 UserInfoService 中的 getValidUser 方法，获取系统所有合法客户。然后通过@ResponseBody 注解，将方法返回结果进行 JSON 格式转换，再发送到前端页面，为客户名称组合框控件提供绑定数据。

在页面 createorder.jsp 中，通过 JavaScript 对 id 为 odbox 的创建为 Easy UI 的 Datagrid 控件的<table>标签进行初始化，代码如下：

```
<script type="text/javascript">
    var $odbox = $('#odbox');
```

```javascript
$(function() {
    $odbox.datagrid({
        rownumbers : true,
        singleSelect : false,
        fit : true,
        toolbar : '#ordertb',
        header : '#divOrderInfo',
        columns : [ [ {
            title : '序号',
            field : '',
            align : 'center',
            checkbox : true
        }, {
            field : 'pid',
            title : '商品名称',
            width : 300,
            editor : {
                type : 'combobox',
                options : {
                    valueField : 'id',
                    textField : 'name',
                    url : 'productinfo/getOnSaleProduct',
                    onChange: function (newValue, oldValue) {
                    var rows = $odbox.datagrid('getRows');
                    var orderprice=0;
                    for (var i = 0; i < rows.length; i++) {
                        var pidEd = $('#odbox').datagrid('getEditor', {
                            index: i,
                            field: 'pid'
                        });
                        var priceEd = $('#odbox').datagrid('getEditor', {
                            index: i,
                            field: 'price'
                        });
                        var totalpriceEd = $('#odbox').datagrid('getEditor', {
                            index: i,
                            field: 'totalprice'
                        });
                        var numEd = $('#odbox').datagrid('getEditor', {
                            index: i,
                            field: 'num'
                        });
                        if (pidEd != null){
                         var pid= $(pidEd.target).combobox('getValue');
        $.ajax({
                            type: 'POST',
                            url: 'productinfo/getPriceById',
                            data: {pid : pid},
```

```js
                                success: function(result) {
                    $(priceEd.target).numberbox('setValue',result);
                        $(totalpriceEd.target).numberbox('setValue',
result * $(numEd.target).numberbox('getValue'));
                    orderprice=Number(orderprice)+
Number($(totalpriceEd.target).numberbox('getValue'));
                                },
                                dataType: 'json',
                                async : false
                            });
                        }
                    }
$("#create_orderprice").textbox("setValue",orderprice);
                    }
                }
            }
        },{
            field : 'price',
            title : '单价',
            width : 80,
            editor: {
                type : "numberbox",
                options: {
                    editable : false
                }
            }
        } , {
            field : 'num',
            title : '数量',
            width : 50,
            editor : {
                type : 'numberbox',
                options :{
                    onChange: function (newValue, oldValue) {
                        var rows = $odbox.datagrid('getRows');
                        var orderprice=0;
                        for (var i = 0; i < rows.length; i++) {
                            var priceEd = $('#odbox').datagrid('getEditor', {
                                index: i,
                                field: 'price'
                            });
                            var totalpriceEd = $('#odbox').datagrid('getEditor', {
                                index: i,
                                field: 'totalprice'
                            });
                            var numEd = $('#odbox').datagrid('getEditor', {
                                index: i,
                                field: 'num'
                            });
$(totalpriceEd.target).numberbox('setValue',
```

```
                $(priceEd.target).numberbox('getValue') *
                $(numEd.target).numberbox('getValue'));
                orderprice=Number(orderprice)+
                Number($(totalpriceEd.target).numberbox('getValue'));
                            }
                $("#create_orderprice").textbox("setValue",orderprice);
                            }
                        }
                    }
                }, {
                    field : 'totalprice',
                    title : '小计',
                    width : 100,
                    editor: {
                        type : "numberbox",
                        options: {
                            editable : false
                        }
                    }
                } ] ]
        });
    });
……
</script>
```

Datagrid 控件的初始化是通过相关属性设置来完成的，rownumbers 属性用于设置是否显示行号，这里为 true，表示显示行号；singleSelect 属性用于设置是否单选，这里为 false，表示允许多选；fit 属性用于设置是否自适应显示数据，这里为 true，允许自适应显示；toolbar 属性用于设置工具栏，这里为#ordertb，表示要将 id 为 ordertb 的<div>标签作为 Datagrid 控件的工具栏；header 属性用于设置表头，这里为#divOrderInfo，表示要将 id 为 divOrderInfo 的<div>标签作为 Datagrid 控件的标题头；columns 属性用于设置 Datagrid 控件显示的列。

其中，商品名称列中使用了 Easy UI 的 ComboBox 控件，通过 url 属性指定其绑定的数据源为 productinfo/getOnSaleProduct，这个 url 请求将映射到控制器类 ProductInfoController 中一个方法。接下来，在 ProductInfoController 类中编写 getOnSaleProduct 方法，代码如下：

```
// 获取在售商品列表
@ResponseBody
@RequestMapping("/getOnSaleProduct")
public List<ProductInfo> getOnSaleProduct() {
    List<ProductInfo> piList = productInfoService.getOnSaleProduct();
    return piList;
}
```

通过在 getOnSaleProduct 方法上标注@RequestMapping 注解，从而将用户对 productinfo/getOnSaleProduct 这个 url 的请求映射到 getOnSaleProduct 方法。在 getOnSaleProduct 方法中，调用业务接口 ProductInfoService 的 getOnSaleProduct 方法获取在售商品列表，并通过@ResponseBody 注解，将 List<ProductInfo>类型的返回值 piList 进行 JSON 格式转换，再发送到前端页面。

此外，在商品名称组合框控件中，还添加了 onChange 事件处理代码，实现根据所选择的商品，显示商品价格列数据，并更新小计列数据和订单金额文本域值。根据商品 id 获取商品单价是通过 AJAX 向后台服务器发送请求获得的，请求的地址为 productinfo/getPriceById，这个 url 请求将映射到控制器类 ProductInfoController 中的 getPriceById 方法，其代码如下：

```java
// 根据商品 id 获取商品单价
@RequestMapping("/getPriceById")
@ResponseBody
public String getPriceById(@RequestParam(value = "pid") String pid) {
    if (pid != null && !"".equals(pid)) {
        ProductInfo pi =
productInfoService.getProductInfoById(Integer.parseInt(pid));
        return pi.getPrice() + "";
    } else {
        return "";
    }
}
```

在数量列的 Easy UI NumberBox 控件中，也添加了 onChange 处理代码，实现根据所填写的数量更新小计列数据和订单金额文本域值。

在创建订单页 createorder.jsp 的工具栏上，包括添加订单明细、删除订单明细和保存订单三个按钮，接下来具体讲解这些按钮功能的实现过程。

1. 添加订单明细

单击添加订单明细按钮时，会触发一个 onclick 事件，通过 JavaScript 函数 addOrderDetail 来处理该事件，以实现在 Easy UI 的 Datagrid 控件上增加一个新行，函数 addOrderDetail 的代码如下：

```javascript
// datagrid 中添加记录行
function addOrderDetail() {
    $odbox.datagrid('appendRow', {
        num : '1',
        price : '0',
        totalprice : '0'
    });
    var rows = $odbox.datagrid('getRows');
// 让添加的行处于可编辑状态
    $odbox.datagrid('beginEdit', rows.length - 1);
}
```

在函数 addOrderDetail 中，使用了 Datagrid 控件的 appendRow 方法，在 Datagrid 控件的尾部添加一个记录行，同时对新增行的 num、price 和 totalprice 三列进行初始化。然后使用 Datagrid 控件的 beginEdit 方法，将新增行设置为可编辑状态。

2. 删除订单明细

在 Datagrid 控件中，选中要删除的记录(支持多选)，单击删除订单明细按钮，可将选择的记录行从控件上清除。单击删除订单明细按钮时，会触发一个 onclick 事件，通过 JavaScript

函数 removeOrderDetail 来处理该事件。函数 removeOrderDetail 的代码如下：

```javascript
//在 datagrid 中删除记录行
function removeOrderDetail() {
    // 获取所选择的行记录
    var rows = $odbox.datagrid('getSelections');
    if (rows.length > 0) {
        // 获取"订单金额"文本域的值
        var create_orderprice =
$("#create_orderprice").textbox("getValue");
        // 遍历选中的行记录，以更新订单金额
        for (var i = 0; i < rows.length; i++) {
            var index = $odbox.datagrid('getRowIndex', rows[i]);
            var totalpriceEd = $('#odbox').datagrid('getEditor', {
                index: index,
                field: 'totalprice'
            });
            create_orderprice = create_orderprice -
Number($(totalpriceEd.target).numberbox('getValue'));
            $odbox.datagrid('deleteRow', index);
        }
        $("#create_orderprice").textbox("setValue",create_orderprice);
    } else {
        $.messager.alert('提示', '请选择要删除的行', 'info');
    }
}
```

在函数 removeOrderDetail 中，首先使用 Datagrid 控件的 getSelections 方法获取所选择的行记录，然后获取订单金额文本域的值，再遍历选中的行记录，以更新订单金额。

3. 保存订单

填写订单和订单明细信息后，单击保存订单按钮，会触发一个 onclick 事件，通过 JavaScript 函数 saveorder 来处理该事件。函数 saveorder 的代码如下：

```javascript
// 保存订单
function saveorder() {
    // 获取订单客户
    var uid = $("#create_uid").combobox("getValue");
    if(uid==0){
        $.messager.alert('提示', '请选择客户名称', 'info');
    } else {
        // 取消 datagrid 控件的行编辑状态
        create_endEdit();
        // 定义 orderinfo 存放订单主表数据
        var orderinfo = [];
        // 获取订单时间
        var ordertime = $("#create_ordertime").datebox("getValue");
        // 获取订单状态
        var status = $("#create_status").combobox("getValue");
        // 获取订单金额
```

```javascript
            var orderprice = $("#create_orderprice").textbox("getValue");
            orderinfo.push({
                ordertime : ordertime,
                uid : uid,
                status : status,
                orderprice : orderprice
            });
            // 获取订单明细(即 datagrid 控件中的行记录)
            if ($odbox.datagrid('getChanges').length) {
                // 获取 datagrid 控件中插入的记录行
                var inserted = $odbox.datagrid('getChanges', "inserted");
                // 获取 datagrid 控件中删除的记录行
                var deleted = $odbox.datagrid('getChanges', "deleted");
                // 获取 datagrid 控件中更新的记录行
                var updated = $odbox.datagrid('getChanges', "updated");
                // 定义 effectRow,保存 inserted 和 orderinfo
                var effectRow = new Object();
                if (inserted.length) {
                    effectRow["inserted"] = JSON.stringify(inserted);
                }
                effectRow["orderinfo"] = JSON.stringify(orderinfo);
                // 提交请求
                $.post(
                "orderinfo/commitOrder",
                effectRow,
                function(data) {
                    if (data == 'success') {
                        $.messager.alert("提示", "创建成功!");
                        $odbox.datagrid('acceptChanges');
                        if ($('#tabs').tabs('exists', '创建订单')) {
                            $('#tabs').tabs('close', '创建订单');
                        }
                        $("#orderDg").datagrid('reload');
                    } else {
                        $.messager.alert("提示", "创建失败!");
                    }
                });
            }
        }
```

在函数 saveorder 中,首先调用自定义的 JavaScript 函数 create_endEdit 取消 Datagrid 控件的行可编辑状态。函数 create_endEdit 的代码如下:

```javascript
// 取消 datagrid 控件的行编辑状态
function create_endEdit() {
    var rows = $odbox.datagrid('getRows');
    for (var i = 0; i < rows.length; i++) {
        $odbox.datagrid('endEdit', i);
```

```
        }
    }
```

在函数 saveorder 中，定义了一个变量 orderinfo，用于存放订单客户、订单时间、订单状态和订单金额等订单数据。接着调用 Datagrid 控件的 getChanges 方法，获取控件上发生改变的情况，包括 inserted、updated 和 deleted 三种类型。由于 Datagrid 控件初始时没有任何记录行，是通过单击添加订单明细按钮创建的，因此即便是删除或者修改某个订单明细，Datagrid 控件上发生的改变都属于 inserted 类型，根据这个类型可获取插入的行数据，并赋值给变量 inserted。然后定义一个对象 effectRow，用于保存 inserted 和 orderinfo，再通过 $.post 将请求提交到 orderinfo/commitOrder，这个 url 请求将映射到一个控制器类中的一个方法。最后在 com.ecpbm.controller 包中，创建一个 OrderInfoController 类，在类中编写一个 commitOrder 方法，代码如下：

```java
package com.ecpbm.controller;
......
@Controller
@RequestMapping("/orderinfo")
public class OrderInfoController {
    @Autowired
    OrderInfoService orderInfoService;
    @Autowired
    UserInfoService userInfoService;
    @Autowired
    ProductInfoService productInfoService;
    // 保存订单
    @ResponseBody
    @RequestMapping(value = "/commitOrder")
    public String commitOrder(String inserted, String orderinfo)
            throws JsonParseException, JsonMappingException, IOException {
        try {
            // 创建 ObjectMapper 对象，实现 JavaBean 和 JSON 的转换
            ObjectMapper mapper = new ObjectMapper();
            // 设置输入时忽略在 JSON 字符串中存在但 Java 对象实际没有的属性
            mapper.disable(DeserializationFeature.FAIL_ON_UNKNOWN_PROPERTIES);
            mapper.configure(SerializationFeature.FAIL_ON_EMPTY_BEANS, false);
            // 将 json 字符串 orderinfo 转换成 JavaBean 对象(订单信息)
            OrderInfo oi = mapper.readValue(orderinfo, OrderInfo[].class)[0];
            // 保存订单信息
            orderInfoService.addOrderInfo(oi);
            // 将 json 字符串转换成 List<OrderDetail>集合(订单明细信息)
            List<OrderDetail> odList = mapper.readValue(inserted, new TypeReference<ArrayList<OrderDetail>>() {
            });
            // 给订单明细对象的其他属性赋值
            for (OrderDetail od : odList) {
                od.setOid(oi.getId());
                // 保存订单明细
                orderInfoService.addOrderDetail(od);
            }
            return "success";
        } catch (Exception e) {
```

```
            return "failure";
        }
    }
    ......
}
</script>
```

在 OrderInfoController 类中，通过@Controller 注解指示该类是一个控制器，通过@RequestMapping 注解将用户对 orderinfo/commitOrder 这个 url 的请求映射到 commitOrder 方法。commitOrder 方法有两个 String 类型的参数 inserted 和 orderinfo，用于封装前端页面传递来的参数。

在 commitOrder 方法中，首先创建 ObjectMapper 对象，用于实现 JavaBean 和 JSON 的转换，并设置输入时忽略在 JSON 字符串中存在但 Java 对象实际没有的属性。然后将 JSON 字符串 orderinfo 转换为 OrderInfo 对象 oi(对应订单信息)，并调用业务接口 OrderInfoService 中的 addOrderInfo 方法，将订单信息保存到数据表 order_info。

将 JSON 字符串 inserted 转换成 List<OrderDetail>集合对象 odList(对应订单明细)，并对集合 odList 进行遍历。每次遍历时，先为当前订单明细对象设置关联的订单 id 号，再调用业务接口 OrderInfoService 中的 addOrderDetail 方法，将订单明细信息保存到数据表 order_detail。

如果执行成功，则向前端页面发送 success，否则发送 failure。在前端页面 createorder.jsp 中，函数 saveorder 会对返回值进行判断，如果成功，则关闭创建订单标签页，再调用 Datagrid 控件的 reload 方法重新获取数据源以更新数据；如果失败，则提示错误信息。

19.9.2 查询订单

在 admin.jsp 页面中，单击 Tree 控件上的查询订单结点，打开订单查询页 searchorder.jsp，如图 19-12 所示。

图 19-12 订单查询页

在订单查询页中，可根据订单编号、客户名称、订单状态和订单时间进行查询，查询表单布局代码如下：

```html
<!-- 查询表单 -->
<div id="searchOrderTb" style="padding:2px 5px;">
    <form id="searchOrderForm">
        <div style="padding:3px">
            订单编号 <input class="easyui-textbox" name="search_oid"
    id="search_oid" style="width:110px" />
        </div>
        <div style="padding:3px">
            客户名称 <input style="width:115px;" id="search_uid"
                class="easyui-combobox" value="0" name="search_uid"
                data-options="valueField:'id',textField:'userName',
url:'userinfo/getValidUser'">   
            订单状态 <select id="search_status"
class="easyui-combobox" name="search_status" style="width:115px;">
                <option value="请选择" selected>请选择</option>
                <option value="未付款">未付款</option>
                <option value="已付款">已付款</option>
                <option value="待发货">待发货</option>
                <option value="已发货">已发货</option>
                <option value="已完成">已完成</option>
            </select>    订单时间  <input
class="easyui-datebox" name="orderTimeFrom" id="orderTimeFrom"
style="width:115px;" /> ~ <input class="easyui-datebox" name="orderTimeTo"
id="orderTimeTo"    style="width:115px;" /> <a href="javascript:void(0)"
class="easyui-linkbutton" iconCls="icon-search" plain="true"
    onclick="searchOrderInfo();">查找</a>
        </div>
    </form>
</div>
```

在订单查询页中，使用了 Easy UI Datagrid 控件来显示订单列表。为此，创建了一个 id 为 orderDg 的<table>标签，并将其 easyui-datagrid 属性设置为 easyui-datagrid，如下所示：

```html
<table id="orderDg" class="easyui-datagrid"></table>
```

为了给 Datagrid 控件提供工具栏，创建了一个 id 为 orderTb 的<div>标签。在<div>标签中，定义了两个<a>标签，将它们的 class 属性都设置为 easyui-linkbutton，将这两个<a>标签创建为查看明细和删除订单两个链接按钮，代码如下：

```html
<!-- 工具栏 -->
<div id="orderTb" style="padding:2px 5px;">
  <a href="javascript:void(0)" class="easyui-linkbutton"
     iconCls="icon-edit" plain="true" onclick="editOrder();">查看明细</a>
  <a href="javascript:void(0)" class="easyui-linkbutton"
iconCls="icon-remove" onclick="removeOrder();" plain="true">删除订单
  </a>
</div>
```

通过 JavaScript 对 id 为 orderDg 的创建为 Easy UI 的 Datagrid 控件的<table>标签进行初始化，代码如下：

```html
<script type="text/javascript">
$(function() {
    $('#orderDg').datagrid({
        singleSelect : false,
        url : 'orderinfo/list', //为datagrid设置数据源
        queryParams : {}, //查询条件
        pagination : true, //启用分页
        pageSize : 5, //设置初始每页记录数(页大小)
        pageList : [ 5, 10, 15 ], //设置可供选择的页大小
        rownumbers : true, //显示行号
        fit : true, //设置自适应
        toolbar : '#orderTb', //为datagrid添加工具栏
        header : '#searchOrderTb', //为datagrid标题头添加搜索栏
        columns : [ [ { //编辑datagrid的列
            title : '序号',
            field : 'id',
            align : 'center',
            checkbox : true
        }, {
            field : 'ui',
            title : '订单客户',
            formatter : function(value, row, index) {
                if (row.ui) {
                    return row.ui.userName;
                } else {
                    return value;
                }
            },
            width : 100
        }, {
            field : 'status',
            title : '订单状态',
            width : 80
        }, {
            field : 'ordertime',
            title : '订单时间',
            width : 100
        }, {
            field : 'orderprice',
            title : '订单金额',
            width : 100
        } ] ]
    });
});
......
</script>
```

Datagrid 控件的初始化是通过相关属性设置来完成的,singleSelect 属性用于设置是否单选,这里为 false,表示允许多选;queryParams 属性用于设置传递到后台控制器的参数列表,这里先设置为{};pagination 属性用于设置是否分页,这里为 true,表示允许分页;pageSize

属性用于设置初始每页记录数，这里为 5；pageList 用于设置可供选择的每页记录数，这里为[5, 10, 15]，表示可以选择每页显示 5、10 或 15 条记录；rownumbers 属性用于设置是否显示行号，这里为 true，表示显示行号；fit 属性用于设置是否自适应显示数据，这里为 true，允许自适应显示；toolbar 属性用于设置工具栏，这里为#orderTb，表示要将 id 为 orderTb 的<div>标签作为 Datagrid 控件的工具栏；header 属性用于设置表头，这里为#searchOrderTb，表示要将 id 为 searchOrderTb 的<div>标签作为 Datagrid 控件的标题头；columns 属性用于设置置 Datagrid 控件显示的列；url 属性用于指定 Datagrid 控件的数据源，这里为 orderinfo/list，这个 url 请求将映射到控制器类 OrderInfoController 中的一个方法。在 OrderInfoController 类中编写 list 方法，根据查询条件分页获取订单列表，代码如下：

```
// 分页显示
@RequestMapping(value = "/list")
@ResponseBody
public Map<String, Object> list(Integer page, Integer rows, OrderInfo orderInfo) {
    // 初始化一个分页类对象 pager
    Pager pager = new Pager();
    pager.setCurPage(page);
    pager.setPerPageRows(rows);
    // 创建对象 params，用于封装查询条件
    Map<String, Object> params = new HashMap<String, Object>();
    params.put("orderInfo", orderInfo);
    // 获取满足条件的订单总数
    int totalCount = orderInfoService.count(params);
    // 获取满足条件的订单列表
    List<OrderInfo> orderinfos = orderInfoService.findOrderInfo(orderInfo, pager);
    // 创建 result 对象，保存查询结果数据
    Map<String, Object> result = new HashMap<String, Object>(2);
    result.put("total", totalCount);
    result.put("rows", orderinfos);
    return result;
}
```

通过在 list 方法上标注@RequestMapping 注解，将用户对 orderinfo/list 这个 url 的请求映射到 list 方法。list 方法有三个参数，一个是 OrderInfo 类参数 orderInfo，用于封装前端页面传递来的查询条件。另外两个参数是 page 和 rows，用于接收从 Datagrid 控件传递来的页码和每页显示的记录数。

在 list 方法中，首先初始化一个分页类对象 pager，给其设置 curPage 和 perPageRows 两个属性值。然后创建 Map<String, Object>类型的对象 params，用于封装查询条件。接着依次调用业务接口 OrderInfoService 中的 count 方法获取满足条件的订单总数，调用 findOrderInfo 方法获取满足条件的订单列表。再创建 Map<String, Object>类型的对象 result，保存查询结果数据。最后将返回结果转为 JSON 格式，以字符串的形式发送到前端页面 searchorder.jsp，为 Datagrid 控件提供数据源。

初始时 Datagrid 控件显示所有订单记录，如果在查询表单中输入查询条件，单击查找

按钮,将执行 JavaScript 函数 searchOrderInfo,代码如下:

```javascript
// 查询订单
function searchOrderInfo() {
    var oid = $('#search_oid').val();
    var status = $('#search_status').combobox("getValue");
    var uid = $('#search_uid').combobox("getValue");
    var orderTimeFrom = $("#orderTimeFrom").datebox("getValue");
    var orderTimeTo = $("#orderTimeTo").datebox("getValue");
    $('#orderDg').datagrid('load', {
        id : oid,
        status : status,
        uid : uid,
        orderTimeFrom : orderTimeFrom,
        orderTimeTo : orderTimeTo
    });
}
```

在函数 searchOrderInfo 中,首先获取用户输入的查询条件,然后调用 Datagrid 控件的 load 方法,再次向控制器类 OrderInfoController 中的 list 方法发送请求,并将参数传递过去。以查询客户名称为 john 的订单为例,执行查询后,结果如图 19-13 所示。

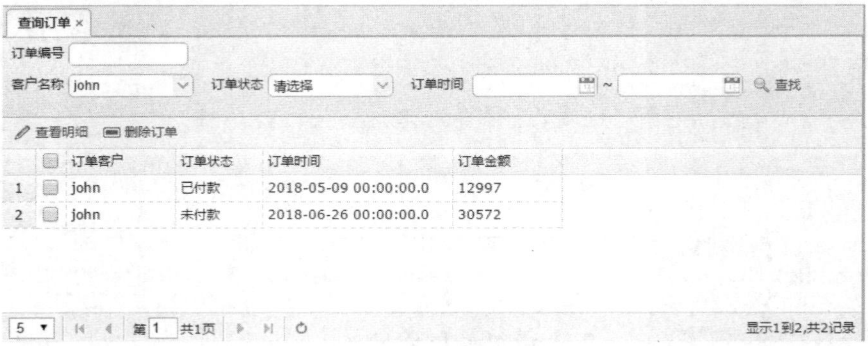

图 19-13　查询客户 john 的订单

19.9.3　删除订单

在订单查询页 searchorder.jsp 中,选中 Datagrid 控件中的某条记录,再单击工具栏中的删除订单按钮,可将订单和关联的明细信息删除。单击删除订单按钮时,将执行 JavaScript 函数 removeOrder,代码如下:

```javascript
// 删除订单
function removeOrder() {
    // 获取选中的订单记录行
    var rows = $("#orderDg").datagrid('getSelections');
    if (rows.length > 0) {
        $.messager.confirm('Confirm', '确认要删除么?', function(r) {
            if (r) {
                var ids = "";
```

```javascript
            // 获取选中订单记录的订单 id, 保存到 ids 中
            for (var i = 0; i < rows.length; i++) {
                ids += rows[i].id + ",";
            }
            // 发送请求
            $.post('orderinfo/deleteOrder', {
                oids : ids
            }, function(result) {
                if (result.success == 'true') {
                    $("#orderDg").datagrid('reload');
                    $.messager.show({
                        title : '提示信息',
                        msg : result.message
                    });
                } else {
                    $.messager.show({
                        title : '提示信息',
                        msg : result.message
                    });
                }
            }, 'json');
        }
    });
} else {
    $.messager.alert('提示', '请选择要删除的行', 'info');
}
}
```

在函数 removeOrder 中，首先获取 Datagrid 控件中选中的订单记录，然后将它们的订单编号以逗号分隔，保存到变量 ids 中，再使用$.post 发送请求 orderinfo/deleteOrder，同时将参数 oids 传递过去，这个 url 请求将映射到控制器类 OrderInfoController 中的一个方法。接下来，在 OrderInfoController 类中编写一个 deleteOrder 方法，代码如下：

```java
// 删除订单
@ResponseBody
@RequestMapping(value = "/deleteOrder", produces = "text/html;charset=UTF-8")
public String deleteOrder(String oids) {
    String str = "";
    try {
        oids = oids.substring(0, oids.length() - 1);
        String[] ids = oids.split(",");
        for (String id : ids) {
            orderInfoService.deleteOrder(Integer.parseInt(id));
        }
        str = "{\"success\":\"true\",\"message\":\"删除成功！\"}";
    } catch (Exception e) {
        str = "{\"success\":\"false\",\"message\":\"删除失败！\"}";
    }
    return str;
}
```

通过在 deleteOrder 方法上标注@RequestMapping 注解，将用户对 orderinfo/deleteOrder 这个 url 的请求映射到 deleteOrder 方法。deleteOrder 方法有一个 String 类型的参数 oids，用于封装从前端页面传递来的以逗号分隔的订单编号。

在 deleteOrder 方法中，循环调用了业务接口 OrderInfoService 中的 deleteOrder 方法，根据订单编号删除订单信息。最后根据执行情况，向前端页面发送成功或失败信息。在前端页面 searchorder.jsp 中，根据结果进行判断，如果执行成功，则调用 Datagrid 控件的 reload 方法，重新向控制器类 OrderInfoController 的 list 方法发送请求，以更新数据源。

19.9.4　查看订单明细

在订单查询页 searchorder.jsp 中，选中 Datagrid 控件中的某条记录，单击工具栏中的"查看明细"按钮，将执行 JavaScript 中的 editOrder 函数，代码如下：

```javascript
// 查看明细
function editOrder() {
    var rows = $("#orderDg").datagrid('getSelections');
    if (rows.length > 0) {
        var row = $("#orderDg").datagrid("getSelected");
        if ($('#tabs').tabs('exists', '订单明细')) {
            $('#tabs').tabs('close', '订单明细');
        }
        $('#tabs').tabs('add', {
            title : "订单明细",
            href : 'orderinfo/getOrderInfo?oid=' + row.id,
            closable : true
        });
    } else {
        $.messager.alert('提示', '请选择要修改的订单', 'info');
    }
}
```

在函数 editOrder 中，首先获取 Datagrid 控件中选中的行，然后发送请求 orderinfo/getOrderInfo，这个 url 请求将映射到控制器类 OrderInfoController 中的一个方法，同时将参数 oid 传递给这个方法。接下来，在 OrderInfoController 类中编写 getOrderInfo 方法，代码如下：

```java
// 根据订单 id 号获取要查看的订单对象，再返回订单明细页
@RequestMapping("/getOrderInfo")
public String getOrderInfo(String oid, Model model) {
    OrderInfo oi =
orderInfoService.getOrderInfoById(Integer.parseInt(oid));
    model.addAttribute("oi", oi);
    return "orderdetail";
}
```

通过在 getOrderInfo 方法上标注@RequestMapping 注解，将用户对 orderinfo/getOrderInfo 这个 url 的请求映射到 getOrderInfo 方法。getOrderInfo 方法有两个参数，一个是 String 类型

的参数 oid，用于封装从前端页面通过 URL 传递来的订单编号，另一个是 Model 类型的参数 model，用于向跳转到的订单明细页传递数据。

在 getOrderInfo 方法中，调用业务接口 OrderInfoService 中的 getOrderInfoById 方法，获取指定编号的订单对象，再将订单对象 oi 添加到 Model 对象中。getOrderInfo 方法执行结束后，跳转到订单明细页 orderdetail.jsp。

在订单明细页 orderdetail.jsp 中，可以从 Model 对象中取出之前存放的订单对象，再将订单信息和订单明细信息显示出来。

订单信息是在一个 id 为 editordertb 的<div>标签中显示的，如下所示：

```html
<div id="editordertb" style="padding: 2px 5px;">
    <div id="editdivOrderInfo">
        <div style="padding: 3px">
            客户名称 <input style="width: 115px;" id="edit_uid"
                class="easyui-textbox" name="edit_uid" readonly="readonly"
value="${requestScope.oi.ui.userName }">
      
            订单金额 <input type="text" name="edit_orderprice"
id="edit_orderprice" value="${requestScope.oi.orderprice }"
    class="easyui-textbox" readonly="readonly" style="width: 115px"
/>  
        </div>
        <div style="padding: 3px">
            订单日期 <input type="text" name="edit_ordertime"
readonly="readonly" id="edit_ordertime"
value="${requestScope.oi.ordertime }" class="easyui-datebox"
style="width: 115px" />   
   订单状态 <input  id="edit_status"
class="easyui-textbox" name="edit_status" style="width: 115px;"
readonly="readonly" value="${requestScope.oi.status }">
    </div>
    </div>
</div>
```

在上述<div>标签中，客户名称、订单金额、订单日期和订单状态这些订单信息都是从 request 域中的 oi 对象获取的。

订单明细信息是通过 Easy UI 的 Datagrid 控件显示的，为此在订单明细页 orderdetail.jsp 中，定义了一个 id 为 editodbox 的<table>标签。

接下来，通过 JavaScript 对 id 为 editodbox 的<table>标签进行初始化，以显示订单明细信息，代码如下：

```html
<script type="text/javascript">
    var $editodbox = $('#editodbox');
    $(function() {
        $editodbox.datagrid({
            url : 'orderinfo/getOrderDetails?oid=${requestScope.oi.id }',
            rownumbers : true,
            singleSelect : false,
```

```
                fit : true,
                toolbar : '#editordertb',
                columns : [ [ {
                    field : 'pid',
                    title : '商品名称',
                    width : 300,
                    formatter : function(value, row, index) {
                        if (row.pi) {
                            return row.pi.name;
                        } else {
                            return value;
                        }
                    }
                }, {
                    field : 'price',
                    title : '单价',
                    width : 80
                }, {
                    field : 'num',
                    title : '数量',
                    width : 50
                }, {
                    field : 'totalprice',
                    title : '小计',
                    width : 100
                } ] ]
            });
        });
    </script>
```

Datagrid 控件的数据源是通过 url 属性指定的,这里设置为 orderinfo/getOrderDetails,这个 url 请求将映射到控制器类 OrderInfoController 中的一个方法,并将参数 oid 传递给该方法。接下来在 OrderInfoController 类中,编写 getOrderDetails 方法,代码如下:

```
// 根据订单id号获取订单明细列表
@RequestMapping("/getOrderDetails")
@ResponseBody
public List<OrderDetail> getOrderDetails(String oid) {
    List<OrderDetail> ods =
orderInfoService.getOrderDetailByOid(Integer.parseInt(oid));
    for (OrderDetail od : ods) {
        od.setPrice(od.getPi().getPrice());
        od.setTotalprice(od.getPi().getPrice() * od.getNum());
    }
    return ods;
}
```

在 getOrderDetails 方法中,调用了业务接口 OrderInfoService 中的 getOrderDetailByOid 方法,根据订单编号获取订单明细信息列表。然后对这个列表进行遍历,将关联的商品编号、价格和小计等信息保存到每一个订单明细对象中。再通过@ResponseBody 注解自动将

List<OrderDetail>类型的 ods 进行 JSON 格式转换，发送到前端页面 orderdetail.jsp，作为页面中 Datagrid 控件的数据源。

在订单查询页 searchorder.jsp 中，选中一条订单记录，再单击查看明细按钮，可以看到这个订单的明细信息，如图 19-14 所示。

图 19-14　查看订单明细

19.10　客户管理

客户管理包括客户列表显示、查询客户、启用和禁用客户功能。

19.10.1　客户列表显示

在 admin.jsp 页面中，单击客户管理下的客户列表结点，打开客户列表页 userlist.jsp，如图 19-15 所示。

图 19-15　客户列表页

在 userlist.jsp 页面中，使用了 Easy UI 的 Datagrid 控件来显示客户列表，该控件是通过 id 为 userListDg 的<table>标签创建的，其定义如下：

```html
<table id="userListDg" class="easyui-datagrid"></table>
```

在页面中,还定义了一个 id 为 userListTb 的<div>标签,并在<div>标签中定义了启用客户和禁用客户两个链接按钮,用作 Datagrid 控件上的工具栏,代码如下:

```html
<!-- 创建工具栏 -->
<div id="userListTb" style="padding:2px 5px;"><a href="javascript:void(0)"
    class="easyui-linkbutton" iconCls="icon-edit" plain="true"
    onclick="SetIsEnableUser(1);">启用客户</a>
    <a href="javascript:void(0)"   class="easyui-linkbutton"
    iconCls="icon-remove" onclick="SetIsEnableUser(0);" plain="true">禁用客户
    </a>
</div>
```

接着定义了一个 id 为 searchUserListTb 的<div>标签,在<div>标签中定义了一个 id 为 searchUserListForm 的表单,在这个表单中定义了一个客户名称文本框和一个查询链接按钮,用作 Datagrid 控件上的搜索栏,代码如下:

```html
<!-- 创建搜索栏 -->
<div id="searchUserListTb" style="padding:4px 3px;">
    <form id="searchUserListForm">
        <div style="padding:3px ">
            客户名称  <input class="easyui-textbox"
name="search_userName" id="search_userName" style="width:110px" /><a
href="javascript:void(0)" class="easyui-linkbutton" iconCls="icon-search"
plain="true"       onclick="searchUserInfo();">查找</a>
        </div>
    </form>
</div>
```

最后通过 JavaScript 对 id 为 userListDg 的创建为 Easy UI 的 Datagrid 控件的<table>标签进行初始化,代码如下:

```javascript
<script type="text/javascript">
$(function() {
    $('#userListDg').datagrid({
        singleSelect : false,
        url : 'userinfo/list',
        queryParams : {}, //查询条件
        pagination : true, //启用分页
        pageSize : 5, //设置初始每页记录数(页大小)
        pageList : [ 5, 10, 15 ], //设置可供选择的页大小
        rownumbers : true, //显示行号
        fit : true, //设置自适应
        toolbar : '#userListTb', //为datagrid添加工具栏
        header : '#searchUserListTb', //为datagrid标题头添加搜索栏
        columns : [ [ { //编辑datagrid的列
            title : '序号',
            field : 'id',
            align : 'center',
            checkbox : true
```

```
            }, {
                field : 'userName',
                title : '登录名',
                width : 100
            }, {
                field : 'realName',
                title : '真实姓名',
                width : 80
            }, {
                field : 'sex',
                title : '性别',
                width : 100
            }, {
                field : 'address',
                title : '住址',
                width : 200
            } , {
                field : 'email',
                title : '邮箱',
                width : 150
            } , {
                field : 'regDate',
                title : '注册日期',
                width : 100
            } , {
                field : 'status',
                title : '客户状态',
                width : 100,
                formatter : function(value, row, index) {
                    if (row.status==1) {
                        return "启用";
                    } else {
                        return "禁用";
                    }
                }
            } ] ]
        });
    });
    ……
</script>
```

Datagrid 控件的初始化是通过相关属性设置来完成的，singleSelect 属性用于设置是否单选，这里为 false，表示允许多选；queryParams 属性用于设置传递到后台控制器的参数列表，这里先设置为{}；pagination 属性用于设置是否分页，这里为 true，表示允许分页；pageSize 属性用于设置每页初始记录数，这里为 5；pageList 用于设置可供选择的每页记录数，这里为[5, 10, 15]，表示可以选择每页显示 5、10 或 15 条记录；rownumbers 属性用于设置是否显示行号，这里为 true，表示显示行号；fit 属性用于设置是否自适应显示数据，这里为 true，允许自适应显示；toolbar 属性用于设置工具栏，这里为#userListTb，表示要将 id 为 userListTb 的<div>标签作为 Datagrid 控件的工具栏；header 属性用于设置 Datadrid 控件的标题头，这里为#searchUserListTb，表示要将 id 为 searchUserListTb 的<div>标签作为 Datagrid 控件的标题头；columns 属性用于设置 Datagrid 控件显示的列；url 属性用于指定 Datagrid 控件的数据源，这里为 userinfo/list，这个 url 请求将映射到控制器类 UserInfoController 中的一个方法。

接下来，在 UserInfoController 类中编写 userlist 方法，根据查询条件分页获取客户列表，代码如下：

```java
@RequestMapping("/list")
@ResponseBody
public Map<String, Object> userlist(Integer page, Integer rows, UserInfo userInfo) {
    // 创建分页类对象
    Pager pager = new Pager();
    pager.setCurPage(page);
    pager.setPerPageRows(rows);
    // 创建对象 params，封装查询条件
    Map<String, Object> params = new HashMap<String, Object>();
    params.put("userInfo", userInfo);
    // 根据查询条件，获取客户记录数
    int totalCount = userInfoService.count(params);
    // 根据查询条件，分页获取客户列表
    List<UserInfo> userinfos = userInfoService.findUserInfo(userInfo, pager);
    // 创建对象 result，保存查询结果数据
    Map<String, Object> result = new HashMap<String, Object>(2);
    result.put("total", totalCount);
    result.put("rows", userinfos);
    return result;
}
```

通过在 userlist 方法上标注@RequestMapping 注解，将用户对 userinfo/list 这个 url 的请求映射到 userlist 方法。userlist 方法有三个参数，一个是 UserInfo 类参数 userInfo，用于封装前端页面传递来的查询条件。另外两个参数是 page 和 rows，用于接收从 Datagrid 控件传递来的页码和每页显示的记录数。

在 userlist 方法中，首先初始化一个分页类对象 pager，给其设置 curPage 和 perPageRows 两个属性值。然后创建 Map<String, Object>类型的对象 params，用于封装查询条件。接着依次调用业务接口 UserInfoService 中的 count 方法获取满足条件的客户总数，调用 findUserInfo 方法获取满足条件的客户列表。再创建 Map<String, Object>类型的对象 result，保存查询结果数据。最后将返回结果转为 JSON 格式，以字符串的形式发送到前端页面 userlist.jsp，为 Datagrid 控件提供数据源。

19.10.2　查询客户

在 userlist.jsp 页面的搜索栏中，输入客户名称，单击"查找"按钮，会根据客户名称进行模糊查询，并将查询结果显示在 Datagrid 控件中。例如，在客户名称文本框中输入 j，单击"查找"按钮，查询结果如图 19-16 所示。

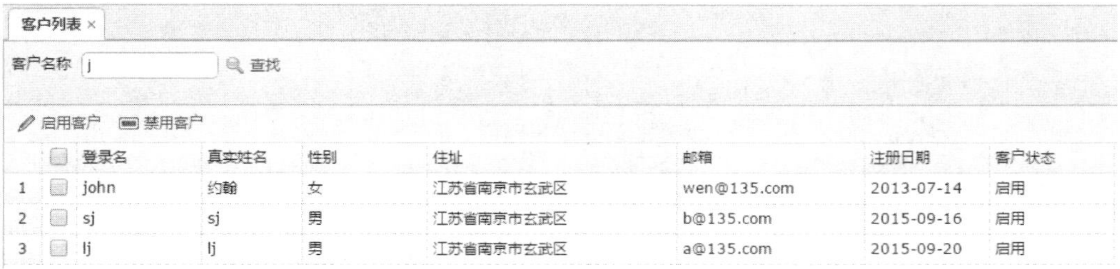

图 19-16　按客户名模糊查询

单击"查找"按钮时，会执行 JavaScript 函数 searchUserInfo，代码如下：

```javascript
function searchUserInfo() {
    var userName = $('#search_userName').textbox("getValue");
    $('#userListDg').datagrid('load', {
        userName : userName
    });
}
```

在函数 searchUserInfo 中，首先获取输入的客户名称，然后调用 Datagrid 控件的 load 方法，这时会向后台服务器重新提交 userinfo/list 请求，即再次执行 UserInfoController 类中的 userlist 方法。由于指定了参数 userName，这个参数会被添加到 queryParams 中，并传递给 userlist 方法。在 userlist 方法中，根据查询条件重新获取数据，再将结果显示在 Datagrid 控件上。

19.10.3　启用和禁用客户

在 userlist.jsp 页面中，选中 Datagrid 控件中的若干条记录，单击启用客户或禁用客户按钮，可以修改客户的状态。单击启用客户或禁用客户按钮后，将执行 JavaScript 函数 SetIsEnableUser，代码如下：

```javascript
// 设置启用或禁用客户
function SetIsEnableUser(flag) {
    var rows = $("#userListDg").datagrid('getSelections');
    if (rows.length > 0) {
        $.messager.confirm('Confirm', '确认要设置么?', function(r) {
            if (r) {
                var uids = "";
                for (var i = 0; i < rows.length; i++) {
                    uids += rows[i].id + ",";
                }
                $.post('userinfo/setIsEnableUser', {
                    uids : uids,
                    flag : flag
                }, function(result) {
                    if (result.success == 'true') {
                        $("#userListDg").datagrid('reload');
                        $.messager.show({
                            title : '提示信息',
```

```
                    msg : result.message
                });
            } else {
                $.messager.show({
                    title : '提示信息',
                    msg : result.message
                });
            }
        }, 'json');
    }
});
    } else {
        $.messager.alert('提示', '请选择要启用或禁用的客户', 'info');
    }
}
```

setIsEnableUser 函数中有一个参数 flag，代表执行启用还是禁用操作。如果 flag 为 1，则执行启用操作；如果为 0，则执行禁用操作。

在 setIsEnableUser 函数中，首先获取选中行的客户编号，并将其以逗号分隔保存到变量 uids 中，然后通过$.post 方法提交请求 userinfo/setIsEnableUser，这个 url 请求将映射到控制器类 UserInfoController 中的一个方法，并将参数 uids 和 flag 传递给这个方法。接下来，在 UserInfoController 类中编写 setIsEnableUser 方法，代码如下：

```
// 更新客户状态
@RequestMapping(value = "/setIsEnableUser", produces = 
"text/html;charset=UTF-8")
@ResponseBody
public String setIsEnableUser(@RequestParam(value = "uids") String 
uids,@RequestParam(value = "flag") String flag) {
    try {
        userInfoService.modifyStatus(uids.substring(0, uids.length() - 1), 
Integer.parseInt(flag));
        return "{\"success\":\"true\",\"message\":\"更改成功\"}";
    } catch (Exception e) {
        return "{\"success\":\"false\",\"message\":\"更改失败\"}";
    }
}
```

通过在 setIsEnableUser 方法上标注@RequestMapping 注解，将用户对 userinfo/setIsEnableUser 这个 url 的请求映射到 setIsEnableUser 方法。setIsEnableUser 方法有两个 String 类型的参数 uids 和 flag，用来绑定前端页面传递来的参数。

在 setIsEnableUser 方法中，调用业务接口 UserInfoService 中的 modifyStatus 方法，将数据表 user_info 中的 Status 字段值设置为 1(启用)或 0(禁用)。最后根据执行结果，向前端页面发送信息。

在前端页面 userlist.jsp 中，JavaScript 函数 setIsEnableUser 根据服务器的返回信息进行判断。如果执行成功，则调用 Datagrid 控件的 reload 方法，重新执行控制器类 UserInfoController 中的 userlist 方法，重新获取数据以更新客户列表。

19.11　小　　结

　　本章基于 Spring、Spring MVC 与 MyBatis 整合框架，采用注解方式并结合前端 Easy UI 框架，详细讲解了一个典型的电商平台后台管理系统的实现过程，系统的主要功能包括商品管理、订单管理和客户管理，按照三层架构开发每个功能模块。

　　通过本章的学习，希望读者能够熟练掌握 Spring、Spring MVC 与 MyBatis 框架整合开发的基本步骤、方法和技巧。

第 20 章 校园通讯管理系统

在第 17 章中，以用户登录功能为例，详细讲解了 Spring、Spring MVC 与 MyBatis 框架 (SSM) 整合的流程。本章将使用 SSM 整合，并结合前端基于 Bootstrap 的 H+框架实现校园通讯管理系统。

20.1 需求与系统分析

在日常生活中，对于一名学生来说，宿舍的设备报修，需要上报给宿管员。宿管员手写报修单，再送至后勤，然后才能派出专业人员进行修理。再比如，院系给教师发送通知，学校给院系发送通知，没有统一的平台供多个不同的群体或者个体通讯。学校对于某一决定让学生个体投票，又或者以院系为单位投票时，最耗费时间的莫过于清点票数，因此，以往的消息传递方式和投票方式与时代的发展已不匹配。于是，一种新式的通讯方式应运而生，这就是校园通讯管理系统。

根据需求分析，使用校园通讯管理系统的人员包括平台管理员、院校管理员和单位用户三类用户，各自拥有的功能权限如下。

1. 平台管理员功能权限

平台管理员的功能权限包括院校管理、院校账户管理和菜单管理。
(1) 院校管理：对各个地区的院校进行增加、删除、修改和查询。
(2) 院校账户(或院校管理员)管理：对已有的院校添加、删除、修改和查询账户。

(3) 菜单(或系统功能模块)管理：对已有的菜单进行管理(区域管理员角色权限的来源)。

2．院校管理员功能权限

院校管理员功能权限包括单位管理、Excel 批量导入单位、单位类别管理、角色管理和用户管理。

(1) 单位管理：对学校的单位(包括学生、教师、宿舍楼、院系、后勤部门等某一个体或群体)进行增删改查。

(2) Excel 批量导入单位：选择角色，导入后自动生成账号密码(账号形式为学校编号_单位编号，默认密码为单位编号)，同时不存在的单位类别会自动生成。

(3) 单位类别管理：单位所属类别的增加、删除、修改和查询。

(4) 角色管理：对角色的增加、删除、修改、查询，以及角色权限的配置。

(5) 用户管理：为已有的单位生成账号和密码，并绑定角色。

3．单位用户功能权限

单位用户功能权限包括通知推送、通知查看、投票和查看投票数据。

(1) 通知推送：筛选单位或者群体，发送消息。

(2) 通知查看：查看通知消息，点击查收，确认收到。

(3) 投票：查看投票消息，并选择赞成或反对票。

(4) 查看投票数据：查看发出的消息或数据，数据分析。

根据上述分析，可以得到系统的模块结构，如图 20-1 所示。

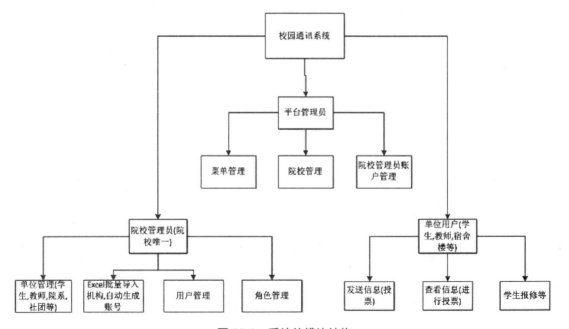

图 20-1　系统的模块结构

20.2 数据库设计

根据系统需求，创建名称为 school 的数据库，创建 10 张数据表，如下所示。
(1) 系统模块(菜单)表 sys_module，用于记录系统功能信息。
(2) 区域/院校表 sys_area，用于记录区域或院校信息。
(3) 用户信息表 sys_user，用于记录用户信息。
(4) 参数信息表 pro_paraminfo，用于记录参数信息。
(5) 角色表 sys_role，用于记录角色信息。
(6) 角色模块表 sys_role_module，用于记录角色对应的模块。
(7) 用户角色表 sys_user_role，用于记录用户对应的角色。
(8) 单位信息表 pro_unitinfo，用于记录学校的单位信息，包括学生、教师、宿舍楼、院系、后勤部门等某一个体或群体。
(9) 通知信息表 notice，用于记录通知信息。
(10) 通知/投票回复表 answer，用于记录消息接收状态或投票计数信息。
其中，客户信息表 user_info 的字段说明如表 20-1 所示。

表 20-1 系统模块(菜单)表 sys_module

字段名	类型	主外键	说明
moduleCode	varchar(36)	PK	模块编号，主键
moduleName	varchar(50)		模块名称
modulePath	varchar(1000)		模块访问路径
parentCode	varchar(36)		父结点编号
isLeaf	int(11)		是否叶子结点
sortNumber	int(11)		排序号

区域/院校表 sys_area 的字段说明如表 20-2 所示。

表 20-2 区域/院校表 sys_area

字段名	类型	主外键	说明
areaNumber	varchar(50)	PK	区域/院校编号，主键
name	varchar(50)		区域/院校名称
type	int(11)		地域类型(1：省；2：市；3：院校)
parentId	varchar(36)		父结点编号
isLeaf	int(11)		是否子结点(0：否；1：是)
sortNum	int(11)		排序号
delState	int(11)		删除状态(1：未删除；2：删除)

用户信息表 sys_user 的字段说明如表 20-3 所示。

表 20-3 用户信息表 sys_user

字段名	类　型	主外键	说　明
userCode	varchar(36)	PK	用户编号，主键
name	varchar(50)		用户名
psw	varchar(50)		密码
operatorId	varchar(36)		操作人
operatorTime	timestamp		操作时间
delState	int(11)		删除状态(1：未删除；2：删除)
unitId	varchar(36)		绑定单位主键
userType	int(11)		用户类型(1：院校人员；2：学校单位用户；3：平台管理人员，可分配不同院校账号；4：各院校管理员)
areaId	varchar(36)		所属地域 ID

参数信息表 pro_paraminfo 的字段说明如表 20-4 所示。

表 20-4　参数信息表 pro_paraminfo

字段名	类　型	主外键	说　明
id	varchar(36)	PK	参数 id，主键
name	varchar(50)		参数名称
parent_id	varchar(36)		父结点编号
type	varchar(10)		参数类型(01：通知类型；02：单位类型)
outside_code	varchar(50)		外部编码
sortNum	int(11)		排序号
delState	int(11)		删除状态(1：未删除；2：删除)
areaId	varchar(36)		所属院校 ID

角色表 sys_role 的字段说明如表 20-5 所示。

表 20-5　角色表 sys_role

字段名	类　型	主外键	说　明
roleCode	varchar(36)	PK	角色编号，主键
roleName	varchar(50)		角色名称
areaId	varchar(36)		所属地域 ID

角色模块表 sys_role_module 的字段说明如表 20-6 所示。

表 20-6　角色模块表 sys_role_module

字段名	类　型	主外键	说　明
rmId	varchar(36)	PK	角色模块编号，主键
roleCode	varchar(36)		角色编号

续表

字段名	类型	主外键	说明
moduleCode	varchar(50)		模块编号
areaId	varchar(36)		所属地域 ID

用户角色表 sys_user_role 的字段说明如表 20-7 所示。

表 20-7 用户角色表 sys_user_role

字段名	类型	主外键	说明
urId	varchar(36)	PK	用户角色编号，主键
userCode	varchar(36)		用户编号
roleCode	varchar(50)		角色编号
areaId	varchar(36)		所属地域 ID

单位信息表 pro_unitinfo 的主要字段说明如表 20-8 所示。

表 20-8 单位信息表 pro_unitinfo

字段名	类型	主外键	说明
id	varchar(36)	PK	单位编号，主键
name	varchar(50)		单位名称
unitTypeId	varchar(36)		单位类型编号，与表 pro_paraminfo 的 id 字段关联
unitGradeId	varchar(36)		等级名称
outside_code	varchar(50)		外部编码
delState	int(11)		删除状态(1：未删除；2：删除)
areaId	varchar(36)		所属地域 ID

通知信息表 notice 的字段说明如表 20-9 所示。

表 20-9 通知信息表 notice

字段名	类型	主外键	说明
id	varchar(50)		通知编号
userId	varchar(50)		用户编号
title	varchar(50)		通知标题
content	varchar(5000)		通知内容
operatetime	varchar(50)		通知时间
type	int(10)		通知类型

通知/投票回复表 answer 的字段说明如表 20-10 所示。

表 20-10 通知/投票回复表 answer

字段名	类型	主外键	说明
id	char(50)		通知/投票回复编号
nid	char(50)		通知编号

续表

字段名	类 型	主外键	说 明
uid	char(50)		用户对应的单位编号
flag	int(10)		接收状态(0:未接收;1:已接收)
vote	int(10)		投票计数
reason	varchar(500)		理由

20.3 环境搭建与配置文件

可以参照第 17 章 Spring 整合 MyBatis，完成校园通讯管理系统的框架搭建及相关配置文件的编写，项目最终的目录结构如图 20-2 所示。

图 20-2 系统的目录结构

com.ccms.controller 包用于存放控制器类，com.ccms.service 包用于存放业务逻辑层接口，com.ccms.service.impl 包用于存放业务逻辑层接口的实现类，com.ccms.dao 包用于存放数据访问层接口，com.ccms.dao.provider 包用于存放 DynaSqlProvider 类，com.ccms.pojo 包用于存放实体类，com.ccms.tools 包用于存放工具类，com.ccms.config 包用于存放包含 SQL 语句字符串的类。dbconfig.properties 为存储数据库连接信息的属性文件，applicationContext.xml 为 Spring 框架的配置文件，dispatcherServlet-servlet.xml 为 Spring MVC 框架的配置文件。

在 src/main/webapp 目录下，commons 子目录中存放与基于 Bootstrap 的 H+框架、jQuery 树插件 ztree、百度图表插件 echarts、HTML 可视化编辑器 kindeditor、jQuery 弹出层插件 layer 等相关的 CSS 和 js 文件；views 子目录及其包含的子目录中存放页面文件。

20.4 创建实体类

在系统开发中，实体类常用于封装数据。在 com.ccms.pojo 包中，依次创建实体类 Answer、Notice、ProParamInfo、ProUnitinfo、SysArea、SysModule、SysRole、SysRoleModule 和 SysUser。

实体类 Answer 用于封装通知/投票回复表 answer 的数据，代码如下：

```java
package com.ccms.pojo;
public class Answer {
    private String id;
    private String nid;
    private String uid;
    private String flag;
    private String reason;
    // 省略上述属性的 getter 和 setter 方法
}
```

实体类 Notice 用于封装通知信息表 notice 的数据，代码如下：

```java
package com.ccms.pojo;
public class Notice {
    private String id;
    private String userId;
    private String content;
    private String title;
    private String operatetime;
    private int type;
    private String typeName;
    private String noticebelong;
    private String flag;
    // 省略上述属性的 getter 和 setter 方法
}
```

实体类 ProParamInfo 用于封装参数信息表 pro_paraminfo 的数据，代码如下：

```java
package com.ccms.pojo;
public class ProParamInfo {
    private String id;
    private String name;
    private String parent_id;
    private String type;
    private String outside_code;
    private int sortNum;
    private String parent_name;
    private String areaId;
    // 省略上述属性的 getter 和 setter 方法
}
```

实体类 ProUnitinfo 用于封装单位信息表 pro_unitinfo 的数据，代码如下：

```java
package com.ccms.pojo;
public class ProUnitinfo {
    private String id;
    private String name;
    private String unitTypeId;
    private String unitGradeId;
    private String outside_code;
    private String delState;
    private String areaId;
    private String unitType;
    private String unitGrade;
    // 省略上述属性的getter和setter方法
}
```

实体类 SysArea 用于封装区域/院校表 sys_area 的数据，代码如下：

```java
package com.ccms.pojo;
public class SysArea {
    private String areaNumber;
    private String name;
    private String type;
    private String parentId;
    private String isLeaf;
    private String sortNum;
    private String delState;
    private String parentName;
// 省略上述属性的getter和setter方法
}
```

实体类 SysModule 用于封装系统模块(菜单)表 sys_module 的数据，代码如下：

```java
package com.ccms.pojo;
import java.util.ArrayList;
import java.util.List;
public class SysModule {
    /**模板编码*/
    private String moduleCode;
    /**模板名称*/
    private String moduleName;
    /**模板路径*/
    private String modulePath;
    /**父级模板编号*/
    private String parentCode;
    /**是否为叶子结点*/
    private int isLeaf;
    /**同级排序编号*/
    private int sortNumber;
    private List<SysModule> children = new ArrayList<SysModule>();
    /**父结点 name*/
    private String parentModuleName;
```

```
        // 省略上述属性的getter和setter方法
}
```

实体类SysRole用于封装角色表sys_role的数据，代码如下：

```java
package com.ccms.pojo;
public class SysRole{
    /**角色编号*/
    private String roleCode;
    /**角色名称*/
    private String roleName;
    /**区域id*/
    private String areaID;
// 省略上述属性的getter和setter方法
}
```

实体类SysRoleModule用于封装角色模块表sys_role_module的数据，代码如下：

```java
package com.ccms.pojo;
public class SysRoleModule {
    private String rmid;
    /**角色编号*/
    private String roleCode;
    /**模板编号*/
    private String moduleCode;
    // 省略上述属性的getter和setter方法
}
```

实体类SysUser用于封装用户信息表sys_user的数据，代码如下：

```java
package com.ccms.pojo;
public class SysUser{
    private String userCode;
    private String name;
    private String psw;
    private String operatorId;
    private String operatorTime;
    private int delState;
    private String unitId;
    private String userType;
    private String unitName;
    private String operator;
    private String roleCodes;
    private String areaId;
    private String areaType;
    private String areaName;
    private String roleNames;
    private String province;
    private String city;
    private String county;
    // 省略上述属性的getter和setter方法
}
```

此外,还创建了分页类 Pager,用于封装分页信息,代码如下:

```java
package com.ccms.pojo;
public class Pager {
    private int curPage;// 待显示页
    private int perPageRows;// 每页显示的记录数
    private int rowCount; // 记录总数
    private int pageCount; // 总页数
    // 根据 rowCount 和 perPageRows 计算总页数
    public int getPageCount() {
        return (rowCount + perPageRows - 1) / perPageRows;
    }
    // 分页显示时,获取当前页的第一条记录的索引
    public int getFirstLimitParam() {
        return (this.curPage - 1) * this.perPageRows;
    }
}
// 省略其他属性的 getter 和 setter 方法
}
```

图表图列类 Echarts,用于封装图表中的图列信息,代码如下:

```java
package com.ccms.pojo;
public class Echarts {
    // 图列名称
    private String name ;
    // 图列值
    private int value ;
    // 省略上述属性的 getter 和 setter 方法
}
```

20.5　后　台　登　录

系统后台登录页为 login.jsp,页面效果如图 20-3 所示。

图 20-3　后台登录页

在 login.jsp 页面中，使用了基于 Bootstrap 的 H+框架进行布局，为了使用 Bootstrap 的 H+框架，在页面的<head></head>标签中，需要引入相关的 css 和 js 文件。在 id 为 loginForm 的<form>标签中，定义了用户名、密码文本域和一个登录按钮，登录表单代码如下：

```html
<form id='loginForm' method="post" action="/ccms/user/login">
    <p class="m-t-md" id="title" style="position: absolute; width: 237px; height: 26px; z-index: 1; top: 8%; left: 13%;">登录到后台</p>
    <div style="position: absolute; width: 237px; height: 26px; z-index: 1; top: 28%; left: 13%;">
        <input type="text" class="form-control uname" placeholder="用户名" name="name" id="userName" style="color:#030303;"/>
    </div>
    <div style="position: absolute; width: 237px; height: 26px; z-index: 2; top: 49%; left: 13%;">
        <input type="password" class="form-control pword m-b" placeholder="密码" name="psw" id="pwd" style="color:#030303;"/>
    </div>
    <div style="position: absolute; width: 185px; height: 26px; z-index: 2; top: 65%; left: 13%;">
        <b><DIV id="div_msg" style="color: #FF9C00;"></DIV></b>
    </div>
    <div style="position: absolute; width:237px; height: 40px; z-index: 3; top: 75%; left: 13%; ">
        <button type="submit" class="btn btn-primary btn-block">登  录</button>
    </div>
</form>
```

在登录表单中，填写用户名和密码，单击"登录"按钮，将请求提交到/ccms/user/login。这个 url 请求被映射到控制器类 UserController 中的 login 方法。

UserController 类位于 com.ccms.controller 包中，login 方法的代码如下：

```java
package com.ccms.controller;
……
@Controller
@RequestMapping("/user")
public class UserController {
    @Autowired
    UserService userService;
    JsonUtil<SysUser> json = new JsonUtil<SysUser>();
    JsonObject jObject = null;
    // 登录
    @RequestMapping("/login")
    public String login(@RequestParam("name") String name, HttpServletRequest req, HttpServletResponse res,
            @RequestParam("psw") String psw, HttpSession session, ModelAndView mv) throws IOException {
        SysUser user = userService.login(name, MD5Util.MD5(psw));
        if (user != null && user.getName() != null) {
```

```java
            req.getSession().setAttribute(CommonValue.USERID, user.getUserCode());
            session.setAttribute(CommonValue.USERNAME, user.getName());
            session.setAttribute(CommonValue.UNITNAME, user.getUnitName());
            session.setAttribute(CommonValue.USERTYPE, user.getUserType());
            session.setAttribute(CommonValue.UNITINFOID, user.getUnitId());
            session.setAttribute(CommonValue.AREANUMBER, user.getAreaId());
            session.setAttribute(CommonValue.AERATYPE, user.getAreaType());
            session.setAttribute(CommonValue.AERANAME, user.getAreaName());
            res.sendRedirect(req.getContextPath() + "/views/index.jsp");
        } else {
            session.setAttribute(CommonValue.USERID, null);
            session.setAttribute(CommonValue.UNITINFOID, null);
            session.setAttribute(CommonValue.USERNAME, null);
            session.setAttribute(CommonValue.UNITNAME, null);
            session.setAttribute(CommonValue.USERTYPE, null);
            session.setAttribute(CommonValue.AREANUMBER, null);
            session.setAttribute(CommonValue.AERATYPE, null);
            session.setAttribute(CommonValue.AERANAME, null);
            res.sendRedirect(req.getContextPath() + "/login.jsp?rtnCode=500");
        }
        return null;
    }
}
```

在 login 方法中，调用业务接口 UserService 中的 login 方法，根据用户名和密码进行登录验证。验证通过后，将登录用户的相关信息存入 HttpSession 对象，再重定向到系统首页面 index.jsp。如果登录失败，则重定向到登录页。CommonValue 类中定义了一些公共变量，该类位于 com.ccms.tools 包中。

UserService 接口位于 com.ccms.service 包中，login 方法的声明如下：

```java
package com.ccms.service;
import java.util.List;
……
public interface UserService {
    // 登录
    public SysUser login(String name, String psw);
}
```

在 com.ccms.service.impl 包中，创建 UserService 接口的实现类 UserServiceImpl，实现 login 方法，如下所示：

```java
package com.ccms.service.impl;
import java.util.ArrayList;
```

```java
……
@Service("userService")
@Transactional(propagation = Propagation.REQUIRED, isolation = Isolation.DEFAULT)
public class UserServiceImpl implements UserService {
    @Autowired
    UserDao userDao;
    @Autowired
    ModuleDao moduleDao;
    @Override
    public SysUser login(String name, String psw) {
        Map<String, Object> params = new HashMap<String, Object>();
        params.put("name", name);
        params.put("psw", psw);
        return UserDao.selectByNameAndPwd(params);
    }
}
```

在 UserServiceImpl 类的 login 方法中，将传入的用户名和密码封装到 Map<String, Object>类型的 params 对象中，再调用数据访问层接口 UserDao 中的 selectByNameAndPwd 方法。

接口 UserDao 位于 com.ccms.dao 包中，selectByNameAndPwd 方法的代码如下：

```java
package com.ccms.dao;
……
import com.ccms.dao.provider.UserDynaSqlProvider;
import com.ccms.pojo.SysUser;
public interface UserDao {
    // 根据登录名和密码查询合法用户
    @SelectProvider(type = UserDynaSqlProvider.class, method = "login")
    public SysUser selectByNameAndPwd(Map<String, Object> params);
}
```

selectByNameAndPwd 方法标注了@SelectProvider 注解，指定由 UserDynaSqlProvider 类中定义的 login 方法提供需要执行的 SELECT 语句。UserDynaSqlProvider 类位于 com.ccms.dao.provider 包中，login 方法的代码如下：

```java
package com.ccms.dao.provider;
import java.util.Map;
import com.ccms.config.SysUserConfig;
import com.ccms.pojo.SysUser;
import com.ccms.tools.CommonValue;
public class UserDynaSqlProvider {
    // 登录
    public String login(Map<String, Object> params) {
        String name = (String) params.get("name");
        String psw = (String) params.get("psw");
        String sql = "select " + " su.*," + " (case when su.userType=1 then '院系账号'" + " when su.userType=3 then '平台管理员'" + " when su.userType=4 then '院校管理员'"    + " else pui.name end) as unitName," + " sa_weiLeaf.type
```

```
                as areaType," + "  (case "+ " when sa_weiLeaf.type=3 then 
CONCAT(sa_province.name,'.',sa_city.name,'.',sa_weiLeaf.name)"
                + " when sa_weiLeaf.type=2 then 
CONCAT(sa_city.name,'.',sa_weiLeaf.name)"
                + " when sa_weiLeaf.type=1 then CONCAT(sa_weiLeaf.name)" +
" else '未知' end" + " ) as areaName"
                + "  from sys_user as su " + " left join pro_unitinfo as pui"
+ " on su.unitId=pui.id"
                + " left join sys_area as sa_weiLeaf" + " on 
su.areaId=sa_weiLeaf.areaNumber"
                + " left join sys_area as sa_city" + " on 
sa_weiLeaf.parentId=sa_city.areaNumber"
                + " left join sys_area as sa_province" + " on 
sa_city.parentId=sa_province.areaNumber"
                + " where 1=1 ";
        sql += " and su.delState=1";
        sql += " and su.name='" + name + "' and su.psw='" + psw + "'";
        return sql;
    }
}
```

在这个 login 方法中，通过数据表 sys_user 对用户名和密码进行验证，同时还关联了数据表 pro_unitinfo 和 sys_area，以获取该用户关联的其他数据。

在登录页 login.jsp 中，如果填写用户名 sysadmin、密码 123456，单击登录按钮，则以平台管理员的身份进入系统首页面 index.jsp，如图 20-4 所示。

如果填写用户名 yzd、密码 123456，单击"登录"按钮，则以院校管理员的身份进入系统首页面 index.jsp，如图 20-5 所示。

如果填写用户名 1404_yx001、密码 123456，单击"登录"按钮，则以单位用户的身份进入系统首页面 index.jsp，如图 20-6 所示。

图 20-4　平台管理员界面

图 20-5　院校管理员界面

图 20-6　单位用户界面

平台管理员界面、院校管理员界面和单位用户界面的主要区别在于左侧导航菜单不同，在系统首页面 index.jsp 中，导航布局代码如下：

```
<!--左侧导航开始-->
<nav class="navbar-default navbar-static-side" role="navigation">
<div class="nav-close">
    <i class="fa fa-times-circle"></i>
</div>
<div class="sidebar-collapse">
    <ul class="nav" id="side-menu">
        <li class="nav-header" style="height: 158px;">
            <div class="dropdown profile-element">
                <span><img alt="image"
                        src="/ccms/commons/images/head-index.jpg"
class="img-circle" height="40%"width="40%" /> </span>
```

```html
                    <a data-toggle="dropdown" class="dropdown-toggle"
href="#"> <span class="clear"> <span class="block m-t-xs"><strong
                        class="font-bold"> <c:choose>
                            <c:when
test="${sessionScope.USERTYPE!='3'}">院校：${sessionScope.AERANAME}
                            </c:when>
                            <c:otherwise>
                            </c:otherwise>
                        </c:choose> <br /></strong> </span> <span
                    class="text-muted text-xs
block">${sessionScope.UNITNAME}:${sessionScope.USERNAME}<b
                        class="caret"></b> </span> </span> </a>
                <ul class="dropdown-menu animated fadeInRight m-t-xs">
                    <li>
                        <a data-toggle="modal" data-keyboard="true"
                            data-backdrop="true" id="btn_ResetPsw">
<b>修改密码</b></a>
                    </li>
                    <li>
                        <a href="/ccms/user/loginOut"><b>安全退出</b>
                        </a>
                    </li>
                </ul>
            </div>
            <br/><br/>
            <div class="logo-element">
            </div>
        </li>
        <c:if test="${sessionScope.USERTYPE=='3'}">
        <li>
            <a class="J_menuItem"
                href="area/areaAdminManager.jsp"><i
                class="fa fa-check"></i> <span class="nav-label">院校管理员
管理</span>
            </a>
        </li>
        <li>
            <a class="J_menuItem"
                href="menu/menuManager.jsp"><i
                class="fa fa-check"></i> <span class="nav-label">
系统功能模块管理</span>
            </a>
        </li>
        <li>
            <a class="J_menuItem"
                href="area/schoolManager.jsp"><i
                class="fa fa-check"></i> <span class="nav-label">
院校管理</span>
            </a>
        </li>
        </c:if>
```

```html
            <c:if test="${sessionScope.USERTYPE=='4'}">
                <li>
                    <a href="#"> <i class="fa fa-check"></i> <span
                        class="nav-label">单位管理</span>
                        <span class="fa arrow"></span>
                    </a>
                    <ul class="nav nav-second-level">
                        <li>
                            <a class="J_menuItem"
                                href="unit/unitInfo.jsp">单位设置</a>
                        </li>
                        <li>
                            <a class="J_menuItem"
                                href="param/orgTypeParam.jsp">单位类别</a>
                        </li>
                    </ul>
                </li>
                <li>
                    <a href="#"> <i class="fa fa-check"></i> <span
                        class="nav-label">用户权限管理</span>
                        <span class="fa arrow"></span>
                    </a>
                    <ul class="nav nav-second-level">
                        <li>
                            <a class="J_menuItem"
                                href="user/userManager.jsp">用户管理</a>
                        </li>
                        <li>
                            <a class="J_menuItem"
                                href="role/roleManager.jsp">角色管理</a>
                        </li>
                    </ul>
                </li>
            </c:if>
        </ul>
    </div>
</nav>
<!--左侧导航结束-->
```

当 USERTYPE=3 时，显示平台管理员菜单；当 USERTYPE=4 时，显示各院校管理员菜单；当 USERTYPE=1 或 2 时，显示的菜单是在自执行函数$(function(){}中指定的，相关代码如下：

```javascript
var userType="<%=session.getAttribute("USERTYPE")%>";
var userId="<%=session.getAttribute("USERID")%>";
if(userType==1||userType==2){
    $.ajax({
        url:'/ccms/user/loadpermissions',
        type:'post',
        async: false,
        cache:false,
```

```javascript
            data:{userId:userId},
            dataType:'json',
            success: function(data){
                for(var i=0,len=data.length;i<len;i++){
                    var ps = data[i];
                    if(ps.children.length>0){
                        var result = RecMenu(ps);
                        $('#side-menu').append(result);
                    }
                }
            },
            error: function (aa, ee, rr) {
            }
        });
```

在页面 index.jsp 加载时，如果用户类型 USERTYPE=1 或 2，则通过 jQuery AJAX 方式向后台服务器发送请求，url 参数用于指定发送请求的地址，这里为/ccms/user/loadpermissions，这个请求将映射到控制器类 UserController 中的 loadPermiss 方法；dataType 参数用于指定预期服务器返回的数据类型，这里指定返回 JSON 数据；type 参数用于指定请求方式，这里为 post 方式；async 参数用于指定是否发送异步请求，这里为 false，表示同步请求，此时将锁住浏览器，用户的其他操作必须等待请求完成才可以执行；cache 属性用于设置是否启用 AJAX 数据缓存，这里为 false，表示禁用；data 参数用于指定发送到服务器的数据，这里传递的参数为用户编号 userId；success 参数的类型为 Function，表示请求成功后的回调函数，服务器返回的数据将传给该函数的 data 参数，前端页面 index.jsp 使用 data 中的数据生成单位用户的功能菜单。

在控制器类 UserController 中，loadPermiss 方法的代码如下：

```java
// 根据用户 id 获取 module 列表
@RequestMapping(value = "/loadpermissions", method = { RequestMethod.GET,
RequestMethod.POST })
@ResponseBody
public List<SysModule> loadPermiss(String userId, HttpServletRequest req,
HttpServletResponse res)
        throws Exception {
    List<SysModule> modules = userService.selectModuleByUserId(userId);
    if (modules == null) {
        modules = new ArrayList<SysModule>();
    }
    String json = new JsonUtil<SysModule>().objectsToJSON(modules);
    res.resetBuffer();
    res.setContentType("text/html;charset=utf-8");
    res.getOutputStream().write(json.getBytes("utf-8"));
    res.getOutputStream().flush();
    return null;
}
```

loadPermiss 方法有 3 个参数，第一个是 String 类型的参数 userId，用于封装前端页面传递来的用户编号；第二个是 HttpServletRequest 类型的参数 req，代表客户端的请求；第三个是 HttpServletResponse 类型的参数 res，代表服务器的响应。

在 loadPermiss 方法中，首先调用业务接口 UserService 的 selectModuleByUserId 方法，根据用户 id 获取系统模块列表；然后通过 JSON 转换工具类 JsonUtil 的 objectsToJSON 方法将系统模块列表转换成 JSON 格式的字符串，并将其发送到前端页面。

在业务接口 UserService 中，声明 selectModuleByUserId 方法，如下所示：

```java
// 根据用户 id 获取 module 列表
public List<SysModule> selectModuleByUserId(String userId);
```

在接口 UserService 的实现类 UserServiceImpl 中，实现这个方法，如下所示：

```java
@Override
public List<SysModule> selectModuleByUserId(String userId) {
    List<SysModule> result = new ArrayList<SysModule>();
    // 查询所有的模块
    List<SysModule> modules = moduleDao.getAllModule();
    if (modules == null) {
        modules = new ArrayList<SysModule>();
    }
    // 查询用户可以操作的模块编号集合
    List<String> mids = moduleDao.getmoduleCodes(userId);
    if (mids == null) {
        mids = new ArrayList<String>();
    }
    // 去除没有权限的叶子结点
    clear(modules, mids);
    // 建立模块层次结构
    for (int i = 0; i < modules.size(); i++) {
        SysModule m = modules.get(i);
        if (m.getParentCode() != null && m.getParentCode().equals("0")) {
            resetModules2(m, modules);
            result.add(m);
        }
    }
    return result;
}
```

在实现类 UserServiceImpl 的 selectModuleByUserId 方法中，首先调用 ModuleDao 接口的 getAllModule 方法查询所有的模块；然后调用 ModuleDao 接口的 getmoduleCodes 方法查询该用户可以操作的模块编号集合；接着调用自定义方法 clear 根据 mids 去除没有权限的叶子结点；最后建立模块层次结构。

在 ModuleDao 接口中，getAllModule 和 getmoduleCodes 方法的代码如下：

```java
package com.ccms.dao;
import java.util.List;
import org.apache.ibatis.annotations.Param;
import org.apache.ibatis.annotations.Select;
import com.ccms.pojo.SysModule;
public interface ModuleDao {
    // 查询所有模块
```

```
@Select("select * from sys_module order by sortNumber asc")
public List<SysModule> getAllModule();
// 查询用户可以操作的模块编号集合
@Select("select distinct moduleCode from sys_role_module where roleCode "
    + "in (select roleCode from sys_user_role where userCode=#{userCode})")
List<String> getmoduleCodes(@Param("userCode") String userCode);
}
```

20.6 平台管理员功能

平台管理员功能包括院校管理员管理、院校管理和系统功能模块管理,这里主要讲解院校管理员管理、院校管理的实现过程。

20.6.1 院校管理员管理

在图 20-4 所示的平台管理员界面中,单击院校管理员管理菜单,打开院校管理员管理页面 areaAdminManager.jsp,如图 20-7 所示。

图 20-7 院校管理员管理页面

在 areaAdminManager.jsp 页面中,需要实现院校管理员列表的显示、编辑、重置密码、新增、删除,以及根据省、市、院校和用户名搜索等功能。

1. 显示院校管理员列表

为了显示院校管理员列表,在页面中定义了一个 id 为 table 的<table>标签,如下所示:

```
<!-- 院校管理员列表显示 -->
<div class="ibox-content">
    <table id="table">
    </table>
</div>
```

然后在$(function() { })代码段中,通过JavaScript初始化这个<table>标签,如下所示:

```
$(function() {
loadCity("/ccms/area/getShengAreaList", "sheng_search", null);
loadCity("/ccms/area/getShengAreaList", "sheng_add", null);
loadCity("/ccms/area/getShengAreaList", "sheng_up", null);
$("#username_add").val('');
$("#userpsw_add").val('');
$("#unitId").val('');
$("#userType").val('');
vform('upform', upUser);
vform('addform', addUser);
vform('resetform', resetPsw);
var $table = $('#table');
// 初始化table
    $table.bootstrapTable({
        url : "/ccms/user/getAreaUser",
        method : 'post',
        contentType : "application/x-www-form-urlencoded",
        dataType : "json",
        pagination : true, //分页
        pageSize : 10,
        pageNumber : 1,
        singleSelect : false,
        queryParamsType : "undefined",
        queryParams : function queryParams(params) { //设置查询参数
            var param = {
                page : params.pageNumber,
                rows : params.pageSize,
                userType : 4
            };
            return param;
        },
        cache : false,
        sidePagination : "server", //服务端处理分页
        columns : [
            {
                checkbox : true
            },
            {
                title : '用户名',
                field : 'name',
                valign : 'middle'
            },
            {
                title : '所属区域',
                field : 'areaName',
                valign : 'middle'
            },
            {
                title : '用户类型',
```

```
                    field : 'userType',
                    valign : 'middle',
                    formatter : function(value, row, index) {
                        if (value == '1') {
                            return '院方';
                        } else if (value == '2') {
                            return '部门';
                        } else if (value == '3') {
                            return '平台管理员';
                        } else if (value == '4') {
                            return '院校管理员';
                        }
                    }
                },
                {
                    title : '操作人',
                    field : 'operator',
                    valign : 'middle'
                },
                {
                    title : '操作时间',
                    field : 'operatorTime',
                    valign : 'middle',
                    formatter : function(value, row, index) {
                        return value.substring(0,
                                value.length - 2);
                    }
                },
                {
                    title : '操作',
                    field : 'id',
                    formatter : function(value, row, index) {
                        var e = '<a href="#" class="btn btn-gmtx-define1" onclick="edit(\''
                                + row.userCode
                                + '\',\''
                                + row.delState
                                + '\',\''
                                + row.name
                                + '\',\''
                                + row.areaName
                                + '\',\''
                                + row.province
                                + '\',\''
                                + row.city
                                + '\',\''
                                + row.county + '\')">编辑</a> ';
                        var d = '<a href="#" class="btn btn-gmtx-define1" onclick="resetPsw(\''
                                + row.userCode
                                + '\')">重置密码</a> ';
                        return e + d;
                    }
```

```
            } ]
    });
});
```

通过调用 bootstrapTable 方法，将<table>标签创建为 Bootstrap 的 table 控件，并通过设置一系列属性来初始化这个 table 控件。url 属性用于设置 table 控件的数据源，这里为 /ccms/user/getAreaUser，这个 url 请求将映射到后台控制器类 UserController 中的 getAreaUser 方法；method 属性用于设置请求方式，这里为 post 方式；contentType 属性用于设置发送到服务器的数据编码类型，这里为 application/x-www-form-urlencoded；dataType 属性用于设置服务器返回的数据类型，这里为 json；pagination 属性用于设置是否显示分页，这里为 true，表示允许分页；pageSize 属性用于设置每页的记录行数，这里为 10；pageNumber 属性用于设置首页页码，这里为 1；singleSelect 属性用于设置是否单选，这里为 false，表示可以多选；queryParamsType 属性用于设置参数格式，这里为 undefined；queryParams 属性用于设置查询参数，Bootstrap 的 table 控件会将 page、rows 和 userType 三个参数传递到控制器类 UserController 中的 getAreaUser 方法；cache 属性用于设置是否启用 AJAX 数据缓存，这里为 false，表示禁用；sidePagination 属性用于设置在哪里进行分页，可选值为 client 或者 server，这里为 server，表示由服务端处理分页；columns 用于设置列。

Bootstrap 的 table 控件的数据源来自控制器类 UserController 中的 getAreaUser 方法的返回结果，代码如下：

```
package com.ccms.controller;
......
import com.google.gson.JsonObject;
@Controller
@RequestMapping("/user")
public class UserController {
    @Autowired
    UserService userService;
    JsonUtil<SysUser> json = new JsonUtil<SysUser>();
    JsonObject jObject = null;
    ......
    // 获取区域管理员
    @RequestMapping("/getAreaUser")
    @ResponseBody
    public Map<String, Object> getAreaUser(Integer page, Integer rows, HttpServletRequest req, HttpServletResponse rep, @ModelAttribute SysUser user) {
        // 初始化分页类对象
        Pager pager = new Pager();
        pager.setCurPage(page);
        pager.setPerPageRows(rows);
        // 创建对象 params，用于封装查询条件
        Map<String, Object> params = new HashMap<String, Object>();
        params.put("user", user);
        // 获取满足条件的区域管理员总数
        int totalCount = userService.count(params);
        // 根据查询条件获取当前页的区域管理员列表
```

```
        List<SysUser> users = userService.findAreaUser(user, pager);
        // 创建对象result,用于保存返回结果
        Map<String, Object> result = new HashMap<String, Object>(2);
        result.put("total", totalCount);
        result.put("rows", users);
        return result;
    }
}
```

通过@RequestMapping 注解，将用户对/ccms/user/getAreaUser 这个 url 的请求映射到 getAreaUser 方法。getAreaUser 方法有 5 个参数，一个是 SysUser 类型的参数 user，用于封装表单传递来的查询条件；有两个Integer类型的参数page 和rows，用于接收从前端Bootstrap 的 table 控件传递来的页码和每页显示的记录数；有一个 HttpServletRequest 类型的参数 req，代表客户端的请求；有一个 HttpServletResponse 类型的参数 rep，代表服务器的响应。

在 getAreaUser 方法中，首先初始化一个分页类对象 pager，给其设置 curPage 和 perPageRows 两个属性值；然后创建 Map<String, Object>类型的对象 params，用于封装查询条件；接着依次调用业务接口 UserService 的 count 方法，获取满足条件的区域管理员总数，调用 findAreaUser 方法，获取满足条件的区域管理员列表；再创建 Map<String, Object>类型的对象 result，保存查询结果数据；最后将返回结果转为 JSON 格式，以字符串的形式发送到前端页面 areaAdminManager.jsp，为 Bootstrap 的 table 控件提供数据源。

在业务接口 UserService 中，声明两个方法，如下所示：

```
// 获取满足条件的区域管理员总数
Integer count(Map<String, Object> params);
// 分页获取区域管理员
List<SysUser> findAreaUser(SysUser user, Pager pager);
```

在接口 UserService 的实现类 UserServiceImpl 中，实现这两个方法，如下所示：

```
public Integer count(Map<String, Object> params) {
    return userDao.count(params);
}
public List<SysUser> findAreaUser(SysUser user, Pager pager) {
    // / 创建对象params
    Map<String, Object> params = new HashMap<String, Object>();
    // 将封装有查询条件的 user 对象放入 params
    params.put("user", user);
    // 根据条件计算区域管理员总数
    int recordCount = userDao.count(params);
    // 给 pager 对象设置 rowCount 属性值(记录总数)
    pager.setRowCount(recordCount);
    if (recordCount > 0) {
        // 将 page 对象放入 params
        params.put("pager", pager);
    }
    // 分页获取区域管理员信息
    return userDao.selectByPage(params);
}
```

在实现类 UserServiceImpl 的 count 方法中，直接调用了接口 UserDao 中的 count 方法。

在 UserDao 接口中，count 方法的代码如下：

```java
// 根据条件查询区域管理员总数
@SelectProvider(type = UserDynaSqlProvider.class, method = "count")
Integer count(Map<String, Object> params);
```

在 UserDao 接口的 count 方法上，通过@SelectProvider 注解指定由 UserDynaSqlProvider 类中定义的 count 方法提供需要执行的 SELECT 语句。

在 UserDynaSqlProvider 类中，count 方法的代码如下：

```java
// 动态查询区域管理员总记录数
public String count(Map<String, Object> params) {
    String sql = "select  count(*) from sys_user as su where su.delState=1";
    if (params.get("user") != null) {
        SysUser user = (SysUser) params.get("user");
        if (user.getName() != null && !"".equals(user.getName())) {
            sql += " and su.name LIKE '%" + user.getName() + "%' ";
        }
        if (user.getAreaId() != null && !user.getAreaId().isEmpty()) {
            sql += " and su.areaId in (select areaNumber from sys_area as sa where sa.areaNumber= '" + user.getAreaId() + "'";
            sql += " union all select areaNumber from sys_area as sa where sa.parentId='" + user.getAreaId() + "'";
            sql += " union all select areaNumber from sys_area as sa where sa.parentId in (select areaNumber from sys_area as sa where sa.parentId='" + user.getAreaId() + "')" + ") ";
        }
        sql += " and su.userType=" + user.getUserType() + "";
        sql += " order by operatorTime desc";
    }
    return sql;
}
```

实现类 UserServiceImpl 的 findAreaUser 方法有两个参数，一个是 SysUser 类型的参数 user，用于封装前端页面传递来的查询条件；另一个是 Pager 类型的参数 pager，用于封装从前端页面传递来的页码和每页记录数。在 findAreaUser 方法中，创建了一个 Map<String, Object>类型的对象 params，用来存放两个对象，一个是 user 对象，另一个是 pager 对象。对于 pager 对象来说，放入 params 前，还需设置 pager 对象的 rowCount 属性值，这个值是通过调用 UserDao 接口的 count 方法获得的。设置好 params 对象后，再调用 UserDao 接口的 selectByPage 方法获得当前页的区域管理员列表。

在 UserDao 接口中，selectByPage 方法的代码如下：

```java
// 根据条件，分页动态查询区域管理员
@SelectProvider(type = UserDynaSqlProvider.class, method = "selectWithParam")
List<SysUser> selectByPage(Map<String, Object> params);
```

在 UserDao 接口的 selectByPage 方法上，通过@SelectProvider 注解指定由 UserDynaSqlProvider 类中定义的 selectWithParam 方法提供需要执行的 SELECT 语句。

在 UserDynaSqlProvider 类中，selectWithParam 方法的代码如下：

```java
// 分页动态查询区域管理员
public String selectWithParam(Map<String, Object> params) {
```

```
        String sql = SysUserConfig.getAllAreaAdmin;
        if (params.get("user") != null) {
            SysUser user = (SysUser) params.get("user");
            if (user.getName() != null && !"".equals(user.getName())) {
                sql += " and su.name LIKE '%" + user.getName() + "%' ";
            }
            if (user.getAreaId() != null && !user.getAreaId().isEmpty()) {
                sql += " and su.areaId in (select areaNumber from sys_area as sa where sa.areaNumber= '" + user.getAreaId() + "'";
                sql += " union all select areaNumber from sys_area as sa where sa.parentId='" + user.getAreaId() + "'";
                sql += " union all select areaNumber from sys_area as sa where sa.parentId in (select areaNumber from sys_area as sa where sa.parentId='" + user.getAreaId() + "')" + ") ";
            }
            sql += " and su.userType=" + user.getUserType() + "";
            sql += " order by operatorTime desc";
        }
        if (params.get("pager") != null) {
            sql += " limit #{pager.firstLimitParam},#{pager.perPageRows}";
        }
        return sql;
    }
```

2．编辑院校管理员

在院校管理员列表中，每个记录行中都有一个编辑按钮。单击"编辑"按钮，打开"用户修改"对话框，如图 20-8 所示。

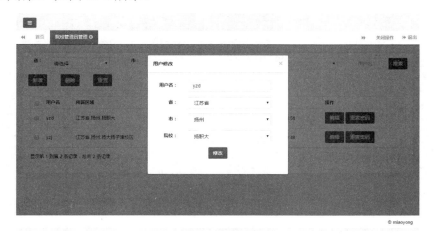

图 20-8 "用户修改"对话框

用户修改对话框的布局如下：

```
<div class="modal fade" id="upwin">
    <div class="modal-dialog" style="width: 400px">
        <div class="modal-content">
            <div class="modal-header">
                <button type="button" class="close" data-dismiss="modal" aria-hidden="true">&times;</button>
                <h4 class="modal-title" id="upwinlable">用户修改</h4>
```

```html
            </div>
            <div class="modal-body">
                <div class="row">
                    <form method="post" class="form-horizontal"id="upform">
                        <div class="form-group">
                            <label class="col-sm-3 control-label">用户名：</label>
                            <div class="col-sm-8 controls">
                                <input type="text" value="" class="form-control" name="name" id="username_up" tabindex="0" />
                            </div>
                        </div>
                        <div class="form-group">
                            <label class="col-sm-3 control-label">省：</label>
                            <div class="col-sm-8 controls">
                                <select data-placeholder="请选择所属省" onchange="changeSheng('_up')" class="form-control" name="sheng_up" id="sheng_up"
                                    tabindex="3">
                                    <option value=''>请选择</option>
                                </select>
                            </div>
                        </div>
                        <div class="form-group">
                            <label class="col-sm-3 control-label"> 市：</label>
                            <div class="col-sm-8 controls">
                                <select data-placeholder="请选择所属市" onchange="changeShi('_up')" class="form-control" name="shi_up" id="shi_up" tabindex="3">
                                    <option value=''>请选择</option>
                                </select>
                            </div>
                        </div>
                        <div class="form-group">
                            <label class="col-sm-3 control-label"> 院校：</label>
                            <div class="col-sm-8 controls">
                                <select data-placeholder="请选择院校" class="form-control" name="xian_up" id="xian_up" tabindex="3">
                                    <option value=''>请选择</option>
                                </select>
                            </div>
                        </div>
                        <div class="form-group">
                            <div class="controls">
                                <button type="submit" class="btn btn-gmtx-define1 center-block">
                                    修改</button>
                            </div>
                        </div>
```

```
                </form>
            </div>
        </div>
    </div>
</div>
```

单击"编辑"按钮,将执行 JavaScript 函数 edit。在函数 edit 中,首先显示用户修改对话框,然后将选中行的用户数据绑定到对话框中的表单元素上,代码如下:

```
function edit(uCode, delState, uname, areaName, province, city, country) {
    clearWin("_up");
    ucode_up = uCode;
    del_State = delState;
    $("#upwin").modal('show');
    $("#username_up").val(uname);
    $("#showArea_up").val(areaName);
    $("#sheng_up").val(province);
    $("#sheng_up").trigger("change");
    $("#shi_up").val(city);
    $("#shi_up").trigger("change");
    $("#xian_up").val(country);
}
```

在用户修改对话框中,"省份"下拉列表控件的数据源是在$(function() { })代码块中定义的,如下所示:

```
loadCity("/ccms/area/getShengAreaList", "sheng_up", null);
```

loadCity 函数定义在 CommonValue.js 文件中,该文件位于 src/main/webapp/commons/jslib 目录下,loadCity 函数如下所示:

```
// 省市院校下拉框
function loadCity(url, idStr, pid) {
    $.ajax({
        url : url,
        dataType : 'json',
        async : false,
        data : {
            parentId : pid
        },
        type : 'post',
        success : function(result) {
            var options = "<option value=''>请选择</option>";
            if (result.count > 0) {
                $.each(result.CityList, function(key, val) {
                    options += '<option value=' + val.areaNumber + '>'
                        + val.name + '</option>';
                });
            }
            $('#' + idStr).empty();
            $('#' + idStr).append(options);
```

```
            },
            error : function() {
            }
        });
    }
```

在 loadCity 函数中，通过 jQuery AJAX 方式向后台服务器发送请求，url 参数用于指定发送请求的地址，这里为 /ccms/area/getShengAreaList，这个请求将映射到控制器类 AreaController 中的 getShengAreaList 方法，方法的返回结果作为省份下拉列表的数据源；dataType 参数用于指定预期服务器返回的数据类型，这里指定返回 JSON 数据；async 参数用于指定是否发送异步请求，这里为 false，表示同步请求，此时将锁住浏览器，用户的其他操作必须等待请求完成才可以执行；data 参数用于指定发送到服务器的数据，将自动转换为请求字符串格式；type 参数用于指定请求方式，这里为 post 方式；success 参数的类型为 Function，表示请求成功后的回调函数，服务器返回的数据将传给该函数的 result 参数，然后根据接收的数据生成省份下拉列表的选项。

在 AreaController 类中，getShengAreaList 方法的代码如下：

```java
package com.ccms.controller;
......
import com.ccms.pojo.SysArea;
import com.ccms.service.AreaService;
@Controller
@RequestMapping(value = "/area", method = { RequestMethod.GET, RequestMethod.POST })
public class AreaController {
    @Autowired
    private AreaService areaService;
    // 获取省份列表
    @RequestMapping(value = "/getShengAreaList", method = { RequestMethod.GET, RequestMethod.POST })
    @ResponseBody
    public Map<String, Object> getShengAreaList(SysArea area, HttpServletRequest req, HttpServletResponse resp) throws IOException {
        area.setType("1");
        List<SysArea> list = areaService.getAreaList(area);
        int count = 0;
        if (list != null && list.size() > 0) {
            count = list.size();
        }
        Map<String, Object> result = new HashMap<String, Object>(2);
        result.put("count", count);
        result.put("CityList", list);
        return result;
    }
    ......
}
```

在 AreaController 类上标注 @Controller 注解，指示该类是一个控制器。通过

@RequestMapping 注解将用户对 /ccms/area/getShengAreaList 这个 url 的请求映射到 getShengAreaList 方法。

在 getShengAreaList 方法中，将 SysArea 类型的对象 area 的 type 属性值设置为 1，再调用业务接口 AreaService 中的 getAreaList 方法，以获取省份信息列表。

AreaService 接口位于 com.ccms.service 包中，getAreaList 方法的声明如下：

```java
package com.ccms.service;
import java.util.List;
import com.ccms.pojo.SysArea;
public interface AreaService {
    // 获取区域列表
    public List<SysArea> getAreaList(SysArea area);
}
```

在 com.ccms.service.impl 包中，创建 AreaService 接口的实现类 AreaServiceImpl，实现 getAreaList 方法，代码如下：

```java
package com.ccms.service.impl;
……
import com.ccms.pojo.SysArea;
import com.ccms.service.AreaService;
@Service("areaService")
@Transactional(propagation = Propagation.REQUIRED, isolation = Isolation.DEFAULT)
public class AreaServiceImpl implements AreaService {
    @Autowired
    AreaDao areaDao;
    @Override
    public List<SysArea> getAreaList(SysArea area) {
        List list = null;
        Map<String, Object> params = new HashMap<String, Object>();
        params.put("area", area);
        list = areaDao.selectAreaList(params);
        return list;
    }
}
```

在实现类 AreaServiceImpl 的 getAreaList 方法中，调用了接口 AreaDao 中的 selectAreaList 方法。在 AreaDao 接口中，selectAreaList 方法的代码如下：

```java
package com.ccms.dao;
……
import com.ccms.dao.provider.AreaDynaSqlProvider;
import com.ccms.pojo.SysArea;
public interface AreaDao {
    @SelectProvider(type = AreaDynaSqlProvider.class, method = "selectWithParam")
    public List<SysArea> selectAreaList(Map<String, Object> params);
}
```

selectAreaList 方法标注了@SelectProvider 注解，指定由 AreaDynaSqlProvider 类中定义

的 selectWithParam 方法提供需要执行的 SELECT 语句。在 AreaDynaSqlProvider 类中，selectWithParam 方法的代码如下：

```java
package com.ccms.dao.provider;
import java.util.Map;
import org.apache.ibatis.jdbc.SQL;
import com.ccms.pojo.SysArea;
public class AreaDynaSqlProvider {
    public String selectWithParam(Map<String, Object> params) {
        SysArea area = (SysArea) params.get("area");
        String sql = new SQL() {
            {
                SELECT("*");
                FROM("sys_area");
                WHERE(" type = #{area.type} ");
                if (area.getType() == "1") {
                    WHERE("parentId is null");
                }
                if (area.getType() != "1") {
                    WHERE("parentId = #{area.parentId} ");
                }
            }
        }.toString();
        return sql;
    }
}
```

在用户修改对话框中，如果省份下拉列表选项发生变化，会触发 onchange 事件。处理这个事件的 JavaScript 函数为 changeSheng('_up')，代码如下：

```javascript
function changeSheng(suffix) {
    var shengid = $('#sheng' + suffix).val();
    $('#shi' + suffix).val('');
    loadCity("/ccms/area/getShiAreaList", "shi" + suffix, shengid);
    $('#xian' + suffix).val('');
}
```

在函数 changeSheng 中，会根据选择的省份，向后台服务器发送一个请求，请求的 url 为 /ccms/area/getShiAreaList。这个 url 请求将映射到控制器类 AreaController 中的 getShiAreaList 方法，根据省份获取城市列表，作为市下拉列表的数据源。在 AreaController 类中，getShiAreaList 方法的代码如下：

```java
// 获取城市列表
@RequestMapping(value = "/getShiAreaList", method = { RequestMethod.GET, RequestMethod.POST })
@ResponseBody
public Map<String, Object> getShiAreaList(SysArea area, HttpServletRequest req, HttpServletResponse resp) throws IOException {
    area.setType("2");
    List list = areaService.getAreaList(area);
    int count = 0;
```

```java
    if (list != null && list.size() > 0) {
        count = list.size();
    }
    Map<String, Object> result = new HashMap<String, Object>(2);
    result.put("count", count);
    result.put("CityList", list);
    return result;
}
```

在 getShiAreaList 方法中,将 SysArea 类型的对象 area 的 type 属性值设置为 2,再调用业务接口 AreaService 中的 getAreaList 方法,以获取城市列表。

同样,在用户修改对话框中,如果市下拉列表选项发生变化,会触发 onchange 事件。处理这个事件的 JavaScript 函数为 changeShi('_up'),如下所示:

```javascript
function changeShi(suffix) {
    var shiid = $('#shi' + suffix).val();
    $('#xian' + suffix).val('');
    loadCity("/ccms/area/getXianAreaList", "xian" + suffix, shiid);
}
```

在函数 changeShi 中,会根据选择的城市,向后台服务器发送一个请求,请求的 url 为 /ccms/area/getXianAreaList。这个 url 请求将映射到控制器类 AreaController 中的 getXianAreaList 方法,根据城市获取院校列表,作为院校下拉列表的数据源。getXianAreaList 方法的代码如下:

```java
// 获取院校(县)列表
@RequestMapping(value = "/getXianAreaList", method = { RequestMethod.GET,
RequestMethod.POST })
@ResponseBody
public Map<String, Object> getXianAreaList(SysArea area, HttpServletRequest
req, HttpServletResponse resp) throws
IOException {
    area.setType("3");
    List list = areaService.getAreaList(area);
    int count = 0;
    if (list != null && list.size() > 0) {
        count = list.size();
    }
    Map<String, Object> result = new HashMap<String, Object>(2);
    result.put("count", count);
    result.put("CityList", list);
    return result;
}
```

在 getXianAreaList 方法中,将 SysArea 类型的对象 area 的 type 属性值设置为 3,再调用业务接口 AreaService 中的 getAreaList 方法,以获取院校列表。这样,就可以实现用户修改对话框中省、市、院校三个下拉列表的联动了。

在用户修改对话框中修改数据,单击修改按钮,将执行 JavaScript 函数 upUser,其代码如下:

```javascript
function upUser() {
    var name = $("#username_up").val();
    var sheng = $("#sheng_up").val();
    var shi = $("#shi_up").val();
    var xian = $("#xian_up").val();
    //未选择区域,提示
    if (isNull(sheng) && isNull(shi) && isNull(xian)) {
        swal({
            title : "系统提示",
            text : "请选择区域",
            type : "warning"
        });
        return;
    }
    var areaId = sheng;
    if (!(shi == undefined || shi == null || shi == '')) {
        areaId = shi
    }
    if (!(xian == undefined || xian == null || xian == '')) {
        areaId = xian
    }
    var userType = 4;
    $.ajax({
        url : '/ccms/user/upAreaAdmin',
        type : 'post',
        async : 'true',
        cache : false,
        data : {
            userCode : ucode_up,
            name : name,
            userType : userType,
            areaId : areaId,
            delState : del_State
        },
        dataType : 'json',
        success : function(data) {
            if (data.isExist) {
                swal({
                    title : "系统提示",
                    text : "已存在该用户名",
                    type : "warning"
                }, function() {
                    $("#userName").val('');
                    $("#addwin").modal('hide');
                });
            } else if (data.existAdmin) {
                swal({
                    title : "系统提示",
                    text : "该区域已绑定管理员",
                    type : "warning"
                }, function() {
```

```
                    $("#sheng_up").val('');
                    $("#shi_up").val('');
                    $("#xian_up").val('');
                });
            } else if (data.success) {
                swal({
                    title : "系统提示",
                    text : "修改成功",
                    type : "success"
                }, function() {
                    $("#userName_edit").val('');
                    $("#upwin").modal('hide');
                    $('#table').bootstrapTable("refresh");
                });
            } else {
                swal({
                    title : "系统提示",
                    text : "修改失败",
                    type : "warning"
                }, function() {
                    $("#upwin").modal('hide');
                    $('#table').bootstrapTable("refresh");
                });
            }
        },
        error : function(aa, ee, rr) {
            swal({
                title : "系统提示",
                text : "请求服务器失败,请稍候再试",
                type : "warning"
            }, function() {
            });
        }
    });
}
```

在 upUser 方法中，首先获取要修改的区域管理员信息，然后通过 jQuery AJAX 方式向后台服务器发送请求，请求的地址为/ccms/user/upAreaAdmin，这个 url 请求将映射到控制器类 UserController 中的 upAreaAdmin 方法；async 用于指定是否发送异步请求，这里为 true，表示异步请求，用户的其他操作不必等待请求完成；data 用于指定发送到服务器的数据；success 参数的类型为 Function，表示请求成功后的回调函数，服务器返回的数据将传给该函数的 data 参数，然后根据接收的数据给出提示信息；error 用于指定请求失败时调用的函数。

在控制器类 UserController 中，upAreaAdmin 方法的代码如下：

```
// 修改区域管理员
@RequestMapping(value = "/upAreaAdmin", method = { RequestMethod.GET,
RequestMethod.POST })
```

```java
public String upAreaAdmin(SysUser user, HttpServletRequest req,
HttpServletResponse res) {
    jObject = new JsonObject();
    boolean result = false;
    String operateId =
req.getSession().getAttribute(CommonValue.USERID).toString();
    user.setOperatorId(operateId);
    // 修改时检验用户是否已存在
    String exist = userService.isExistAreaAdminName(user);
    if (exist.equals("exit")) {
        jObject.addProperty("isExist", true);
    } else if (exist.equals("exitAdmin")) {
        jObject.addProperty("existAdmin", true);
    } else if (exist.equals("no")) {
        result = userService.upAreaAdmin(user);
        jObject.addProperty("success", result);
    }
    try {
        ServletOutputStream jos = res.getOutputStream();
        jos.write(jObject.toString().getBytes("utf-8"));
        jos.flush();
        jos.close();
    } catch (IOException e) {
        e.printStackTrace();
    }
    return null;
}
```

在 upAreaAdmin 方法上，通过@RequestMapping 注解，将用户对/ccms/user/upAreaAdmin 这个 url 的请求映射到 upAreaAdmin 方法。upAreaAdmin 方法有三个参数，其中，SysUser 类型的参数 user 用于封装表单传递来的查询条件。在 upAreaAdmin 方法中，首先调用业务接口 UserService 中的 isExistAreaAdminName 方法，检验用户是否已存在；然后调用 upAreaAdmin 方法，更新区域管理员信息。通过 JsonObject 类的 write 方法，可向客户端浏览器发送 JSON 格式的数据。

在 UserService 接口中，声明两个方法，代码如下：

```java
// 查询是否存在该区域管理员
public String isExistAreaAdminName(SysUser user);
// 修改区域管理员
public boolean upAreaAdmin(SysUser user);
```

在 UserService 接口的实现类 UserServiceImpl 中，实现这两个方法，代码如下：

```java
// 查询是否存在该区域管理员
public String isExistAreaAdminName(SysUser user) {
    String result;
    // 查询该管理员名是否存在
    int i = (Integer) userDao.isExistAreaAdminName(user);
    // 查询该区域是否绑定管理员
    int j = (Integer) userDao.isExistAreaAdmin(user);
    if (j > 0) {
```

```
            result = "exitAdmin";
        } else if (i > 0) {
            result = "exit";
        } else {
            result = "no";
        }
        return result;
    }
    // 修改区域管理员
    public boolean upAreaAdmin(SysUser user) {
        int result = userDao.upAreaAdmin(user);
        return result > 0;
    }
```

在 isExistAreaAdminName 方法中，首先调用接口 UserDao 中的 isExistAreaAdminName 方法，查询该管理员名是否存在；然后调用 isExistAreaAdmin 方法，查询该区域是否绑定管理员。

在 UserDao 接口中，isExistAreaAdminName 和 isExistAreaAdmin 方法的代码如下：

```
// 查询是否存在该区域管理员
@Select("select count(userCode) from sys_user where name = #{name} and userCode <> #{userCode} and userType=#{userType} and delState =1")
public int isExistAreaAdminName(SysUser user);
// 检验该区域是否绑定管理员
@Select("select count(userCode) from sys_user where areaId = #{areaId} and userCode <> #{userCode} and userType=#{userType} and delState =1")
public int isExistAreaAdmin(SysUser user);
```

在 upAreaAdmin 方法中，直接调用接口 UserDao 中的 upAreaAdmin 方法。在 UserDao 接口中，upAreaAdmin 方法的代码如下：

```
// 更新区域管理员
@Update("update sys_user set name=#{name},areaId=#{areaId}, operatorId=#{operatorId},operatorTime=now() " + "where userCode= #{userCode} and delState=1")
public int upAreaAdmin(SysUser user);
```

3．新增院校管理员

在 areaAdminManager.jsp 页面中，单击新增按钮，打开区域管理员添加对话框，如图 20-9 所示。

区域管理员添加对话框与修改对话框的布局基本类似，此处不再列出。填写管理员和密码，选择省、市和院校，单击"添加"按钮，将执行 JavaScript 函数 addUser，其代码如下：

图 20-9　区域管理员添加对话框

```
function addUser() {
    var name = $("#username_add").val();
    var psw = $("#userpsw_add").val();
    var sheng = $("#sheng_add").val();
    var shi = $("#shi_add").val();
```

```javascript
            var xian = $("#xian_add").val();
            if (isNull(sheng) && isNull(shi) && isNull(xian)) {
                swal({
                    title : "系统提示",
                    text : "请选择区域",
                    type : "warning"
                });
                return;
            }
            var areaId = sheng;
            if (!(shi == undefined || shi == null || shi == '')) {
                areaId = shi
            }
            if (!(xian == undefined || xian == null || xian == '')) {
                areaId = xian
            }
            var userType = $("#userType_add").val();
            $.ajax({
                url : '/ccms/user/addAreaAdmin',
                type : 'post',
                async : 'true',
                cache : false,
                data : {
                    name : name,
                    psw : psw,
                    userType : userType,
                    areaId : areaId
                },
                dataType : 'json',
                success : function(data) {
                    if (data.isExist) {
                        swal({
                            title : "系统提示",
                            text : "已存在该管理员名",
                            type : "warning"
                        }, function() {
                            $("#username_add").val('');
                        });
                    } else if (data.existAdmin) {
                        swal({
                            title : "系统提示",
                            text : "该区域已绑定管理员",
                            type : "warning"
                        }, function() {
                            $("#sheng_add").val('');
                            $("#shi_add").val('');
                            $("#xian_add").val('');
                        });
                    } else if (data.success) {
                        swal({
                            title : "系统提示",
```

```javascript
                        text : "添加成功",
                        type : "success"
                    }, function() {
                        clearWin("_add");
                        $("#addwin").modal('hide');
                        $('#table').bootstrapTable("refresh");
                    });
                } else {
                    swal({
                        title : "系统提示",
                        text : "添加失败",
                        type : "warning"
                    }, function() {
                        $("#name").val('');
                        $("#psw").val('');
                        clearWin("_add");
                    });
                }
            },
            error : function(aa, ee, rr) {
                swal({
                    title : "系统提示",
                    text : "请求服务器失败,请稍候再试",
                    type : "warning"
                }, function() {
                    clearWin("_add");
                    $("#addwin").modal('hide');
                });
            }
        });
    }
```

在函数 addUser 中，首先获取表单中填写的管理员、密码、省、市和院校等数据，然后通过 jQuery AJAX 方式向后台服务器发送请求，请求的地址为/ccms/user/addAreaAdmin，这个 url 请求将映射到控制器类 UserController 中的 addAreaAdmin 方法；并将 name、psw、userType 和 areaId 四个参数传递过去。

在 UserController 类中，addAreaAdmin 方法的代码如下：

```java
// 添加区域管理员
@RequestMapping(value = "/addAreaAdmin", method = { RequestMethod.GET, RequestMethod.POST })
@ResponseBody
public String addAreaAdmin(SysUser user, HttpServletRequest req, HttpServletResponse res) throws IOException {
    jObject = new JsonObject();
    String operateId = req.getSession().getAttribute(CommonValue.USERID).toString();
    user.setOperatorId(operateId);
    user.setPsw(MD5Util.MD5(user.getPsw()));
    user.setUserCode("");
```

```java
        user.setUserType("4");
        // 添加时检验该区域管理员是否已存在
        String exist = userService.isExistAreaAdminName(user);
        if (exist.equals("exit")) {
            jObject.addProperty("isExist", true);
        } else if (exist.equals("exitAdmin")) {
            jObject.addProperty("existAdmin", true);
        } else if (exist.equals("no")) {
            boolean result = userService.addUser(user);
            jObject.addProperty("success", result);
        }
        ServletOutputStream jos = res.getOutputStream();
        jos.write(jObject.toString().getBytes("utf-8"));
        return null;
    }
```

与区域管理员修改功能类似，在 addAreaAdmin 方法中，首先调用业务接口 UserService 中的 isExistAreaAdminName 方法，检验该区域管理员是否已存在。如果不存在，则调用业务接口 UserService 中的 addUser 方法，执行添加操作。

在 UserService 接口中，addUser 方法的声明如下：

```java
// 添加用户
public boolean addUser(SysUser user);
```

在实现类 UserServiceImpl 中，实现 addUser 方法，代码如下：

```java
@Override
public boolean addUser(SysUser user) {
    Object[] params = null;
    String code = UUIDGenerator.getUUID();
    user.setUserCode(code);
    user.setDelState(1);
    int result = userDao.insertUser(user);
    return result > 0;
}
```

在 addUser 方法中，首先调用 UUIDGenerator 类的 getUUID 方法，生成一个新的用户编号，将其保存在 user 对象中。然后直接调用 UserDao 接口中的 insertUser 方法，将区域管理员信息插入数据表 sys_user 中。

在 UserDao 接口中，insertUser 方法的代码如下：

```java
// 添加用户
@Insert("insert into sys_user(userCode,name,psw,operatorId,delState,unitId,userType,areaId) " + "values(#{userCode},#{name},#{psw},#{operatorId},#{delState},#{unitId},#{userType},#{areaId})")
public int insertUser(SysUser user);
```

输入图 20-9 所示的数据，添加成功后，在区域管理员列表上，显示出新增的区域管理员记录，如图 20-10 所示。

图 20-10　新增的区域管理员记录

4．删除院校管理员

在区域管理员列表上，选中一条记录的复选框，单击删除按钮，将执行 JavaScript 函数 delUser，其代码如下：

```javascript
function delUser() {
    var ids = '';
    var selects = $('#table').bootstrapTable('getSelections');
    if (selects.length <= 0) {
        swal('系统提示', '请选择要删除的用户！', 'warning');
        return;
    }
    ids = "'" + selects[0].userCode + "'";
    for (var i = 1; i < selects.length; i++) {
        ids += ",'" + selects[i].userCode + "'";
    }
    swal({
        title : "您确定要删除这条信息吗",
        text : "删除后将无法恢复，请谨慎操作！",
        type : "warning",
        showCancelButton : true,
        confirmButtonColor : "#DD6B55",
        confirmButtonText : "删除",
        cancelButtonText : "取消",
        closeOnConfirm : false,
        closeOnCancel : false
    }, function(isConfirm) {
        if (isConfirm) {
            $.ajax({
                url : '/ccms/user/delUser',
                type : 'post',
                async : 'true',
                cache : false,
                data : {
                    ids : ids
                },
                dataType : 'json',
                success : function(data) {
                    if (data.success) {
                        swal("删除成功！","您已经删除了这条信息。","success");
                        $('#table').bootstrapTable("refresh");
                    } else {
```

```
                    swal({
                        title : "系统提示",
                        text : "删除失败",
                        type : "warning"
                    });
                }
            },
            error : function(aa, ee, rr) {
                swal({
                    title : "系统提示",
                    text : "请求服务器失败,请稍候再试",
                    type : "warning"
                });
            }
        });
    } else {
        swal("已取消", "您取消了删除操作！", "error")
    }
});
```

在函数 delUser 中，首先获取 Bootstrap 的 table 控件上选中的记录行的用户编号，并将这些编号以逗号分隔；然后通过 jQuery AJAX 方式向后台服务器发送请求，请求的地址为 /ccms/user/delUser，这个 url 请求将映射到控制器类 UserController 中的 delUser 方法，并将参数 ids 传递过去。

在 UserController 类中，delUser 方法的代码如下：

```
@RequestMapping(value = "/delUser", method = { RequestMethod.GET,
RequestMethod.POST })
public String delUser(HttpServletRequest req, HttpServletResponse res) {
    String ids = req.getParameter("ids");
    boolean result = userService.delUser(ids, 2);
    jObject = new JsonObject();
    jObject.addProperty("success", result);
    try {
        ServletOutputStream jos = res.getOutputStream();
        jos.write(jObject.toString().getBytes("utf-8"));
        jos.flush();
        jos.close();
    } catch (IOException e) {
        e.printStackTrace();
    }
    return null;
}
```

在 UserController 类的 delUser 方法中，调用业务接口 UserService 中的 delUser 方法，这个方法有两个参数，第一个参数 ids 是以逗号分隔的用户编号，第二个参数 2 表示删除状态。

在 UserService 接口中，delUser 方法的声明如下：

```
// 删除用户
public boolean delUser(String ids, int flag);
```

在实现类 UserServiceImpl 中，实现 delUser 方法，代码如下：

```
public boolean delUser(String ids, int flag) {
    int result = userDao.delUser(ids, flag);
    return result > 0;
}
```

在 UserServiceImpl 类的 delUser 方法中，直接调用了接口 UserDao 中的 delUser 方法。在接口 UserDao 中，delUser 方法的代码如下：

```
// 删除用户(即更新用户状态)
@Update("update sys_user set delState=#{flag} where userCode in (${ids})")
public int delUser(@Param("ids") String ids, @Param("flag") int flag);
```

删除区域管理员的操作，实质上是修改区域管理员的状态，即将数据表 sys_user 中指定编号的用户记录的 delState 字段值设置为 2。

5. 根据省、市、院校和用户名搜索

在 areaAdminManager.jsp 页面中，提供了根据省、市、院校和用户名查询区域管理员的功能。首先需要为省份下拉列表控件指定数据源，这个数据源是在$(function() { })代码块中定义的，代码如下：

```
loadCity("/ccms/area/getShengAreaList", "sheng_search", null);
```

在区域管理员修改功能开发时，已经对 loadCity 函数做过介绍。同样，之前也介绍过省、市和院校三个下拉列表的联动。

选择省、市和院校，或输入用户名，单击搜索按钮，将执行 JavaScript 函数 search，其代码如下：

```
function search() {
    var _name = $('#name_search').val();
    var sheng = $("#sheng_search").val();
    var shi = $("#shi_search").val();
    var xian = $("#xian_search").val();
    var areaId = sheng;
    if (!(shi == undefined || shi == null || shi == '')) {
        areaId = shi
    }
    if (!(xian == undefined || xian == null || xian == '')) {
        areaId = xian
    }
    var userType = $('#userType_search').val();
    $('#table').bootstrapTable('refresh', {
        query : {
            name : _name,
            areaId : areaId,
            userType : '4',
        }
    });
```

}

在函数 search 中，首先获取省、市和院校，或用户名信息，然后调用 Bootstrap 的 table 控件的 refresh 方法。这时会重新发送请求/ccms/user/getAreaUser，即再次执行控制器类 UserController 中的 getAreaUser 方法，并将参数 name、areaId 和 userType 传递过去，根据条件重新获取区域管理员列表，并显示在 table 控件上。以查询江苏省扬州市扬大扬子津校区为例，查询结果如图 20-11 所示。

图 20-11　区域管理员查询

20.6.2　院校管理

在图 20-4 所示的平台管理员界面中，单击院校管理菜单，打开院校管理页 schoolManager.jsp，该页面位于 src/main/webapp/views/area 目录下，如图 20-12 所示。

图 20-12　院校管理页

在 schoolManager.jsp 页面中，需要实现院校的显示、修改、添加和删除等功能。

1．显示院校

为了以树形结构显示院校信息，使用了 zTree 插件，它是一个依靠 jQuery 实现的多功能的树插件。在页面中，它的布局如下：

```
<div class="col-sm-3">
    <ul id="schooltree" class="ztree" style="background:#fff"></ul>
```

```
</div>
```

在页面中使用 zTree 插件时,需要引用 jquery.ztree.all-3.5.js 文件,如下所示:

```
<script type="text/javascript"
    src="/ccms/commons/jslib/ztreeV3.5.15/jquery.ztree.all-3.5.js">
</script>
```

在页面 schoolManager.jsp 中,定义了一个自执行函数,如下所示:

```
$(function(){
    vform('addform',addModule);
    vform('upform',upModule);
    loadSchoolTree();
});
```

在页面加载时,函数中的代码会被执行。loadSchoolTree 方法用来给 zTree 插件绑定数据,代码如下:

```
function loadSchoolTree(){
    $.ajax({
        url:'/ccms/school/getTreeList',
        type:'post',
        async:'true',
        cache:false,
        data:{},
        dataType:'json',
        success: function(data){
            $.fn.zTree.init($("#schooltree"), setting, data.zNodes);
        }
    });
}
```

在 loadSchoolTree 方法中,通过 jQuery AJAX 方式向后台服务器发送请求,请求的地址为/ccms/school/getTreeList,这个 url 请求将映射到控制器类 SchoolController 中的 getTreeList 方法; success 参数的类型为 Function,表示请求成功后的回调函数,服务器返回的数据将传给该函数的 data 参数,然后通过$.fn.zTree.init 填充 zTree 插件。绑定数据源后,zTree 插件就变成了一棵院校树。

在 SchoolController 类中,getTreeList 方法的代码如下:

```
package com.ccms.controller;
……
import com.ccms.pojo.SysArea;
import com.ccms.service.SchoolService;
import com.ccms.tools.Tree;
@Controller
@RequestMapping(value = "/school", method = { RequestMethod.GET,
RequestMethod.POST })
public class SchoolController {
    @Autowired
    private SchoolService schoolService;
    // 获取院校树
```

```java
    @RequestMapping(value = "/getTreeList", method = {
RequestMethod.GET, RequestMethod.POST })
    @ResponseBody
    public Map<String, Object> getTreeList(HttpServletRequest req,
HttpServletResponse resp)
            throws UnsupportedEncodingException, IOException {
        List<Tree> tree = schoolService.getSchoolTree();
        Map<String, Object> result = new HashMap<String, Object>();
        result.put("zNodes", tree);
        return result;
    }
    ……
}
```

在 SchoolController 类上标注@Controller 注解，指示该类是一个控制器。并通过@RequestMapping 注解将用户对/ccms/school/getTreeList 这个 url 的请求映射到 getTreeList 方法。在 getTreeList 方法中，调用业务接口 SchoolService 中的 getSchoolTree 方法，以获取院校列表。

SchoolService 接口位于 com.ccms.service 包中，getSchoolTree 方法的声明如下：

```java
package com.ccms.service;
import java.util.List;
import com.ccms.pojo.SysArea;
import com.ccms.tools.Tree;
public interface SchoolService {
    // 获取院校树
    public List<Tree> getSchoolTree();
}
```

在接口 SchoolService 的实现类 SchoolServiceImpl 中，实现 getSchoolTree 方法，代码如下：

```java
package com.ccms.service.impl;
import com.ccms.service.SchoolService;
import com.ccms.tools.Tree;
……
@Service("schoolService")
@Transactional(propagation = Propagation.REQUIRED, isolation = Isolation.DEFAULT)
public class SchoolServiceImpl implements SchoolService {
    @Autowired
    SchoolDao schoolDao;
    @Override
    public List<Tree> getSchoolTree() {
        List<String> attributesList = new ArrayList<String>();
        attributesList.add("isLeaf");
// 定义了一个 Tree 类型的对象 rootTree,
// 并设置了相关属性，作为 zTree 插件的根结点
        Tree rootTree = new Tree();
        rootTree.setId(null);
        rootTree.setName("院校配置");
```

```java
            rootTree.setOpen(true);
            rootTree.setAttributes("{\"isLeaf\":\"0\"}");
            // 获取院校/区域的所有父结点
            List<SysArea> parentAreaLst = schoolDao.getParentArea();
            List<Tree> parentTreesLst = new ArrayList<Tree>();
            // 将List<SysArea>类型转换成List<Tree>
            for (SysArea area : parentAreaLst) {
                Tree rootTreeData = new Tree();
                rootTreeData.setId(area.getAreaNumber());
                rootTreeData.setName(area.getName());
                rootTreeData.setOpen(true);
                rootTreeData.setChecked(false);
                rootTreeData.setParent(false);
                rootTreeData.setNocheck(false);
                JSONObject attributesJO = new JSONObject();
                attributesJO.put("isLeaf", area.getIsLeaf());
                rootTreeData.setAttributes(attributesJO.toString());
                parentTreesLst.add(rootTreeData);
            }
            if (parentTreesLst != null) {
                // 遍历List<Tree>类型对象parentTreesLst
                for (Tree parentData : parentTreesLst) {
                    // 获取当前父结点区域编号
                    String areaNumber = parentData.getId();
                // 根据当前父结点区域编号,获取其所包含的区域信息(子结点列表)
                    List<SysArea> list = schoolDao.getChildArea(areaNumber);
                    List<Tree> treeList = new ArrayList<Tree>();
                    // 将List<SysArea>类型转换成List<Tree>类型
                    for (SysArea sysArea : list) {
                        Tree tempTreeData = new Tree();
                        tempTreeData.setId(sysArea.getAreaNumber());
                        tempTreeData.setName(sysArea.getName());
                        tempTreeData.setOpen(true);
                        tempTreeData.setChecked(false);
                        tempTreeData.setParent(false);
                        tempTreeData.setNocheck(false);
                        JSONObject attributesJO = new JSONObject();
                        attributesJO.put("isLeaf", sysArea.getIsLeaf());
                    tempTreeData.setAttributes(attributesJO.toString());
                        List childList = new ArrayList<Tree>();
//调用自定义方法getChildTreeById,根据当前区域Id(子结点区
//域编号)获取其包含的区域信息(子子结点数据列表)
                        childList = getChildTreeById(sysArea.getAreaNumber());
                        // 将子结点列表赋值到父结点的children属性
                        tempTreeData.setChildren(childList);
                        treeList.add(tempTreeData);
                    }
// 给parentData结点设置children属性
                    parentData.setChildren(treeList);
                }
            }
```

```java
        // 将 parentTreesLst 设置到根结点对象 rootTree 的 children 属性
        rootTree.setChildren(parentTreesLst);
        List<Tree> resultTree = new ArrayList<Tree>();
        resultTree.add(rootTree);
        return resultTree;
    }
}
```

在 getSchoolTree 方法中，首先定义了一个 Tree 类型的对象 rootTree，并设置了相关属性，作为 zTree 插件的根结点。Tree 这个类位于 com.ccms.tools 包中，用于描述 zTree 插件上的一个结点，Tree 类的定义如下：

```java
package com.ccms.tools;
import java.util.List;
public class Tree {
    private String id;
    private String pId;
    private String name;
    private boolean isParent;
    private boolean open;
    private boolean checked;
    private boolean nocheck;
    private List<Tree> children;
    private String attributes;
    // 省略了上述属性的 getter 和 setter 方法
    // 省略了 hashCode 和 equals 方法
}
```

然后调用接口 SchoolDao 中的 getParentArea 方法获取院校或区域的父结点列表，返回 List<SysArea>类型的对象 parentAreaLst。接着定义了 List<Tree>类型的对象 parentTreesLst，并将 List<SysArea>类型的对象 parentAreaLst 转换成 List<Tree>类型的对象 parentTreesLst，此时 parentTreesLst 中每个元素的 children 属性还没有赋值。之后遍历对象 parentTreesLst，每次遍历都是先获取当前父结点区域编号，再调用接口 schoolDao 中的 getChildArea 方法，根据当前父结点区域编号获取其所包含的区域信息(即子结点列表)，并将作为返回结果的 List<SysArea>类型的对象 list 转换成 List<Tree>类型的对象 treeList。在这个转换过程中会调用自定义方法 getChildTreeById，根据当前区域编号(子结点区域编号)获取其所包含的区域信息(子子结点列表)。在自定义方法 getChildTreeById 中，会递归调用 getChildTreeById 方法，从而为每个 Tree 结点都设置 children 属性值。转换完成后，将对象 treeList 设置到当前遍历元素的 children 属性。这样遍历结束时，parentTreesLst 列表中每个元素的 children 属性都有了值。最后将 parentTreesLst 设置到根结点对象 rootTree 的 children 属性，再将这个根结点对象 rootTree 添加到 List<Tree>类型的对象 resultTree 并返回。

在接口 SchoolDao 中，getParentArea 方法从数据表 sys_area 中获取区域的父结点列表，代码如下：

```java
// 获取区域的父结点列表
@Select("select * from sys_area where parentId is null or parentId = '' order by sortNum asc")
```

```
public List<SysArea> getParentArea();
```

在接口 SchoolDao 中，getChildArea 方法根据当前父结点区域编号，获取其所包含的区域信息，如下所示：

```
// 获取所有子结点
@Select("select * from sys_area where parentId=#{areaNumber} order by sortNum asc")
public List<SysArea> getChildArea(@Param("areaNumber") String areaNumber);
```

绑定数据源后，zTree 插件就变成了一个院校菜单树。

2．修改院校

在院校菜单树上，单击一个结点，打开修改院校界面，如图 20-13 所示。

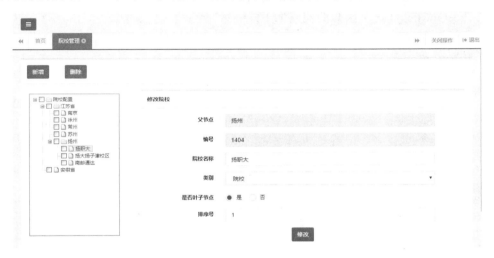

图 20-13　修改院校界面

在页面 schoolManager.jsp 的 loadSchoolTree 方法中，使用$.fn.zTree.init 填充 zTree 插件时，init 方法还指定了一个 JSON 格式的参数 setting，这个参数的内容如下：

```
var setting = {
    check: {
        enable: true,
        autoCheckTrigger: true,
        chkStyle: "checkbox",
        chkboxType: { "Y": "s", "N": "ps" }
    },
    data: {
        simpleData: {
            enable: false,
        }
    },
    edit:{
        enable: true,
        showRemoveBtn: false,
```

```
                showRenameBtn: false,
                drag: {
                    autoExpandTrigger: true,
                    prev: false,
                    inner: true,
                    next: false
                }
            },
            callback:{
            onClick: zTreeOnClick,
            onDrop:zTreeOnDrop,
            beforeDrop: zTreeBeforeDrop
            }
        };
```

因此，在院校菜单树上单击结点时，会触发 onClick 事件，处理这个事件的 JavaScript 函数为 zTreeOnClick，这个函数的代码如下：

```
//结点点击事件
function zTreeOnClick(event, treeId, treeNode) {
    $('#addwin').hide();
    if(isSelected("schooltree"))
    {
        if(getSelected("schooltree").id!="null")
        {
            $('#upwin').show();
            loadSingleData(getSelected("schooltree").id);
        }else
        {
            $("#addwin").hide();
            $("#upwin").hide();
        }
    }
}
```

在页面 schoolManager.jsp 中，有添加院校和修改院校两个表单，添加院校表单定义在 id 为 addwin 的<div>标签中的，修改院校表单定义在 id 为 upwin 的<div>标签中。在函数 zTreeOnClick 中，首先隐藏添加院校表单，然后显示修改院校表单，再调用 JavaScript 函数 loadSingleData，其代码如下：

```
//加载单条数据
function loadSingleData(id) {
    $.ajax({
        url : '/ccms/school/getSingleData',
        dataType : 'json',
        data : {
            id : id
        },
        type : 'post',
        success : function(data) {
            $("#pname_edit").val(data.moduleData.parentName);
            $("#id_up").val(data.moduleData.areaNumber);
```

```javascript
            $("#moduleCode_edit").val(data.moduleData.areaNumber);
            $("#moduleName_edit").val(data.moduleData.name);
            $("#parentCode_edit").val(data.moduleData.parentId);
            $("#isLeaf_edit>input[name='isLeaf']:checked").prop(
                    'checked', false);
            $("#isLeaf" + data.moduleData.isLeaf + "_edit").prop(
                    'checked', true);
            $("#sortNumber_edit").val(data.moduleData.sortNum);
            $("#areatype_up").val(data.moduleData.type);
            if (data.moduleData.areaNumber == '') {
                $("#id_up").attr('disabled', true);
                $("#moduleName_edit").attr('disabled', true);
                $("#moduleCode_edit").attr('disabled', true);
                $("#modulePath_edit").attr('disabled', true);
                $("#parentCode_edit").attr('disabled', true);
                $("#areatype_up").attr('disabled', true);
                $("#isLeaf_edit input[name='isLeaf']").attr(
                        'disabled', true);
                $("#sortNumber_edit").attr('disabled', true);
            } else {
                $("#id_up").attr('disabled', true);
                $("#moduleName_edit").attr('disabled', false);
                $("#areatype_up").attr('disabled', false);
                $("#moduleCode_edit").attr('disabled', false);
                $("#modulePath_edit").attr('disabled', false);
                $("#parentCode_edit").attr('disabled', false);
                $("#isLeaf_edit input[name='isLeaf']").attr(
                        'disabled', false);
                $("#sortNumber_edit").attr('disabled', false);
            }
        },
        error : function() {
            swal('系统提示', '抱歉,数据加载失败。', 'info');
        }
    });
}
```

在函数 loadSingleData 中,通过 jQuery AJAX 方式向后台服务器发送请求,请求的地址为 /ccms/school/getSingleData,这个 url 请求将映射到控制器类 SchoolController 中的 getSingleData 方法,并将表示区域或院校编号的参数 id 传递过去;success 参数的类型为 Function,表示请求成功后的回调函数,服务器返回的数据将传给该函数的参数 data,再将接收的数据绑定到修改院校表单中的父结点、编号、院校名称、类别、是否叶子结点和排序号六个元素中。

在 SchoolController 类中,getSingleData 方法的代码如下:

```java
// 根据 id 获取单条数据
@RequestMapping(value = "/getSingleData", method = { RequestMethod.GET,
        RequestMethod.POST })
@ResponseBody
```

```java
public Map<String, Object> getSingleData(String id, HttpServletRequest req,
HttpServletResponse resp)
        throws UnsupportedEncodingException, IOException {
    SysArea info = schoolService.getSingleDate(id);
    Map<String, Object> result = new HashMap<String, Object>();
    result.put("moduleData", info);
    return result;
}
```

在 getSingleData 方法中，调用业务接口 SchoolService 中的 getSingleDate 方法，根据选中结点编号获取单个结点数据。

在接口 SchoolService 中，getSingleDate 方法的代码声明如下：

```java
// 根据 id 获取院校
public SysArea getSingleDate(String areaNumber);
```

在接口 SchoolService 的实现类 SchoolServiceImpl 中，实现 getSingleDate 方法，代码如下：

```java
@Override
public SysArea getSingleDate(String areaNumber) {
    return schoolDao.getSingleData(areaNumber);
}
```

在上述 getSingleDate 方法中，直接调用了接口 SchoolDao 中的方法 getSingleData。在 SchoolDao 接口中，getSingleData 方法的代码如下：

```java
// 获取单条数据
@Select("select sac.*,(case when sac.parentId is null then '院校配置' else sap.name end) as parentName "
        + "from sys_area as sac left join sys_area as sap on sac.parentId = sap.areaNumber "
        + "where sac.areaNumber = #{areaNumber} order by sortNum asc")
public SysArea getSingleData(@Param("areaNumber") String areaNumber);
```

至此，在修改院校表单中，数据绑定就完成了。接下来，修改院校名称、类别、是否叶子结点和排序号这四项信息，再单击修改按钮，会执行 JavaScript 函数 upModule，代码如下：

```javascript
function upModule() {
    var id = $("#id_up").val();
    var moduleName = $("#moduleName_edit").val();
    var moduleCode = $("#moduleCode_edit").val();
    var modulePath = $("#modulePath_edit").val();
    var parentCode = $("#parentCode_edit").val();
    var isLeaf = $("#isLeaf_edit input[name='isLeaf']:checked")
            .val();
    var sortNumber = $("#sortNumber_edit").val();
    var areatype = $("#areatype_up").val();
    if (areatype == 0) {
        swal({
            title : "系统提示",
```

```javascript
                text : "请选择类别",
                type : "warning"
            });
        }
        $.ajax({
            url : '/ccms/school/upSchool',
            type : 'post',
            async : 'true',
            cache : false,
            data : {
                areaNumber : id,
                areaNumber : moduleCode,
                name : moduleName,
                parentId : parentCode,
                isLeaf : isLeaf,
                sortNum : sortNumber,
                type : areatype
            },
            dataType : 'json',
            success : function(data) {
                if (data.exist == 'existName') {
                    swal({
                        title : "系统提示",
                        text : "已存在该名称",
                        type : "warning"
                    }, function() {
                        $("#moduleName_edit").val('');
                    });
                } else if (data.success) {
                    swal({
                        title : "系统提示",
                        text : "修改成功",
                        type : "success"
                    }, function() {
                        loadSchoolTree();
                    });
                } else {
                    swal({
                        title : "系统提示",
                        text : "修改失败",
                        type : "warning"
                    });
                }
            },
            error : function(aa, ee, rr) {
                swal({
                    title : "系统提示",
                    text : "请求服务器失败,请稍候再试",
                    type : "warning"
                });
            }
        });
```

}

在函数 upModule 中，首先获取修改后的结点信息，通过 jQuery AJAX 方式向后台服务器发送请求，请求的地址为/ccms/school/upSchool，这个 url 请求将映射到控制器类 SchoolController 中的 upModule 方法；data 参数用于指定发送到服务器的数据；success 参数的类型为 Function，表示请求成功后的回调函数，服务器返回的数据将传给该函数的参数 data，然后根据接收的数据给出提示信息。

在控制器类 ModuleController 中，upModule 方法的代码如下：

```java
@RequestMapping(value = "/upSchool", method = { RequestMethod.GET,
RequestMethod.POST })
public String upModule(SysArea area, HttpServletRequest req,
HttpServletResponse resp)
        throws UnsupportedEncodingException, IOException {
    JSONObject jsonObject = new JSONObject();
    int result = schoolService.upSchool(area);
    if (result > 0) {
        jsonObject.put("success", "true");
    }
    resp.getOutputStream().write(jsonObject.toString().getBytes("utf-8")
);
    resp.getOutputStream().flush();
    return null;
}
```

在 upModule 方法中，调用业务接口 SchoolService 中的 upSchool 方法执行更新操作。在接口 SchoolService 中，upSchool 方法的声明如下：

```java
// 修改
public int upSchool(SysArea area);
```

在实现类 SchoolServiceImpl 中，实现 upSchool 方法，代码如下：

```java
@Override
public int upSchool(SysArea area) {
    return schoolDao.updateSchool(area);
}
```

在上述 upSchool 方法中，直接调用接口 SchoolDao 中的 updateSchool 方法，代码如下：

```java
// 更新
@Update("update sys_area set name = #{name},type= #{type},
sortNum=#{sortNum},isLeaf=#{isLeaf} where areaNumber=#{areaNumber}")
public int updateSchool(SysArea area);
```

至此，修改院校功能就讲解完了。接下来，介绍添加院校功能的实现过程。

3．添加院校

在院校管理页 schoolManager.jsp 中，单击一个非叶子结点(如扬州)，再单击新增按钮，打开添加院校界面，如图 20-14 所示。

图 20-14　添加院校界面

单击新增按钮时，将执行 JavaScript 函数 showAdd，其代码如下：

```javascript
//显示添加窗口
function showAdd() {
    if (isSelected("schooltree")) {
        // 判断院校树上选择的结点是否为叶子结点
        var node = jsonToObj(getSelected("schooltree").attributes);
        if (node.isLeaf == '1') {
            swal('系统提示', '该结点为叶子结点!', 'warning');
            return;
        }
        // 获取选择的结点中的区域/院校名称
        $("#pname_add").val(getSelected("schooltree").name);
        // 获取选择的结点中的区域/院校编号
        $("#parentCode").val(getSelected("schooltree").id);
        // 隐藏修改院校表单
        $('#upwin').hide();
        // 显示添加院校表单
        $('#addwin').show();
    } else {
        swal("系统提示", "请选择父结点", "warning");
    }
}
```

在函数 showAdd 中，首先判断院校树上选择的结点是否为叶子结点。如果不是叶子结点，就获取选择的结点中的区域/院校名称和编号，然后隐藏修改院校表单，并显示添加院校表单。

在添加院校表单中，填写编号、名称、类别、是否叶子结点和排序号这 5 项信息，单击添加按钮，将执行 JavaScript 函数 addModule，其代码如下：

```javascript
//添加
function addModule() {
    var id = $("#id").val();
    var name = $("#moduleName").val();
    var areatype = $("#areatype").val();
    if (areatype == 0) {
```

```javascript
            swal({
                title : "系统提示",
                text : "请选择类别",
                type : "warning"
            });
        }
        var parentId = $("#parentCode").val();
        var isLeaf = $("#isLeaf_add input[name='isLeaf']:checked")
                .val();
        var sortNumber = $("#sortNumber").val();
        if (sortNumber == '') {
            sortNumber = 0;
        }
        $.ajax({
            url : '/ccms/school/addSchool',
            type : 'post',
            async : 'true',
            cache : false,
            data : {
                areaNumber : id,
                name : name,
                parentId : parentId,
                isLeaf : isLeaf,
                sortNum : sortNumber,
                type : areatype
            },
            dataType : 'json',
            success : function(data) {
                if (data.exist == 'existName') {
                    swal({
                        title : "系统提示",
                        text : "已存在该名称",
                        type : "warning"
                    }, function() {
                        $("#moduleName").val('');
                    });
                } else if (data.success) {
                    swal({
                        title : "系统提示",
                        text : "添加成功",
                        type : "success"
                    }, function() {
                        $("#id").val('');
                        $("#moduleName").val('');
                        $("#modulePath").val('');
                        $("#parentCode").val('');
                        $("#isLeaf_add input[name='isLeaf']:checked")
                                .attr('checked', false);
                        $("#sortNumber").val('');
                        $("#addwin").modal('hide');
                        loadSchoolTree();
```

```
            });
        } else {
            swal({
                title : "系统提示",
                text : "添加失败",
                type : "warning"
            });
        }
    },
    error : function(aa, ee, rr) {
        swal({
            title : "系统提示",
            text : "请求服务器失败,请稍候再试",
            type : "warning"
        }, function() {
        });
    }
});
}
```

在函数 addModule 中,首先获取表单中填写的院校信息,然后通过 jQuery AJAX 方式向后台服务器发送请求,请求的地址为/ccms/school/addSchool,这个 url 请求将映射到控制器类 SchoolController 中的 addModule 方法;data 参数用于指定发送到服务器的数据;success 参数的类型为 Function,表示请求成功后的回调函数,服务器返回的数据将传给该函数的 data 参数,然后根据接收的数据给出提示信息;error 用于指定请求失败时调用的函数。

在控制器类 SchoolController 中,addModule 方法的代码如下:

```
// 添加院校
@RequestMapping(value = "/addSchool", method = { RequestMethod.GET,
RequestMethod.POST })
public String addModule(SysArea area, HttpServletRequest req,
HttpServletResponse resp)
        throws UnsupportedEncodingException, IOException {
    JSONObject jsonObject = new JSONObject();
    // 添加操作
    int result = schoolService.addSchool(area);
    if (result > 0) {
        jsonObject.put("success", "true");
    }
    resp.getOutputStream().write(jsonObject.toString().
getBytes("utf-8"));
    resp.getOutputStream().flush();
    return null;
}
```

在 addModule 方法中,调用业务接口 SchoolService 中的 addSchool 方法。在 SchoolService 接口中,addSchool 方法的声明如下:

```
// 添加
public int addSchool(SysArea area);
```

在实现类 SchoolServiceImpl 中，实现 addSchool 方法，代码如下：

```java
@Override
public int addSchool(SysArea area) {
    return schoolDao.insertSchool(area);
}
```

在 SchoolServiceImpl 类的 addSchool 方法中，直接调用了接口 SchoolDao 中的 insertSchool 方法，向数据表 sys_area 中插入一条记录。insertSchool 方法的代码如下：

```java
// 添加
@Insert("INSERT INTO sys_area (areaNumber, name, type, parentId, isLeaf, sortNum, delState) VALUES "+ "(#{areaNumber}, #{name}, #{type}, #{parentId}, #{isLeaf}, #{sortNum}, 1)")
public int insertSchool(SysArea area);
```

以在扬州市这个结点下添加院校扬州工职院为例，测试一下添加院校功能。在添加院校表单中，首先填写编号 yzgzy，名称为扬州工职院，类别为院校，选中叶子结点，排序号为 4，然后单击添加按钮。此时，可以看到院校树上，扬州这个结点下多了一个子结点扬州工职院，如图 20-15 所示。

至此，添加院校功能就讲解完了。接下来，介绍删除院校功能的实现过程。

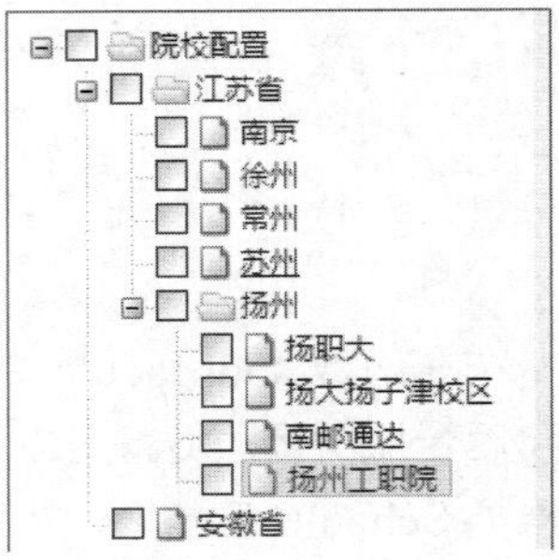

图 20-15　院校树

4．删除院校

在院校管理页 schoolManager.jsp 中，单击一个叶子结点(如扬州工职院)，再单击删除按钮，将执行 JavaScript 函数 delModule，其代码如下：

```javascript
function delModule() {
    var ids = '';
    var selects = getCheckeds('schooltree', true);
    if (selects == null) {
        swal('系统提示', '请选择要删除的模板！', 'warning');
        return;
    }
    ids = "'" + selects[0].id + "'";
    for (var i = 1; i < selects.length; i++) {
        if (selects[i].id == 0) {
            swal('系统提示', '请勿删除根结点！', 'warning');
            return;
        }
        ids += ",'" + selects[i].id + "'";
    }
    swal(
    {
        title : "您确定要删除吗",
        text : "删除后将无法恢复，请谨慎操作！",
        type : "warning",
        showCancelButton : true,
```

```
                confirmButtonColor : "#DD6B55",
                confirmButtonText : "删除",
                cancelButtonText : "取消",
                closeOnConfirm : false,
                closeOnCancel : false
            },
            function(isConfirm) {
                if (isConfirm) {
                    $.ajax({
                        url : '/ccms/school/delSchool',
                        type : 'post',
                        async : 'true',
                        cache : false,
                        contentType : "application/x-www-form-urlencoded; charset=utf-8", data : {
                            ids : ids
                        },
                        dataType : 'json',
                        success : function(data) {
                            if (data.success) {
                                swal("系统提示!", "删除成功!",
                                        "success");
                                loadSchoolTree();
                            } else {
                                swal({
                                    title : "系统提示",
                                    text : "删除失败",
                                    type : "warning"
                                });
                            }
                        },
                        error : function(aa, ee, rr) {
                            swal({
                                title : "系统提示",
                                text : "请求服务器失败,请稍候再试",
                                type : "warning"
                            });
                        }
                    });
                } else {
                    swal("已取消", "您取消了删除操作!", "error")
                }
            });
    }
```

在函数 delModule 中，首先获取院校树上选择的叶子结点，并以逗号分隔保存在变量 ids 中；然后通过 jQuery AJAX 方式向后台服务器发送请求，请求的地址为 /ccms/school/delSchool，这个 url 请求将映射到控制器类 SchoolController 中的 deleteBatch 方法；data 参数用于指定发送到服务器的数据，这里为 ids。

在控制器类 SchoolController 中，deleteBatch 方法的代码如下：

```java
// 删除
@RequestMapping(value = "/delSchool", method = { RequestMethod.GET,
RequestMethod.POST })
public String deleteBatch(HttpServletRequest req, HttpServletResponse resp)
        throws UnsupportedEncodingException, IOException {
    JSONObject jsonObject = new JSONObject();
    req.setCharacterEncoding("utf-8");
    String ids = req.getParameter("ids");
    // 删除数据
    boolean result = schoolService.delSchool(ids) > 0;
    jsonObject.put("success", result);
    resp.getOutputStream().write(jsonObject.toString().
getBytes("utf-8"));
    resp.getOutputStream().flush();
    return null;
}
```

在 deleteBatch 方法中，调用了业务接口 SchoolService 中的 delSchool 方法。在 SchoolService 接口中，delSchool 方法的声明如下：

```java
// 删除
public int delSchool(String ids);
```

在实现类 SchoolServiceImpl 中，实现 delSchool 方法，如下所示：

```java
@Override
public int delSchool(String ids) {
    return schoolDao.deleteSchool(ids);
}
```

在 SchoolServiceImpl 类的 delSchool 方法中，直接调用接口 SchoolDao 中的 deleteSchool 方法，从数据表 sys_area 中删除指定编号的记录。

在 SchoolDao 接口中，deleteSchool 方法的代码如下：

```java
// 删除
@Delete("delete from sys_area where areaNumber in (${ids})")
public int deleteSchool(@Param("ids") String ids);
```

以删除院校扬州工职院为例，测试一下删除院校功能。在院校树上，首先选中扬州工职院这个结点的复选框，然后单击删除按钮。此时，院校树上就看不到这个结点了。

至此，实现了院校的显示、修改、添加和删除功能。

20.7 院校管理员功能

在后台登录页中，以用户名 yzd、密码 123456 登录系统，以院校管理员的身份进入系统首页面 index.jsp，如图 20-5 所示。

院校管理员功能包括单位管理和用户权限管理，单位管理包括单位设置，用户权限管理包括用户管理和角色管理。

用户是依附在单位之上的，用户可以绑定由平台管理员提供的一个或多个角色。

20.7.1 单位管理

在校园通讯管理系统中，学校的单位是学生、教师、宿舍楼、院系、后勤部门等某一个体或群体。

在图 20-5 所示的院校管理员界面中，单击单位管理栏目下的单位设置菜单，打开单位设置页 unitInfo.jsp，该页面位于 src/main/webapp/views/unit 目录下，如图 20-16 所示。

图 20-16　单位设置页

在单位设置页 unitInfo.jsp 中，需要实现单位列表的显示、添加、修改、删除，以及根据单位名称和单位类别搜索等功能。

1．显示单位列表

为了显示单位列表，定义了一个 id 为 table 的<table>标签，如下所示：

```
<!-- 单位列表显示 -->
<div class="ibox-content">
    <table id="table" data-click-to-select="true" >    </table>
</div>
```

然后通过 JavaScript 对这个<table>标签进行初始化，代码如下：

```
var $table = $('#table');
  $table.bootstrapTable({
   url: "/ccms/unitinfo/queryUnitInfoList_table",
   dataType: "json",
   method: 'post',
    contentType: "application/x-www-form-urlencoded",
   pagination: true,
   pageSize: 30,
   pageNumber:1,
   singleSelect: false,
           checkboxHeader: true,
           clickToSelect: true,
```

```javascript
            queryParamsType : "undefined",
            queryParams: function queryParams(params) {  // 设置查询参数
            var param = {
                page: params.pageNumber,
                rows: params.pageSize
            };
                return param;
            },
            cache: false,
            sidePagination: "server", // 服务端处理分页
                columns: [{
                    checkbox: true
                },
                {
                    title: '单位名称',
                    field: 'name',
                    valign: 'middle'
                },
                {
                    title: '单位类别',
                    field: 'unitType',
                    //align: 'center',
                    valign: 'middle'
                },
                {
                    title: '编码',
                    field: 'outside_code',
                    //align: 'center',
                    valign: 'middle'
                }
            ]
        });
    });
```

通过调用 bootstrapTable 方法,将<table>标签创建为 Bootstrap 的 table 控件,并通过设置一系列属性来初始化这个 table 控件。url 属性用于设置 table 控件的数据源,这里为 /ccms/unitinfo/queryUnitInfoList_table,这个 url 请求将映射到后台控制器类 UnitInfoController 中的 queryUnitInfoList_Table 方法;method 属性用于设置请求方式,这里为 post 方式;contentType 属性用于设置发送到服务器的数据编码类型,这里为 application/x-www-form-urlencoded;dataType 属性用于设置服务器返回的数据类型,这里为 json;pagination 属性用于设置是否显示分页,这里为 true,表示允许分页;pageSize 属性用于设置每页的记录行数,这里为 30;pageNumber 属性用于设置首页页码,这里为 1;singleSelect 属性用于设置是否单选,这里为 false,表示可以多选;queryParamsType 属性用于设置参数格式,这里为 undefined;queryParams 属性用于设置查询参数,Bootstrap 的 table 控件会将 page 和 rows 两个参数传递到控制器类 UnitInfoController 中的 queryUnitInfoList_Table 方法;cache 属性用于设置是否启用 AJAX 数据缓存,这里为 false,表示禁用;sidePagination 属性用于设置在哪里进行分页,可选值为 client 或者 server,这里为 server,表示由服务端处理分页;columns

属性用于设置列。

在控制器类 UnitInfoController 中，queryUnitInfoList_Table 方法的代码如下：

```java
package com.ccms.controller;
import com.ccms.service.UnitInfoService;
import com.google.gson.JsonObject;
……
@Controller
@RequestMapping("/unitinfo")
public class UnitInfoController {
    @Autowired
    UnitInfoService unitInfoService;
    // 单位列表分页显示
    @RequestMapping("/queryUnitInfoList_table")
    @ResponseBody
    public Map<String, Object> queryUnitInfoList_Table(Integer page,
Integer rows, String nameOrCode, String unitTypeId,
            String unitGradeId, HttpServletRequest req, HttpServletResponse resp) throws IOException {
        // 从session中获取当前登录用户的区域编号
        String areaId = req.getSession().getAttribute("AREANUMBER").toString();
        // 初始化分页类对象
        Pager pager = new Pager();
        pager.setCurPage(page);
        pager.setPerPageRows(rows);
        // 创建对象 params，用于封装查询条件
        Map<String, Object> params = new HashMap<String, Object>();
        params.put("nameOrCode", nameOrCode);
        params.put("unitTypeId", unitTypeId);
        params.put("unitGradeId", unitGradeId);
        params.put("areaId", areaId);
        // 获取满足条件的单位总数
        int totalCount = unitInfoService.count(params);
        // 根据查询条件获取当前页的单位列表
        List<ProUnitinfo> list = unitInfoService.queryUnitInfoList_Table(params, pager);
        // 创建对象 result，用于保存返回结果
        Map<String, Object> result = new HashMap<String, Object>(2);
        result.put("total", totalCount);
        result.put("rows", list);
        return result;
    }
    ……
}
```

在 queryUnitInfoList_Table 方法中，首先从 session 中获取当前登录用户的区域编号；然后初始化一个分页类对象 pager，给其设置 curPage 和 perPageRows 两个属性值；接着创建 Map<String, Object>类型的对象 params，用于封装查询条件；接下来依次调用业务接口 UnitInfoService 的 count 方法获取满足条件的单位总数，调用 queryUnitInfoList_Table 方法

获取满足条件的单位列表；再创建 Map<String, Object>类型的对象 result，保存查询结果数据；最后将返回结果转为 JSON 格式，以字符串的形式发送到前端页面 unitInfo.jsp，为 Bootstrap 的 table 控件提供数据源。

在业务接口 UnitInfoService 中，声明两个方法，如下所示：

```java
// 获取满足条件的单位总数
Integer count(Map<String, Object> params);
// 分页显示单位列表
public List<ProUnitinfo> queryUnitInfoList_Table(Map<String, Object> params,
Pager pager);
```

在接口 UnitInfoService 的实现类 UnitInfoServiceImpl 中，实现 count 方法，如下所示：

```java
@Override
public Integer count(Map<String, Object> params) {
    return unitInfoDao.count(params).size();
}
```

在上述 count 方法中，直接调用了接口 UnitInfoDao 中的 count 方法。在 UnitInfoDao 接口中，count 方法的代码如下：

```java
// 根据条件查询总数
@SelectProvider(type = UnitInfoDynaSqlProvider.class, method = "count")
List<ProUnitinfo> count(Map<String, Object> params);
```

在上述 count 方法中，通过调用 UnitInfoDynaSqlProvider 类中的 count 方法返回需要执行的 SELECT 语句。在 UnitInfoDynaSqlProvider 类中，count 方法的代码如下：

```java
// 动态查询总记录数
public String count(Map<String, Object> params) {
    String sql = "select u.*,pt.name as unitType,pg.name as unitGrade from pro_unitinfo u "+ "left join pro_paraminfo pt on u.unitTypeId = pt.id "
            + "left join pro_paraminfo pg on u.unitGradeId = pg.id " + "where u.delState='1' and 1=1";
    if (params.get("nameOrCode") != null) {
        String nameOrCode = (String) params.get("nameOrCode");
        sql += " and (u.name like '%" + nameOrCode + "%' or u.outside_code like '%" + nameOrCode + "%' )";
    }
    if (params.get("unitTypeId") != null) {
        String unitTypeId = (String) params.get("unitTypeId");
        sql += " and unitTypeId='" + unitTypeId + "'";
    }
    if (params.get("unitGradeId") != null) {
        String unitGradeId = (String) params.get("unitGradeId");
        sql += " and unitGradeId='" + unitGradeId + "'";
    }
    if (params.get("areaId") != null) {
        String areaId = (String) params.get("areaId");
        sql += " and u.areaId='" + areaId + "'";
    }
    sql += " order by outside_code desc";
```

```
        return sql;
    }
```

在实现类 UnitInfoServiceImpl 中,实现 queryUnitInfoList_Table 方法,代码如下:

```
@Override
public List<ProUnitinfo> queryUnitInfoList_Table(Map<String, Object> params,
Pager pager) {
    int recordCount = unitInfoDao.count(params).size();
    pager.setRowCount(recordCount);
    if (recordCount > 0) {
        params.put("pager", pager);
    }
    return unitInfoDao.selectByPage(params);
}
```

queryUnitInfoList_Table 方法有两个参数,一个是 Pager 类型的参数 pager,用于封装从前端页面传递来的页码和每页记录数,另一个是 Map<String, Object>类型的参数 params,用于封装从前端页面传递来的查询条件。在 queryUnitInfoList_Table 方法中,首先调用接口 UnitInfoDao 中的 count 方法,获取满足条件的单位总数,将其赋值给 pager 对象的 rowCount 属性,并将 pager 对象放入 params 中;然后调用接口 UnitInfoDao 中的 selectByPage 方法,分页获取满足条件的单位列表。

在 UnitInfoDao 接口中,selectByPage 方法的代码如下:

```
// 分页显示单位列表
@SelectProvider(type = UnitInfoDynaSqlProvider.class, method =
"selectWithParam")
List<ProUnitinfo> selectByPage(Map<String, Object> params);
```

在上述 selectByPage 方法中,通过调用 UnitInfoDynaSqlProvider 类中的 selectWithParam 方法返回需要执行的 SELECT 语句。在 UnitInfoDynaSqlProvider 类中,selectWithParam 方法的代码如下:

```
public String selectWithParam(Map<String, Object> params) {
    String sql = "select u.*,pt.name as unitType,pg.name as unitGrade from
pro_unitinfo u " + "left join pro_paraminfo pt on u.unitTypeId = pt.id "
            + "left join pro_paraminfo pg on u.unitGradeId = pg.id " + "where
u.delState='1' and 1=1";
    if (params.get("nameOrCode") != null) {
        String nameOrCode = (String) params.get("nameOrCode");
        sql += " and (u.name like '%" + nameOrCode + "%' or u.outside_code
like '%" + nameOrCode + "%' )";
    }
    if (params.get("unitTypeId") != null
&& !params.get("unitTypeId").equals("")) {
        String unitTypeId = (String) params.get("unitTypeId");
        sql += " and unitTypeId='" + unitTypeId + "'";
    }
    if (params.get("unitGradeId") != null) {
        String unitGradeId = (String) params.get("unitGradeId");
```

```
            sql += " and unitGradeId='" + unitGradeId + "'";
        }
        if (params.get("areaId") != null) {
            String areaId = (String) params.get("areaId");
            sql += " and u.areaId='" + areaId + "'";
        }
        sql += " order by outside_code desc";
        if (params.get("pager") != null) {
            sql += " limit #{pager.firstLimitParam} , #{pager.perPageRows} ";
        }
        return sql;
    }
```

至此,单位列表显示功能就讲解完了。接下来,介绍单位添加功能的实现过程。

2. 添加单位

在单位设置页面中,单击新增按钮,打开单位添加对话框,如图 20-17 所示。

图 20-17 单位添加对话框

在单位添加对话框中,单位类别是一个下拉列表控件,其布局如下:

```
<select class="form-control unitTypeId" name="unitTypeId" id="unitTypeId"></select>
```

在$(function() { })代码段中,调用了一个 JavaScript 函数 loadUnitType,为单位类别下拉列表提供数据源并生成下拉列表选项,函数调用如下:

```
loadUnitType("/ccms/unitinfo/getUnitTypeList","unitTypeId");
```

loadUnitType 函数定义在 CommonValue.js 中,该文件位于 src/main/webapp/commons/jslib 目录中,其代码如下:

```
function loadUnitType(url, idStr) {
    $.ajax({
        url : url,
        dataType : 'json',
        data : {},
        type : 'post',
        success : function(data) {
```

```
            var options = "<option value=''>请选择</option>";
            $.each(data.unitTypeList, function(key, val) {
                options += '<option value=' + val.id + '>' + val.name
                    + '</option>';
            });
            $('#' + idStr).empty();
            $('#' + idStr).append(options);
        },
        error : function() {
        }
    });
}
```

在函数 loadUnitType 中，通过 jQuery AJAX 方式向后台服务器发送请求，请求的地址为/ccms/unitinfo/getUnitTypeList，这个 url 请求将映射到控制器类 UnitInfoController 中的 getUnitTypeList 方法，方法的返回结果作为单位类别下拉列表的数据源；success 参数的类型为 Function，表示请求成功后的回调函数，服务器返回的数据将传给该函数的参数 data，然后根据接收的数据生成单位类别下拉列表的选项。

在控制器类 UnitInfoController 中，getUnitTypeList 方法的代码如下：

```
// 获取单位类别列表
@RequestMapping(value = "/getUnitTypeList", method = { RequestMethod.GET,
RequestMethod.POST })
@ResponseBody
public Map<String, Object> getUnitTypeList(HttpServletRequest req,
HttpServletResponse resp) throws IOException {
    // 从session中获取当前区域管理员的区域编号
    String areaId =
req.getSession().getAttribute("AREANUMBER").toString();
    // 根据区域编号获取所有单位类别
    List<ProParamInfo> list = unitInfoService.getUnitTypeList(areaId);
    int count = 0;
    if (list != null && list.size() > 0) {
        count = list.size();
    }
    // 创建对象result，保存返回结果
    Map<String, Object> result = new HashMap<String, Object>(2);
    result.put("count", count);
    result.put("unitTypeList", list);
    return result;
}
```

在上述 getUnitTypeList 方法中，首先从 session 中获取当前区域管理员的区域编号；然后调用业务接口 UnitInfoService 中的 getUnitTypeList 方法，根据区域编号获取所有单位类别，再创建 Map<String, Object>类型的对象 result，用于保存单位类别总数和单位类别列表；最后将结果 result 转为 JSON 格式，发送到前端页面。

在接口 UnitInfoService 中，声明 getUnitTypeList 方法，代码如下：

```
// 根据当前院校管理员的区域编号获取所有单位类别
```

```
List<ProParamInfo> getUnitTypeList(String areaId);
```

在接口 UnitInfoService 的实现类 UnitInfoServiceImpl 中，实现 getUnitTypeList 方法，代码如下：

```
@Override
public List<ProParamInfo> getUnitTypeList(String areaId) {
    return unitInfoDao.getUnitTypeList(areaId);
}
```

在实现类 UnitInfoServiceImpl 的 getUnitTypeList 方法中，直接调用了 UnitInfoDao 接口中的 getUnitTypeList 方法。

在 UnitInfoDao 接口中，getUnitTypeList 方法的代码如下：

```
// 根据当前院校管理员的区域编号获取所有单位类别
@Select("select * from pro_paraminfo where type='02' and areaId = #{areaId} order by sortNum asc")
List<ProParamInfo> getUnitTypeList(@Param("areaId") String areaId);
```

在 UnitInfoDao 接口的 getUnitTypeList 方法中，根据当前院校管理员的区域编号从参数信息表 pro_paraminfo 中查询 type=02，即单位类型记录。

对于区域管理员 yzd 来说，单位类别下拉列表中的选项如图 20-18 所示。

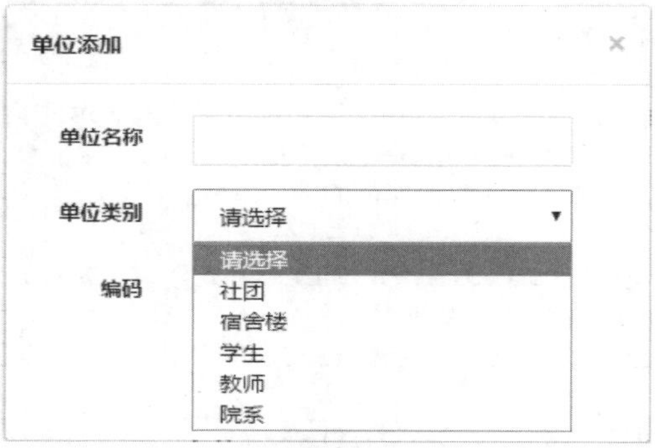

图 20-18　单位类别下拉列表

在单位添加对话框中，填写单位名称和编码，并选择单位类别，再单击添加按钮，将执行 JavaScript 中的 addUnit 函数，其代码如下：

```
function addUnit(){
    var name = $("#name").val();
    var outside = $("#outside_code").val();
    var unitTypeId = $("#unitTypeId").val();
    var unitGradeId = $("#unitGradeId").val();
    $.ajax({
        url:'/ccms/unitinfo/addUnit',
        type:'post',
        async:'true',
        cache:false,
data:{name:name,outside_code:outside, unitTypeId:unitTypeId, unitGradeId:unitGradeId},
```

```javascript
                dataType:'json',
                success: function(data){
                  if(data.exist=='exitOutCode'){
                      swal({
                          title: "系统提示",
                            text: "已存在该编码",
                            type: "warning"
                        },function(){
                            $("#outside_code").val('');
                        });
                    }else if(data.success){
                        swal({
                            title: "系统提示",
                            text: "添加成功",
                            type: "success"
                        });
                    $("#name").val('');
                    $("#outside_code").val('');
                        $("#unitTypeId").val('');
                        $("#unitGradeId").val('');
                    $("#addwin").modal('hide');
                    $('#table').bootstrapTable("refresh");
                    }else{
                      swal({
                          title: "系统提示",
                            text: "添加失败",
                            type: "warning"
                        });
                    }
              },
              error: function (aa, ee, rr) {
                  swal({
                    title: "系统提示",
                      text: "请求服务器失败,请稍候再试",
                      type: "warning"
                    },function(){
                    });
                }
            });
        }
```

在函数 addUnit 中，首先获取表单中填写的内容，然后通过 jQuery AJAX 方式向后台服务器发送请求，请求的地址为/ccms/unitinfo/addUnit，这个 url 请求将映射到控制器类 UnitInfoController 中的 addUnit 方法；data 用于指定发送到服务器的数据；success 用于指定请求成功后调用的回调函数，服务器方法的返回结果将保存在该函数的参数 data 中。

在控制器类 UnitInfoController 中，addUnit 方法的代码如下：

```java
// 添加
@RequestMapping(value = "/addUnit", method = { RequestMethod.GET,
RequestMethod.POST })
```

```java
public String addUnit(ProUnitinfo info, HttpServletRequest req,
HttpServletResponse res) {
    // 从 session 中获取当前区域管理员的区域编号
    String areaId =
req.getSession().getAttribute("AREANUMBER").toString();
    // 将区域编号封装在 ProUnitinfo 类型对象 info 中
    info.setAreaId(areaId);
    JsonObject jObject = new JsonObject();
    // 查询是否存在外部编码
    String exist = unitInfoService.isExistOutCode(info);
    if (exist.equals("no")) {
        // 添加单位
        int result = unitInfoService.addUnit(info);
        jObject.addProperty("success", result);
    } else {
        jObject.addProperty("exist", exist);
    }
    try {
        ServletOutputStream jos = res.getOutputStream();
        jos.write(jObject.toString().getBytes("utf-8"));
        jos.flush();
        jos.close();
    } catch (IOException e) {
        e.printStackTrace();
    }
    return null;
}
```

addUnit 方法有一个 ProUnitinfo 类型的参数 info，用于封装从前端页面传递来的单位信息。在 addUnit 方法中，首先调用业务接口 UnitInfoService 中的 isExistOutCode 方法，根据当前区域管理员的区域编号查询是否存在外部编码。如果不存在该外部编码，则调用接口 UnitInfoService 中的 addUnit 方法，添加这个单位信息。

在接口 UnitInfoService 中，isExistOutCode 方法的声明如下：

```java
// 查询是否存在外部编码
public String isExistOutCode(ProUnitinfo info);
```

在实现类 UnitInfoServiceImpl 中，实现 isExistOutCode 方法，代码如下：

```java
@Override
public String isExistOutCode(ProUnitinfo info) {
    String result;
    if (info.getId() == null) {
        info.setId("");
    }
    String outside_code = info.getOutside_code();
    String id = info.getId();
    String areaId = info.getAreaId();
    int existOutCode = unitInfoDao.unitValidOutCode(outside_code, id, areaId);
    if (existOutCode > 0) {
```

```
            result = "exitOutCode";
        } else {
            result = "no";
        }
        return result;
    }
```

在 isExistOutCode 方法中，调用了接口 UnitInfoDao 中的 unitValidOutCode 方法，以验证单位外部编码是否已经存在。如果已经存在，则返回字符串 exitOutCode，否则返回字符串 no。在 UnitInfoDao 接口中，unitValidOutCode 方法的代码如下：

```
// 验证单位外部编码
@Select("select count(*) from pro_unitinfo where outside_code = 
#{outside_code} and id <> #{id} and areaId = #{areaId} and delState=1")
public int unitValidOutCode(@Param("outside_code") String outside_code,
@Param("id") String id,@Param("areaId") String areaId);
```

在接口 UnitInfoService 中，addUnit 方法的声明如下：

```
// 添加
public int addUnit(ProUnitinfo info);
```

在实现类 UnitInfoServiceImpl 中，实现 addUnit 方法，代码如下：

```
@Override
public int addUnit(ProUnitinfo info) {
    String uid = UUIDGenerator.getUUID();
    info.setId(uid);
    return unitInfoDao.addUnit(info);
}
```

在上述 addUnit 方法中，首先调用 UUIDGenerator 类的 getUUID 方法，生成一个新的单位编号，并保存在 info 对象中；然后调用接口 UnitInfoDao 中的 addUnit 方法，将对象 info 中的数据插入到数据表 pro_unitinfo 中。在接口 UnitInfoDao 中，addUnit 方法的代码如下：

```
// 添加
@Insert("insert into pro_unitinfo(id,name,unitTypeId,unitGradeId,
outside_code,delState,areaId) "+ "values(#{id},#{name},
#{unitTypeId},#{unitGradeId},#{outside_code},1,#{areaId})")
public int addUnit(ProUnitinfo info);
```

以添加一个类型为学生的单位为例，在添加单位对话框中，输入单位名称张山山，编号 180101001，选中单位类别为学生，单击添加按钮。此时，数据表 pro_unitinfo 中会增加一条学生类型的单位记录。

至此，单位添加功能就讲解完了。接下来，介绍根据单位名称和单位类别搜索单位功能的实现过程。

3. 根据单位名称和单位类别搜索

在单位设置页 unitInfo.jsp 中，输入单位名称或选择单位类别，单击搜索按钮，将执行 JavaScript 函数 search，其代码如下：

```
//查询
function search(){
  var _name = $('#_search').val();
    $('#table').bootstrapTable('refresh', {
       query: {
           nameOrCode:$("#unitInfoName").val(),
           unitTypeId:$("#unitTypeSel option:selected").val(),
           unitGradeId:$("#unitGradeSel option:selected").val()
       }
    });
}
```

在函数 search 中，通过调用 Bootstrap 的 table 控件的 refresh 方法，重新发送 /ccms/unitinfo/queryUnitInfoList_table 这个 url 请求，即再次执行控制器类 UnitInfoController 中的 queryUnitInfoList_Table 方法，并将查询参数传递过去。在 queryUnitInfoList_Table 方法中，根据条件重新获取单位列表，并显示在 Bootstrap 的 table 控件中。

单位修改和删除功能的实现思路与院校管理中的院校修改和删除类似，由于篇幅所限，此处不再描述，读者可以参照源代码加以理解。

20.7.2 角色管理

在如图 20-5 所示的院校管理员界面中，单击用户权限管理栏目下的角色管理菜单，打开角色管理页 roleManager.jsp，该页面位于 src/main/webapp/views/role 目录下，如图 20-19 所示。

图 20-19 角色管理页

在角色管理页 roleManager.jsp 中，需要实现角色列表的显示、新增、编辑、删除和权限设置功能。本小节主要就角色列表显示和权限设置功能作详细讲解，由于篇幅所限，其他功能不作介绍，读者可以参照源代码加以理解。

1．角色列表显示

为了显示角色列表，定义了一个 id 为 table 的<table>标签，如下所示：

```
<!-- 角色列表显示 -->
<div class="ibox-content">
    <table id="table">
    </table>
</div>
```

然后通过 JavaScript 对这个<table>标签进行初始化，代码如下：

```
$(function() {
    vform('upform', upRole);
    vform('addform', addRole);
    var $table = $('#table');
    $table.bootstrapTable({
        url : "/ccms/role/getAllRole",
        method : 'post',
        contentType : "application/x-www-form-urlencoded",
        dataType : "json",
        pagination : true, //分页
        pageSize : 3,
        pageNumber : 1,
        singleSelect : false,
        queryParamsType : "undefined",
        queryParams : function queryParams(params) { //设置查询参数
            var param = {
                page : params.pageNumber,
                rows : params.pageSize,
            };
            return param;
        },
        cache : false,
        sidePagination : "server", //服务端处理分页
        columns : [
        {
            title : '角色',
            field : 'roleName',
            width : '50%',
            valign : 'middle'
        },
        {
            title : '操作',
            field : 'id',
            formatter : function(value, row, index) {
                var e = '<a href="#" class="btn btn-gmtx-define1" onclick="edit(\'' + row.roleCode + '\',\''
                    + row.roleName + '\')">编辑</a> ';
                var d = '<a href="#" class="btn btn-gmtx-define1" onclick="delRole(\'' + row.roleCode + '\')">删除</a> ';
```

```
                var f = '<a href="#" class="btn btn-gmtx-define1" 
onclick="accessShow(\'' + row.roleCode + '\')">权限设置</a> ';
                return e + d + f;
            }
        } ]
    });
});
```

通过调用 bootstrapTable 方法,将<table>标签创建为 Bootstrap 的 table 控件,并通过设置一系列属性来初始化这个 table 控件。url 属性用于设置 table 控件的数据源,这里为 /ccms/role/getAllRole,这个 url 请求将映射到后台控制器类 RoleController 中的 getAllRole 方法;queryParams 属性用于设置查询参数,Bootstrap 的 table 控件会将 page 和 rows 两个参数传递到控制器类 RoleController 中的 getAllRole 方法;columns 属性用于设置列。

在控制器类 RoleController 中,getAllRole 方法的代码如下:

```java
// 查询所有 role
@RequestMapping(value = "/getAllRole", method = { RequestMethod.GET, 
RequestMethod.POST })
@ResponseBody
public Map<String, Object> getAllRole(Integer page, Integer rows, 
HttpServletRequest req, HttpServletResponse rep) {
    // 从 session 中获取当前区域管理员的区域编号
    String areaId = 
req.getSession().getAttribute("AREANUMBER").toString();
    // 初始化分页类对象
    Pager pager = new Pager();
    pager.setCurPage(page);
    pager.setPerPageRows(rows);
    // 创建对象 params,用于封装查询条件
    Map<String, Object> params = new HashMap<String, Object>();
    params.put("areaId", areaId);
    // 获取满足条件的角色总数
    int totalCount = roleService.count(params);
    // 根据查询条件获取当前页的角色列表
    List<SysRole> roles = roleService.getAllRole(areaId, pager);
    // 创建对象 result,用于保存返回结果
    Map<String, Object> result = new HashMap<String, Object>(2);
    result.put("total", totalCount);
    result.put("rows", roles);
    return result;
}
```

在 getAllRole 方法中,首先从 session 中获取当前区域管理员的区域编号 areaId,然后初始化一个分页类对象 pager,给其设置 curPage 和 perPageRows 两个属性值;接着创建 Map<String, Object>类型的对象 params,将区域管理员的区域编号放入 params 中;依次调用业务接口 RoleService 中的 count 方法获取满足条件的角色总数,调用 getAllRole 方法获取满足条件的当前页的角色列表;再创建 Map<String, Object>类型的对象 result,保存查询结果数据;最后将返回结果转为 JSON 格式,以字符串的形式发送到前端页面

roleManager.jsp，为 Bootstrap 的 table 控件提供数据源。

在业务接口 RoleService 中，声明两个方法，代码如下：

```java
// 获取满足条件的角色总数
Integer count(Map<String, Object> params);
// 分页查询所有角色
public List<SysRole> getAllRole(String areaId, Pager pager);
```

在接口 RoleService 的实现类 RoleServiceImpl 中，实现 count 方法，代码如下：

```java
package com.ccms.service.impl;
......
import com.ccms.service.RoleService;
import com.ccms.tools.Tree;
import com.ccms.tools.UUIDGenerator;
@Service("roleService")
@Transactional(propagation = Propagation.REQUIRED, isolation = Isolation.DEFAULT)
public class RoleServiceImpl implements RoleService {
    @Autowired
    RoleDao roleDAO;
    @Override
    public Integer count(Map<String, Object> params) {
        return roleDAO.count(params);
    }
    ......
}
```

在实现类 RoleServiceImpl 的 count 方法中，直接调用了接口 RoleDAO 中的 count 方法。在 RoleDAO 接口中，count 方法的代码如下：

```java
// 根据条件查询角色总数
@SelectProvider(type = RoleDynaSqlProvider.class, method = "count")
int count(Map<String, Object> params);
```

在上述 count 方法中，通过 RoleDynaSqlProvider 类的 count 方法获取要执行的 SELECT 语句。在 RoleDynaSqlProvider 类中，count 方法的代码如下：

```java
// 动态查询角色总记录数
public String count(Map<String, Object> params) {
    String areaId = (String) params.get("areaId");
    String sql = "select count(*) from sys_role where areaId='" + areaId + "'";
    return sql;
}
```

在接口 RoleService 的实现类中，实现 getAllRole 方法，代码如下：

```java
@Override
public List<SysRole> getAllRole(String areaId, Pager pager) {
    Map<String, Object> params = new HashMap<String, Object>();
    params.put("areaId", areaId);
    int recordCount = roleDAO.count(params);
    pager.setRowCount(recordCount);
```

```java
    if (recordCount > 0) {
        params.put("pager", pager);
    }
    return roleDAO.selectByPage(params);
}
```

getAllRole 方法有两个参数，一个是 Pager 类型的参数 pager，用于封装从前端页面传递来的页码和每页记录数；另一个是 String 类型的参数 areaId，用于封装当前区域管理员的区域编号。在 getAllRole 方法中，首先调用接口 RoleDAO 中的 count 方法，获取满足条件的角色总数，将其赋值给 pager 对象的 rowCount 属性，并将 pager 对象放入 params 中；然后调用接口 RoleDAO 中的 selectByPage 方法，分页获取满足条件的角色列表。

在 RoleDAO 接口中，selectByPage 方法的代码如下：

```java
// 分页获取所有角色
@SelectProvider(type = RoleDynaSqlProvider.class, method =
"selectWithParam")
List<SysRole> selectByPage(Map<String, Object> params);
```

在 selectByPage 方法中，通过调用 RoleDynaSqlProvider 类中的 selectWithParam 方法返回需要执行的 SELECT 语句。在 RoleDynaSqlProvider 类中，selectWithParam 方法的代码如下：

```java
// 分页动态查询角色
public String selectWithParam(Map<String, Object> params) {
    String areaId = (String) params.get("areaId");
    String sql = "select roleCode,roleName from sys_role where areaId='" +
areaId + "'";
    if (params.get("pager") != null) {
        sql += " limit #{pager.firstLimitParam} , #{pager.perPageRows} ";
    }
    return sql;
}
```

至此，角色列表显示功能讲解完了。接下来，介绍权限设置功能的实现过程。

2．权限设置

权限设置包括权限绑定和权限修改两个部分。

1) 权限绑定

在角色管理页中的用于显示角色列表的 Bootstrap 的 table 控件上，单击某个角色(如学生)的权限设置按钮，打开绑定菜单对话框，如图 20-20 所示。

在绑定菜单对话框中，以树的形式显示该角色所拥有的功能，其布局如下：

```html
<!-- 绑定菜单窗口 -->
<div class="modal fade" id="accesswin">
    <div class="modal-dialog" style="width: 400px; height: 250px">
        <div class="modal-content">
            <div class="modal-header">
                <button type="button" class="close" data-dismiss="modal"
aria-hidden="true">&times;</button>
```

```html
                    <h4 class="modal-title">绑定菜单</h4>
                </div>
                <div class="modal-body">
                    <div class="row">
                        <div class="center-gmtx">
                            <ul id="treeModule" class="ztree" style="height: 280px"></ul>
                        </div>
                    </div>
                </div>
                <div class="modal-footer">
                    <button type="button" class="btn btn-gmtx-define1 center-block" onclick="saveAccess()">保存</button>
                </div>
            </div>
        </div>
    </div>
```

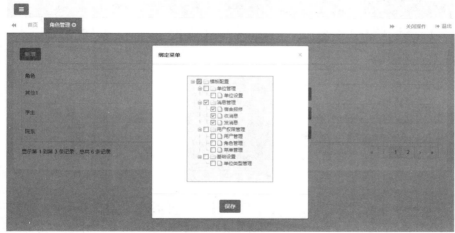

图 20-20 绑定菜单对话框

在这个布局中，定义了一个 zTree 插件，只要为其提供数据源，就可以显示功能权限树了。单击权限设置按钮时，将执行 JavaScript 函数 accessShow，并将当前行的角色编号作为参数，函数 accessShow 的代码如下：

```javascript
/**菜单显示*/
var roleCode_access;
function accessShow(roleCode) {
    roleCode_access = roleCode;
// 显示绑定菜单窗口
    $('#accesswin').modal('show');
    //树
    $.ajax({
        url : '/ccms/role/getModule',
        type : 'post',
        async : 'true',
```

```
            cache : false,
            data : {
                roleCode : roleCode_access
            },
            dataType : 'json',
            success : function(data) {
                $.fn.zTree.init($("#treeModule"), setting, data);
                $("#py").bind("change", setCheck);
                $("#sy").bind("change", setCheck);
                $("#pn").bind("change", setCheck);
                $("#sn").bind("change", setCheck);
            }
        });
    }
    var setting = {
        check : {
            enable : true,
            autoCheckTrigger : true,
            chkStyle : "checkbox",
            chkboxType : {
                "Y" : "ps",
                "N" : "ps"
            }
        },
        data : {
            simpleData : {
                enable : true
            }
        }
    };
```

在函数 accessShow 中，首先显示绑定菜单窗口，然后通过 jQuery AJAX 方式向后台服务器发送请求，请求的地址为/ccms/role/getModule，这个 url 请求将映射到控制器类 RoleController 中的 getModuleListCheckedByRoleId 方法,方法的返回结果作为 zTree 插件(功能权限树)的数据源，getModuleListCheckedByRoleId 方法的代码如下：

```
// 获取所有 module
@RequestMapping(value = "/getModule", method = { RequestMethod.GET,
RequestMethod.POST })
@ResponseBody
public List<Tree> getModuleListCheckedByRoleId(String roleCode,
HttpServletRequest req, HttpServletResponse res) {
    // 从 session 中获取当前区域管理员的区域编号
    String aredId =
req.getSession().getAttribute("AREANUMBER").toString();
    List<Tree> tree = roleService.getModuleListCheckedByRoleId(roleCode,
aredId);
    return tree;
}
```

在 getModuleListCheckedByRoleId 方法中，首先从 session 中获取当前区域管理员的区

域编号，然后调用业务接口 RoleService 中的 getModuleListCheckedByRoleId 方法。在 RoleService 接口中，getModuleListCheckedByRoleId 方法的声明如下：

```java
// 根据 roleid 获取
public List<Tree> getModuleListCheckedByRoleId(String roleCode, String areaId);
```

在实现类 RoleServiceImpl 中，实现 getModuleListCheckedByRoleId 方法，代码如下：

```java
@Override
public List<Tree> getModuleListCheckedByRoleId(String roleCode, String areaId) {
    // 获取所有功能菜单
    List<Tree> allModuleList = roleDAO.getAllModule();
    // 根据角色编号和当前区域管理员区域编号获取功能菜单
    List<Tree> roleModuleList = roleDAO.getModuleByRoleId(roleCode, areaId);
    if (allModuleList != null & roleModuleList != null) {
        for (Tree tree : allModuleList) {
            if (roleModuleList.contains(tree)) {
                tree.setChecked(true);
            }
            tree.setOpen(true);
        }
    }
    return allModuleList;
}
```

在 getModuleListCheckedByRoleId 方法中，首先获取所有功能菜单，然后根据角色编号和当前区域管理员区域编号获取功能菜单，再根据该角色所拥有的权限，将功能权限树上的相应结点选中。这样，学生这个角色的权限绑定就完成了。

2）权限修改

在如图 20-20 所示的绑定菜单对话框中，可以重新设置各个结点的选中状态，单击保存按钮后，将执行 JavaScript 函数 saveAccess，其代码如下：

```javascript
//保存菜单
function saveAccess() {
    var mids = '';
    var treeObj = $.fn.zTree.getZTreeObj("treeModule");
    var nodes = treeObj.getCheckedNodes(true);
    for (i = 0; i < nodes.length; i++) {
        mids = mids + nodes[i].id + ',';
    }
    $.ajax({
        url : '/ccms/role/bindModule',
        type : 'post',
        async : 'true',
        cache : false,
        data : {
            roleCode : roleCode_access,
            mids : mids
```

```javascript
        },
        dataType : 'json',
        success : function(data) {
            if (data.success) {
                swal("系统提示!", "绑定成功。", "success");
            } else {
                swal({
                    title : "系统提示",
                    text : "绑定失败",
                    type : "warning"
                });
            }
        },
        error : function(aa, ee, rr) {
            swal({
                title : "系统提示",
                text : "请求服务器失败,请稍候再试",
                type : "warning"
            });
        }
    });
}
```

在函数 saveAccess 中，首先遍历权限树上选中的结点，将这些结点所代表的功能菜单编号以逗号分隔，保存在变量 mids 中。然后通过 jQuery AJAX 方式向后台服务器发送请求，请求的地址为/ccms/role/bindModule，这个 url 请求将映射到控制器类 RoleController 中的 bindModuleByRoleId 方法；data 用于指定发送到服务器的数据；success 用于指定请求成功后调用的回调函数，控制器类方法的返回结果将传递给该函数的参数 data。

在控制器类 RoleController 中，bindModuleByRoleId 方法的代码如下：

```java
// 绑定 module
@RequestMapping(value = "/bindModule", method = { RequestMethod.GET,
RequestMethod.POST })
public String bindModuleByRoleId(String roleCode, String mids,
HttpServletRequest req, HttpServletResponse res) {
// 从 session 中获取当前区域管理员的区域编号
    String aredId =
req.getSession().getAttribute("AREANUMBER").toString();
    int result = roleService.bindModuleByRoleId(roleCode, mids, aredId);
    jObject = new JsonObject();
    jObject.addProperty("success", result);
    try {
        ServletOutputStream jos = res.getOutputStream();
        jos.write(jObject.toString().getBytes("utf-8"));
        jos.flush();
        jos.close();
    } catch (IOException e) {
        e.printStackTrace();
    }
    return null;
}
```

在 bindModuleByRoleId 方法中，首先从 session 中获取当前区域管理员的区域编号，然后调用业务接口 RoleService 中的 bindModuleByRoleId 方法。在 RoleService 接口中，bindModuleByRoleId 方法的声明如下：

```java
public int bindModuleByRoleId(String roleCode,String mids,String areaId);
```

在实现类 RoleServiceImpl 中，实现 bindModuleByRoleId 方法，代码如下：

```java
@Override
public int bindModuleByRoleId(String roleCode, String mids, String areaId)
{
    if (mids.length() > 0) {
        mids = mids.substring(0, mids.length() - 1);
    }
    String[] midStrings = mids.split(",");
    int resulta = 0, resultd = 0;
    // 不存在时返回 0,不能作为判断失败的标准
    resultd = roleDAO.deleteModuleByRoleCode(roleCode, areaId);
// 当给某个角色删除全部权限时
    if ((resultd > 0) && (mids == null || mids.isEmpty())) {
        return 1;
    }
    for (int i = 0; i < midStrings.length; i++) {
// 当更改为没有权限时 mids 为空,
// resulta 也不能作为判断操作成功与失败的唯一标准
        resulta = roleDAO.insertModuleBuRoleCode(UUIDGenerator.getUUID(), roleCode, midStrings[i], areaId);
    }
    return resulta;
}
```

bindModuleByRoleId 方法有三个参数，roleCode 表示角色编号；mids 封装了从前端页面传递来的以逗号分隔的功能菜单编号；areaId 表示当前区域管理员的区域编号。在 bindModuleByRoleId 方法中，首先调用接口 RoleDAO 中的 deleteModuleByRoleCode 方法，删除该角色编号下的所有功能菜单；然后调用接口 RoleDAO 中的 insertModuleBuRoleCode 方法，根据角色编号重新添加功能菜单。

在 RoleDAO 接口中，deleteModuleByRoleCode 方法的代码如下：

```java
// 删除所有 roleid 下的 module
@Delete("delete from sys_role_module where roleCode = #{roleCode} and areaId=#{areaId}")
public int deleteModuleByRoleCode(@Param("roleCode") String roleCode, @Param("areaId") String areaId);
```

在 RoleDAO 接口中，insertModuleBuRoleCode 方法的代码如下：

```java
// 根据 roleid 添加 module
@Insert("insert into sys_role_module(rmId,roleCode,moduleCode,areaId) values(#{rmId},#{roleCode},#{moduleCode},#{areaId})")
```

```java
public int insertModuleBuRoleCode(@Param("rmId") String rmId,
@Param("roleCode") String roleCode, @Param("moduleCode") String
moduleCode, @Param("areaId") String areaId);
```

至此，权限设置功能就讲解完了。

20.7.3 用户管理

在图 20-5 所示的院校管理员界面中，单击用户权限管理栏目下的用户管理菜单，打开用户管理页 userManager.jsp，该页面位于 src/main/webapp/views/user 目录下，如图 20-21 所示。

图 20-21 用户管理页 userManager.jsp

在用户管理页 userManager.jsp 中，需要实现用户列表的显示、新增、删除、编辑、重置密码、角色分配，以及按用户类型、单位、用户名搜索功能。本小节主要就用户新增和角色分配功能作详细讲解，由于篇幅所限，其他功能不作介绍，读者可以参照源代码加以理解。

1．用户新增

在用户管理页中，单击新增按钮，打开用户添加对话框，如图 20-22 所示。

图 20-22 用户添加对话框

用户添加对话框布局如下：

```html
<div class="modal fade" id="addwin">
    <div class="modal-dialog" style="width: 400px">
        <div class="modal-content">
            <div class="modal-header">
                <button type="button" class="close" data-dismiss="modal" aria-hidden="true">
                    &times;
                </button>
                <h4 class="modal-title" id="addwinlable">
                    用户添加
                </h4>
            </div>
            <div class="modal-body">
                <div class="row">
                    <form method="post" class="form-horizontal" id="addform">
                        <div class="form-group">
                            <label class="col-sm-3 control-label">
                                用户名：
                            </label>
                            <div class="col-sm-8 controls">
                                <input type="text" value="" class="form-control" name="name" id="username_add" tabindex="1" />
                            </div>
                        </div>
                        <div class="form-group">
                            <label class="col-sm-3 control-label">
                                密码：
                            </label>
                            <div class="col-sm-8 controls">
                                <input type="password" value="" class="form-control" name="psw" id="userpsw_add" tabindex="2" />
                            </div>
                        </div>
                        <div class="form-group">
                            <label class="col-sm-3 control-label">
                                用户类型：
                            </label>
                            <div class="col-sm-8 controls">
                                <select onchange="changeSheBao('')" data-placeholder="请选择用户类型" class="form-control" tabindex="4" name="userType" id="userType">
                                    <option value="" hassubinfo="true">
                                        请选择
                                    </option>
                                    <option value="1" hassubinfo="true">
                                        院方
                                    </option>
                                    <option value="2" hassubinfo="true">
```

```html
                                单位
                            </option>
                        </select>
                    </div>
                </div>
                <div class="form-group">
                    <label class="col-sm-3 control-label">
                        单位:
                    </label>
                    <div class="col-sm-8 controls" style="position:relative">
                        <div class="input-group search_div">
                            <input type="text" class="form-control" id="unitId" name="unitId">
                            <ul class="dropdown-menu dropdown-menu-right search_ul" style="position:absolute;left:0px" role="menu"></ul>
                        </div>
                    </div>
                </div>
                <div class="form-group">
                    <div class="controls">
                        <button type="submit" class="btn btn-gmtx-define1 center-block">
                            添加
                        </button>
                    </div>
                </div>
            </form>
        </div>
    </div>
</div>
```

在用户添加对话框中，用户类型有院方和单位两个选项，当用户类型为院方时禁用单位文本框。单位文本框使用了 Bootstrap 的 Search Suggest 插件，相关代码如下：

```javascript
// 添加单位提示框
$("#unitId").bsSuggest('init', {
    clearable : true,
    url : "/ccms/unitinfo/getUnitList",
    showBtn : false,
    idField : "id",
    keyField : "name",
    effectiveFields : [ "name", "outside_code" ],
    effectiveFieldsAlias : {
        "name" : "机构名",
        "outside_code" : "编码"
    },
}).on("onSetSelectValue", function(e, keyword) {
    unitId_add = keyword.id;
```

```
}).on("onUnsetSelectValue", function(e) {
    unitId_add = '';
});
```

在上述代码中，通过 jQuery AJAX 方式向后台服务器发送请求，请求的地址为 /ccms/unitinfo/getUnitList，这个 url 请求将映射到控制器类 UnitinfoController 中的 getUnitList 方法获取单位列表。在返回的结果中，name(单位名称)作为单位输入框显示内容，id(单位编号)作为单位输入框选择的值。

在控制器类 UnitinfoController 中，getUnitList 方法的代码如下：

```java
// 获取单位
@RequestMapping("/getUnitList")
@ResponseBody
public Map<String, Object> getUnitList(HttpServletRequest req,
HttpServletResponse res) {
    // 从session中获取当前登录用户(院校管理员)的区域编号
    String areaId =
req.getSession().getAttribute("AREANUMBER").toString();
    List<ProUnitinfo> list = unitInfoService.getUnitList(areaId);
    int count = 0;
    if (list != null && list.size() > 0) {
        count = list.size();
    }
    Map<String, Object> result = new HashMap<String, Object>();
    result.put("count", count);
    result.put("value", list);
    return result;
}
```

在 getUnitList 方法中，首先从 session 中获取当前登录用户(院校管理员)的区域编号；然后调用业务接口 UnitInfoService 中的 getUnitList 方法，根据该区域编号获取单位列表；接着创建 Map<String, Object>类型的对象 result，以 count 为键保存获取的单位总数，以 value 为键保存单位列表，再转换成 JSON 格式的字符串发送到前端页面。

在接口 UnitInfoService 中，声明 getUnitList 方法，代码如下：

```java
// 获取单位列表
List<ProUnitinfo> getUnitList(String areaId);
```

在接口 UnitInfoService 的实现类 UnitInfoServiceImpl 中，实现 getUnitList 方法，代码如下：

```java
@Override
public List<ProUnitinfo> getUnitList(String areaId) {
    return unitInfoDao.getUnitList(areaId);
}
```

在实现类 UnitInfoServiceImpl 的 getUnitList 方法中，直接调用了接口 UnitInfoDao 中的 getUnitList 方法。

在接口 UnitInfoDao 中，getUnitList 方法的代码如下：

```java
// 获取单位信息
@Select("select * from pro_unitinfo where delState = '1' and areaId = #{areaId} order by outside_code desc")
List<ProUnitinfo> getUnitList(String areaId);
```

填写用户名、密码，选择用户类型为单位，再选择一个单位，然后单击添加按钮，将执行 JavaScript 函数 addUser，其代码如下：

```javascript
/**------添加用户------*/
function addUser() {
    var name = $("#username_add").val();
    var psw = $("#userpsw_add").val();
    var unitId = unitId_add;
    var userType = $("#userType").val();
    console.info(unitId);
    if (userType != 1) {
        if (isNull(unitId)) {
            swal({
                title : "系统提示",
                text : "请选择机构",
                type : "warning"
            });
            return;
        }
    }
    $.ajax({
        url : '/ccms/user/addUser',
        type : 'post',
        async : 'true',
        cache : false,
        data : {
            name : name,
            psw : psw,
            unitId : unitId,
            userType : userType
        },
        dataType : 'json',
        success : function(data) {
            if (data.isExist) {
                swal({
                    title : "系统提示",
                    text : "已存在该用户名",
                    type : "warning"
                }, function() {
                    $("#username_add").val('');
                });
            } else if (data.isExistUnit) {
                swal({
                    title : "系统提示",
                    text : "该机构已绑定用户名",
                    type : "warning"
```

```
            }, function() {
                $("#unitId").val('');
            });
        } else if (data.success) {
            swal({
                title : "系统提示",
                text : "添加成功",
                type : "success"
            }, function() {
                $("#username_add").val('');
                $("#userpsw_add").val('');
                $("#unitId").val('');
                $("#userType").val('');
                $("#addwin").modal('hide');
                $('#table').bootstrapTable("refresh");
            });
        } else {
            swal({
                title : "系统提示",
                text : "添加失败",
                type : "warning"
            }, function() {
                $("#name").val('');
                $("#psw").val('');
                $("#unitId").val('');
                $("#userType").val('');
                $("#addwin").modal('hide');
            });
        }
    },
    error : function(aa, ee, rr) {
        swal({
            title : "系统提示",
            text : "请求服务器失败,请稍候再试",
            type : "warning"
        }, function() {
            $("#username_add").val('');
            $("#userpsw_add").val('');
            $("#unitId").val('');
            $("#userType").val('');
            $("#addwin").modal('hide');
        });
    }
});
}
```

在函数 addUser 中，首先获取填写的用户信息，然后通过 jQuery AJAX 方式向后台服务器发送请求，请求的地址为 /ccms/user/addUser，这个 url 请求将映射到控制器类 UserController 中的 addUser 方法；data 参数用于指定发送到服务器的数据；success 用于指定请求成功后调用的回调函数，控制器类方法的返回结果将传递给该函数的参数 data，然后根据接收的数据给出提示信息。

在控制器类 UserController 中，addUser 方法的代码如下：

```java
// 添加用户
@RequestMapping(value = "/addUser", method = { RequestMethod.GET,
RequestMethod.POST })
public String addUser(SysUser user, HttpServletRequest req,
HttpServletResponse res) {
    // 从 session 中获取当前区域管理员的区域编号
    String areaId =
req.getSession().getAttribute("AREANUMBER").toString();
    // 给对象 user 设置所属区域编号属性
    user.setAreaId(areaId);
    jObject = new JsonObject();
    // 从 session 中获取当前区域管理员用户编号
    String operateId = req.getSession().getAttribute("USERID").toString();
    // 给对象 user 设置操作人属性
    user.setOperatorId(operateId);
    // 给对象 user 设置密码属性
    user.setPsw(MD5Util.MD5(user.getPsw()));
    // 判断该用户名是否存在
    String existName = userService.isExistName(user.getName(), "", areaId);
    // 判断该单位是否存在
    String existUnit = userService.isExistUnit(user.getUnitId(), "",
areaId);
    if (existUnit.equals("exitUnit")) {
        jObject.addProperty("isExistUnit", true);
    } else if (existName.equals("exit")) {
        jObject.addProperty("isExist", true);
    } else if (existName.equals("no")) {
        // 调用业务方法添加用户
        boolean result = userService.addUser(user);
        jObject.addProperty("success", result);
    }
    try {
        ServletOutputStream jos = res.getOutputStream();
        jos.write(jObject.toString().getBytes("utf-8"));
    } catch (IOException e) {
        e.printStackTrace();
    }
    return null;
}
```

addUser 方法有一个 SysUser 类型的参数 user，用于封装从前端页面传递来的新增用户信息。在 addUser 方法中，首先从 session 中获取当前院校管理员的区域编号和用户编号，给对象 user 设置所属区域编号和密码属性；然后依次调用业务接口 UserService 中的 isExistName 方法判断用户名是否已经存在，调用 isExistUnit 方法检验该单位是否已经被绑定；最后调用业务接口 UserService 中的 addUser 方法添加用户。

在接口 UserService 中，isExistName、isExistUnit 和 addUser 方法的声明如下：

```java
// 查询是否存在该用户名
public String isExistName(String name, String userCode, String areaId);
// 检测是否存在该单位
public String isExistUnit(String unitId, String userCode, String areaId);
// 添加用户
public boolean addUser(SysUser user);
```

在实现类 UserServiceImpl 中，实现 isExistName 方法，代码如下：

```java
@Override
public String isExistName(String name, String userCode, String areaId) {
    String result;
    int i = userDao.isExistName(name, userCode, areaId);
    if (i > 0) {
        result = "exit";
    } else {
        result = "no";
    }
    return result;
}
```

在 isExistName 方法中，直接调用了接口 UserDao 中的 isExistName 方法，代码如下：

```java
// 检验是否存在该用户名
@Select("select count(userCode) from sys_user where name = #{name} and userCode <> #{userCode} and areaId=#{areaId} and delState = 1")
public int isExistName(@Param("name") String name, @Param("userCode") String userCode, @Param("areaId") String areaId);
```

在实现类 UserServiceImpl 中，实现 isExistUnit 方法，代码如下：

```java
@Override
public String isExistUnit(String unitId, String userCode, String areaId) {
    String result;
    int i = userDao.isExistUnit(unitId, userCode, areaId);
    if (i > 0) {
        result = "exitUnit";
    } else {
        result = "noUnit";
    }
    return result;
}
```

在 isExistUnit 方法中，直接调用了接口 UserDao 中的 isExistUnit 方法，代码如下：

```java
// 检验该单位是否已经被绑定
@Select("select count(userCode) from sys_user where unitId = #{unitId} and userCode <> #{userCode} and unitId <> '' and areaId =#{areaId} and unitId is not null and delState = 1")
public int isExistUnit(@Param("unitId") String unitId, @Param("userCode") String userCode,    @Param("areaId") String areaId);
```

在实现类 UserServiceImpl 中，实现 addUser 方法，代码如下：

```java
@Override
public boolean addUser(SysUser user) {
    Object[] params = null;
    String code = UUIDGenerator.getUUID();
    user.setUserCode(code);
    user.setDelState(1);
    int result = userDao.insertUser(user);
    return result > 0;
}
```

在 addUser 方法中，直接调用了接口 UserDao 中的 insertUser 方法，代码如下：

```java
// 添加用户
@Insert("insert into sys_user(userCode,name,psw,operatorId,delState,unitId,userType,areaId) " + "values(#{userCode},#{name},#{psw},#{operatorId},#{delState},#{unitId},#{userType},#{areaId})")
public int insertUser(SysUser user);
```

在用户添加对话框中，填写用户名张山山、密码 123456，选择用户类型为单位，选择单位为之前添加的张山山，再单击添加按钮，会在数据表 sys_user 中插入一条记录。此时，在用户管理页中的用户列表上，会出现该用户的记录，如图 20-23 所示。

图 20-23 用户列表

创建新用户张山山后，由于还未为其分配角色，所以用户列表中所属角色一栏显示为空。接下来，将讲解角色分配功能的实现过程。

2．角色分配

在用户管理页中的用户列表上，单击某条用户记录(如张山山)中的角色分配按钮，显示"配置角色"对话框，如图 20-24 所示。

角色分配可以分成原有角色显示和新角色绑定两个阶段来实现。

1) 原有角色显示

在用户管理页中，"配置角色"对话框的布局代码如下：

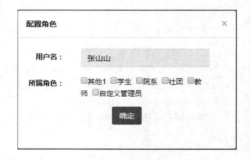

图 20-24 "配置角色"对话框

```html
<!--配置角色对话框-->
<div class="modal fade" id="bindRoleWin">
    <div class="modal-dialog" style="width: 400px">
        <div class="modal-content">
            <div class="modal-header">
                <button type="button" class="close" data-dismiss="modal" aria-hidden="true">&times;</button>
                <h4 class="modal-title">配置角色</h4>
            </div>
            <div class="modal-body">
                <div class="row">
                    <form method="post" class="form-horizontal" id="bindRoleform">
                        <div class="form-group">
                            <label class="col-sm-3 control-label">用户名:</label>
                            <div class="col-sm-8 controls">
                                <input type="text" value="" class="form-control" name="name" id="username_bindRole"
                                disabled tabindex="1" />
                            </div>
                        </div>
                        <div class="form-group">
                            <label class="col-sm-3 control-label">所属角色:</label>
                            <div class="col-sm-8 controls" id='roleAccess'></div>
                        </div>
                        <div class="form-group">
                            <div class="controls">
                                <button class="btn btn-gmtx-define1 center-block" onclick="javascript:bindRole();return false;">确定</button>
                            </div>
                        </div>
                    </form>
                </div>
            </div>
        </div>
    </div>
</div>
```

在配置角色对话框中，原有角色的显示是通过一组复选框的选中与否来实现的，每个复选框代表一个角色。为了实现这一效果，在用户管理页中首先定义一个 id 为 roleAccess 的<div>标签，然后调用 JavaScript 函数 loadRoleChk 来生成多个角色复选框，loadRoleChk 函数的调用代码如下：

```
loadRoleChk("/ccms/role/getRoleList", "roleAccess");
```

函数 loadRoleChk 定义在 src/main/webapp/commons/jslib 目录下的 CommonValue.js 文件中，其代码如下：

```javascript
function loadRoleChk(url, idStr) {
    $.ajax({
        url : url,
        dataType : 'json',
        data : {},
        type : 'post',
        success : function(data) {
            var options = "";
            $.each(
                data.roleList,
                function(key, val) {
                    options += "<input type='checkbox' class='roles' name='roles' id='"              + val.roleCode
                            + "' value='"
                            + val.roleCode
                            + "'>"
                            + val.roleName + "  "
                });
            $('#' + idStr).empty();
            $('#' + idStr).html(options);
        },
        error : function() {
        }
    });
}
```

在 loadRoleChk 函数中，通过 jQuery AJAX 方式向后台服务器发送请求，请求的地址为 /ccms/role/getRoleList，这个 url 请求将映射到控制器类 RoleController 中的 getRoleList 方法；success 用于指定请求成功后调用的回调函数，控制器类方法的返回结果将传递给该函数的参数 data，然后根据接收的数据生成角色复选框。

在控制器类 RoleController 中，getRoleList 方法的代码如下：

```java
// 获取角色列表
@RequestMapping(value = "/getRoleList", method = { RequestMethod.GET,
        RequestMethod.POST })
@ResponseBody
public Map<String, Object> getRoleList(HttpServletRequest req,
        HttpServletResponse resp) throws IOException {
    // 从 session 中获取当前区域管理员的区域编号
    String areaId = req.getSession().getAttribute("AREANUMBER").toString();
    // 根据当前区域管理员的区域编号，获取角色列表
    List<SysRole> list = roleService.getRoleList(areaId);
    int count = 0;
    if (list != null && list.size() > 0) {
        count = list.size();
    }
    Map<String, Object> result = new HashMap<String, Object>(2);
    result.put("count", count);
    result.put("roleList", list);
    return result;
}
```

在 getRoleList 方法中，首先从 session 中获取当前区域管理员的区域编号，然后调用业务接口 RoleService 中的 getRoleList 方法，根据当前区域管理员的区域编号，获取角色列表。

在 RoleService 接口中，getRoleList 方法的声明如下：

```java
// 根据areaId获取角色列表
public List<SysRole> getRoleList(String areaId);
```

在实现类 RoleServiceImpl 中，实现 getRoleList 方法，代码如下：

```java
@Override
public List<SysRole> getRoleList(String areaId) {
    return roleDAO.getRoleList(areaId);
}
```

在实现类 RoleServiceImpl 的 getRoleList 方法中，直接调用了接口 RoleDAO 中的 getRoleList 方法。在 RoleDAO 接口中，getRoleList 方法的代码如下：

```java
@Select("select * from sys_role where areaId=#{areaId}")
public List<SysRole> getRoleList(@Param("areaId") String areaId);
```

角色复选框生成后，可以通过设置复选框的选中状态来显示用户的原有角色，而将哪些角色复选框设置为选中状态，则是在单击角色分配按钮时完成的。单击角色分配按钮，将执行 JavaScript 函数 bindRoleShow，其代码如下：

```javascript
var userCode_bindRole = '';
function bindRoleShow(code, name) {
$("#roleAccess").find("input[type='checkbox']").each(function() {
    $(this).prop('checked', false);
});
    userCode_bindRole = code;
$('#username_bindRole').val(name);
$.ajax({
    url : '/ccms/user/getCheckedRole',
    type : 'post',
    async : false,
    cache : false,
    data : {
        userCode : userCode_bindRole
    },
    dataType : 'json',
    success : function(data) {
        roleIds = data.roleCodes;
        for (var i = 0; i < roleIds.length; i++) {
            if (roleIds[i] != null && roleIds[i] != '') {
                $("#" + roleIds[i]).prop("checked", true);
            }
        }
    },
    error : function(aa, ee, rr) {
    }
});
```

```javascript
$('#bindRoleWin').modal('show');
}
```

bindRoleShow 函数有两个参数，code 表示用户编号；name 表示用户名。在 bindRoleShow 函数中，首先取消所有角色复选框的选中状态，然后通过 jQuery AJAX 方式向后台服务器发送请求，请求的地址为/ccms/user/getCheckedRole，这个 url 请求将映射到控制器类 UserController 中的 getCheckedRole 方法；success 用于指定请求成功后调用的回调函数，控制器类方法的返回结果将传递给该函数的参数 data，然后根据接收的数据来选中相应角色复选框。

在控制器类 UserController 中，getCheckedRole 方法的代码如下：

```java
// 获取已绑定的 rolecodes
@RequestMapping(value = "/getCheckedRole", method = { RequestMethod.GET,
RequestMethod.POST })
@ResponseBody
public Map<String, Object> getCheckedRole(String userCode,
HttpServletRequest req, HttpServletResponse res) {
    Map<String, Object> result = new HashMap<String, Object>();
    List<String> list = userService.getCheckedRole(userCode);
    result.put("roleCodes", list);
    return result;
}
```

在 getCheckedRole 方法中，调用了业务接口 UserService 中的 getCheckedRole 方法。在 UserService 接口中，getCheckedRole 方法的声明如下：

```java
// 根据 Userid 获取用户角色
public List<String> getCheckedRole(String userCode);
```

在实现类 UserServiceImpl 中，实现 getCheckedRole 方法，代码如下：

```java
@Override
public List<String> getCheckedRole(String userCode) {
    return userDao.getCheckedRole(userCode);
}
```

在实现类 UserServiceImpl 的 getCheckedRole 方法中，直接调用了接口 UserDao 中的 getCheckedRole 方法，从 sys_user_role 表中根据用户编号 userCode 获取用户角色编号 roleCode 列表。在 UserDao 接口中，getCheckedRole 方法的代码如下：

```java
// 从 sys_user_role 表中根据 userCode 获取 roleCode 列表
@Select("select roleCode from sys_user_role where userCode=#{userCode}")
public List<String> getCheckedRole(@Param("userCode")
String userCode);
```

至此，作为角色分配功能第一阶段的原有角色显示就实现了。接下来，将实现第二阶段的新角色绑定。

2) 新角色绑定

在配置角色对话框中，选中需要给用户分配的角色复选框，单击确定按钮，将执行 JavaScript 函数 bindRole，为用户绑定新的角色，其代码如下：

```javascript
function bindRole() {
    var roles = "";
    var r = document.getElementsByName("roles");
    for (var i = 0; i < r.length; i++) {
        if (r[i].checked) {
            roles = roles + r[i].value + ",";
        }
    }
    if (roles == "") {
        swal('系统提示', '请至少选择一个角色!', 'warning');
        return false;
    }
    $.ajax({
        url : '/ccms/user/bindRole',
        type : 'post',
        async : 'true',
        cache : false,
        data : {
            userCode : userCode_bindRole,
            roleCodes : roles
        },
        dataType : 'json',
        success : function(data) {
            if (data.success) {
                swal("系统提示!", "绑定成功。", "success");
                $('#table').bootstrapTable("refresh");
            } else {
                swal({
                    title : "系统提示",
                    text : "绑定失败",
                    type : "warning"
                });
            }
            $('#bindRoleWin').modal('hide');
        },
        error : function(aa, ee, rr) {
            swal({
                title : "系统提示",
                text : "请求服务器失败,请稍候再试",
                type : "warning"
            });
        }
    });
}
```

在函数 bindRole 中，首先获取所有选中的角色编号，并以逗号分隔后保存在变量 roles 中，然后通过 jQuery AJAX 方式向后台服务器发送请求，请求的地址为/ccms/user/bindRole，这个 url 请求将映射到控制器类 UserController 中的 bindRole 方法；data 用于指定发送到服务器的数据，userCode 表示用户编号，roleCodes 表示为该用户分配的以逗号分隔的角色编号；success 用于指定请求成功后调用的回调函数，服务器返回的数据将传给该函数的参数 data，然后根据接收的数据给出提示信息。

在控制器类 UserController 中，bindRole 方法的代码如下：

```java
// 绑定新角色
@RequestMapping(value = "/bindRole", method = { RequestMethod.GET,
RequestMethod.POST })
public String bindRole(SysUser user, HttpServletRequest req,
HttpServletResponse res)      throws UnsupportedEncodingException {
    String areaId =
req.getSession().getAttribute(CommonValue.AREANUMBER).toString();
    int result = userService.bindRole(user.getUserCode(),
user.getRoleCodes(), areaId);
    JSONObject jObject = new JSONObject();
    jObject.put("success", result);
    try {
        ServletOutputStream jos = res.getOutputStream();
        jos.write(jObject.toString().getBytes("utf-8"));
        jos.flush();
        jos.close();
    } catch (IOException e) {
        e.printStackTrace();
    }
    return null;
}
```

在上述 bindRole 方法中，调用了业务接口 UserService 中的 bindRole 方法。在 UserService 接口中，bindRole 方法的声明如下：

```java
// 根据Userid绑定role
public int bindRole(String userCode, String roleCodes, String areaId);
```

在实现类 UserServiceImpl 中，实现 bindRole 方法，代码如下：

```java
@Override
public int bindRole(String userCode, String roleCodes, String areaId) {
    if (roleCodes.length() > 0) {
        roleCodes = roleCodes.substring(0, roleCodes.length() - 1);
    }
    String[] roleCode = roleCodes.split(",");
    int resulta = userDao.delRoleByUserId(userCode);
// 当给某个角色删除全部权限时
    if ((resulta > 0) && (roleCode == null || roleCodes.isEmpty())) {
        return 1;
    }
    int result = 0;
    for (int i = 0; i < roleCode.length; i++) {
        result = userDao.bindRoleByUserId(UUIDGenerator.getUUID(),
userCode, roleCode[i], areaId);
    }
    return result;
}
```

在实现类 UserServiceImpl 的 bindRole 方法中，首先调用接口 UserDao 中的 delRoleByUserId 方法删除指定用户编号的角色；然后调用接口 UserDao 中的 bindRoleByUserId 方法给用户绑定新角色。

在接口 UserDao 中，delRoleByUserId 方法的代码如下：

```
// 删除指定用户编号的角色
@Delete("delete from sys_user_role where userCode=#{userCode}")
public int delRoleByUserId(@Param("userCode") String userCode);
```

在接口 UserDao 中，bindRoleByUserId 方法的代码如下：

```
// 给用户绑定新的角色
@Insert("insert into sys_user_role(urId,userCode,roleCode,areaId) values (#{urId},#{userCode},#{roleCode},#{areaId})")
public int bindRoleByUserId(@Param("urId") String urId, @Param("userCode") String userCode, @Param("roleCode") String roleCode, @Param("areaId") String areaId);
```

以用户张山山为例，在"配置角色"对话框中，选中"学生"复选框，单击"确定"按钮。此时，在用户管理页的用户列表中，可以看到，张山山这条记录所属角色一栏显示为学生，如图 20-25 所示。

图 20-25　用户张山山记录

至此，角色分配功能就讲解完了。

20.8　单位用户功能

在后台登录页中，若以用户名张山山、密码 123456 登录系统，则以单位用户的身份进入系统首页面 index.jsp，如图 20-6 所示。单位用户功能包括发送消息、接收消息、投票、查看投票和宿舍报修。

20.8.1　发送消息

在图 20-6 所示的单位用户界面中，单击消息管理栏目下的发消息菜单，打开已发送消息列表页 SendNoticeList.jsp，该页面位于 src/main/webapp/views/notice 目录下，如图 20-26 所示。

在已发送消息列表页中，可以新增消息、查看消息、删除消息。对用户张山山来说，由于还没有发送过任何消息，因此已发送消息列表中没有任何记录。只有新增消息后，才能在列表中显示消息记录，也才可以查看消息和删除消息。

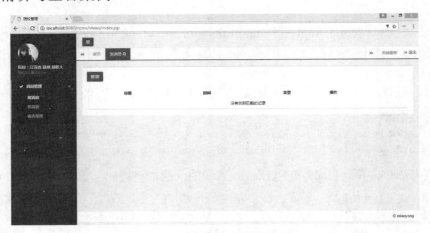

图 20-26　已发送消息列表页

1. 新增消息

在已发送消息列表页中，单击"新增"按钮，将执行 JavaScript 函数 addNotice，其代码如下：

```
function addNotice() {
    window.location.href = "/ccms/views/notice/sendNotice.jsp";
}
```

此时，页面重定向到消息发送页 sendNotice.jsp，该页面位于 src/main/webapp/views/notice 目录下，页面效果如图 20-27 所示。

图 20-27　消息发送页

在消息发送页中，"类型"下拉列表框中有消息和投票两个选项，发送对象右侧有一个"批量添加"按钮，可用来添加消息发送的对象。单击这个按钮，将执行 JavaScript 函数 selectSecond，其代码如下：

```
function selectSecond() {
    layer.open({
        type : 2,
```

```
            title : "批量添加单位",
            shade : 0.8,
            fix : false,
            shadeClose : true,
            area : [ '800px', '90%' ],
            content : '/ccms/views/notice/secondBatch.jsp',
            end : function() {
            }
        });
    }
```

在函数 selectSecond 中使用了 layer,这是一个常用的 Web 弹出框组件。content 用于指定弹出框中显示的内容,这里为/ccms/views/notice/secondBatch.jsp。secondBatch.jsp 页面可用来批量添加单位,效果如图 20-28 所示。

图 20-28 secondBatch.jsp 页面

在 secondBatch.jsp 页面的下方,列出了系统中所有的单位名称。当然,也可以根据单位名称或单位类别搜索所需的单位名称。通过选中复选框,可以指定要将信息发送给哪些单位。这里选中学生 AA 前的复选框,然后单击添加单位按钮。此时,页面返回到消息发送页,如图 20-29 所示。

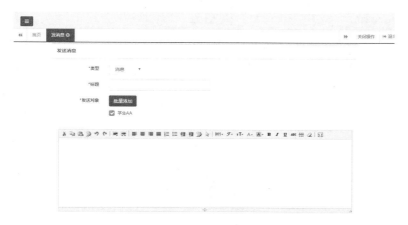

图 20-29 选择发送对象后的消息发送页

选择发送对象后，选择类型为消息，标题为"这是第一条测试通知"，消息内容为"这是第一条测试通知，请学生 AA 查收！"，再单击完成按钮。此时，将执行 JavaScript 函数 addNotice，其代码如下：

```javascript
function addNotice(editor) {
    var type = $("#type").val();
    var title = $("#title").val();
    var mainBody = encodeURIComponent(editor.html());
    if (mainBody == null || mainBody == undefined || mainBody == '') {
        swal({
            title : "请输入内容",
            text : ""
        });
        return false;
    }
    var selectedCount = 0;
    var secondIdsAry = new Array();
    var secondIds = "";
    $('input[name="checkbox_unit_parent"]:checked').each(function() {
        var secondEle = $(this).val().split("|");
        secondIdsAry.push(secondEle[0]);
        selectedCount++;
    });
    if (selectedCount <= 0) {
        swal({
            title : "请选择单位",
            text : ""
        });
        return false;
    }
    secondIds = secondIdsAry.join(",");
    $.ajax({
        url : '/ccms/notice/addNotice',
        type : 'post',
        async : 'true',
        cache : false,
        data : {
            type : type,
            secondIds : secondIds,
            content : mainBody,
            title : title
        },
        dataType : 'json',
        success : function(data) {
            if (data.success) {
                swal({
                    title : "系统提示",
                    text : "添加成功",
                    type : "success"
                }, function() {
```

```
                        forwardListPage();
                    });
                } else {
                    swal({
                        title : "系统提示",
                        text : "添加失败",
                        type : "warning"
                    }, function() {
                    });
                }
            },
            error : function(aa, ee, rr) {
                swal({
                    title : "系统提示",
                    text : "请求服务器失败,请稍候再试",
                    type : "warning"
                }, function() {
                });
            }
        });
    }
```

在函数 addNotice 中,首先获取表单中填写的消息信息,然后通过 jQuery AJAX 方式向后台服务器发送请求,请求的地址为/ccms/notice/addNotice,这个 url 请求将映射到控制器类 NoticeController 中的 addNotice 方法;data 用于指定发送到服务器的数据;success 用于指定请求成功后调用的回调函数,服务器返回的数据将传给该函数的 data 参数,然后根据接收的数据给出提示信息。

在控制器类 NoticeController 中,addNotice 方法的代码如下:

```
package com.ccms.controller;
import com.ccms.tools.CommonTool;
import com.ccms.tools.JsonUtil;
import com.google.gson.JsonObject;
……
@Controller
@RequestMapping(value = "/notice", method = { RequestMethod.GET,
RequestMethod.POST })
public class NoticeController {
    @Autowired
    private NoticeService noticeService;
    JsonUtil<Notice> json = new JsonUtil<Notice>();
    // 添加 notice
    @RequestMapping(value = "/addNotice", method = { RequestMethod.GET,
RequestMethod.POST })
    @ResponseBody
    public Map<String, Object> addNotice(Integer type, String secondIds,
String content, String title, HttpServletRequest req,
            HttpServletResponse resp) throws IOException {
        content = CommonTool.changeImageSrc(req, URLDecoder.decode(content,
"utf-8"));
```

```java
        Notice notice = new Notice();
        notice.setType(type);
        notice.setTitle(title);
        notice.setContent(content);
        String uid = (String) req.getSession().getAttribute("USERID");
        boolean flag = noticeService.addNotice(notice, secondIds, uid);
        Map<String, Object> result = new HashMap<String, Object>();
        if (flag) {
            result.put("success", true);
        } else {
            result.put("success", false);
        }
        return result;
    }
    ......
}
```

在 addNotice 方法中，调用业务接口 NoticeService 中的 addNotice 方法。在 NoticeService 接口中，addNotice 方法的声明如下：

```java
package com.ccms.service;
......
public interface NoticeService {
    // 添加
    public boolean addNotice(Notice notice, String answerids, String uid);
}
```

在 NoticeService 接口的实现类 NoticeServiceImpl 中，实现 addNotice 方法，代码如下：

```java
package com.ccms.service.impl;
......
@Service("noticeService")
@Transactional(propagation = Propagation.REQUIRED, isolation = Isolation.DEFAULT)
public class NoticeServiceImpl implements NoticeService {
    @Autowired
    NoticeDao noticeDao;
    @Override
    public boolean addNotice(Notice notice, String answerids, String uid) {
        // 添加 notice
        String nid = UUIDGenerator.getUUID();
        notice.setId(nid);
        notice.setUserId(uid);
        notice.setOperatetime(CommonTool.getNowDateStr());
        int resultNotice = noticeDao.addNotice(notice);
        // 添加 answer
        String[] ids = answerids.split(",");
        int count = 0;
        for (int i = 0; i < ids.length; i++) {
            count += noticeDao.addAnswer(UUIDGenerator.getUUID(), nid, ids[i]);
```

```
        }
        return (resultNotice > 0) && (count > 0);
    }
    ......
}
```

在实现类 NoticeServiceImpl 的 addNotice 方法中，首先调用接口 NoticeDao 中的 addNotice 方法，向数据表 notice 中插入一条消息；然后调用接口 NoticeDao 中的 addAnswer 方法，向数据表 answer 中插入一条回复信息。

在 NoticeDao 接口中，addNotice 方法的代码如下：

```
// 添加通知
@Insert("insert into notice
values(#{id},#{userId},#{title},#{content},#{operatetime},#{type})")
public int addNotice(Notice notice);
```

在 NoticeDao 接口中，addAnswer 方法的代码如下：

```
//添加通知回复
@Insert("insert into answer(id,nid,uid) values(#{id},#{nid},#{uid})")
    public int addAnswer(@Param("id") String id, @Param("nid") String nid,
@Param("uid") String uid);
```

发送消息时，如果选择类型为投票，标题为"这是一条投票测试"，消息内容为"这是一条投票测试，请学生 AA 投票！"，再单击"完成"按钮，就可以添加一个投票消息。发送消息和投票后，在已发送消息列表页 SendNoticeList.jsp 中，可以看到这两个类型的消息，如图 20-30 所示。

图 20-30　已发送消息列表

2. 已发送消息列表显示

在已发送消息列表页 SendNoticeList.jsp 中，为了显示已发送消息列表，首先定义了一个 id 为 table 的<div>标签，代码如下：

```
<div class="ibox-content">
    <table id="table">  </table>
</div>
```

然后通过 JavaScript 对这个<table>标签进行初始化，代码如下：

```javascript
$(function() {
    var $table = $('#table');
    $table.bootstrapTable({
        url : "/ccms/notice/getAllNotice",
        method : 'post',
        contentType : "application/x-www-form-urlencoded",
        dataType : "json",
        pagination : true, //分页
        pageSize : 50,
        pageNumber : 1,
        singleSelect : false,
        queryParamsType : "undefined",
        queryParams : function queryParams(params) { //设置查询参数
            var param = {
                page : params.pageNumber,
                rows : params.pageSize,
            };
            return param;
        },
        cache : false,
        sidePagination : "server", //服务端处理分页
        columns : [
            {
                title : '标题',
                field : 'title',
                align : 'center',
                valign : 'middle'
            },
            {
                title : '时间',
                field : 'operatetime',
                align : 'center',
                valign : 'middle'
            },
            {
                title : '类型',
                field : 'type',
                align : 'center',
                valign : 'middle',
                formatter : function(value, row, index) {
                    if (value == '1') {
                        return "消息";
                    } else if (value == '2') {
                        return "投票";
                    } else if (value == '3') {
                        return "报修";
                    }
                }
            },
            {
                title : '操作',
```

```
                field : 'id',
                formatter : function(value, row, index) {
                    var e = '<a href="#" class="btn btn-gmtx-define1"
onclick="check(\''       + row.id + '\')">查看</a>';
                    var d = ' <a href="#" class="btn btn-gmtx-define1"
onclick="del(\''          + row.id + '\')">删除</a>';
                    return e + d;
                }
            } ]
    });
});
```

在$(function() { })代码段中，通过调用 bootstrapTable 方法，将<table>标签创建为 Bootstrap 的 table 控件，并通过设置一系列属性来初始化这个 table 控件。url 属性用于设置 table 控件的数据源，这里为/ccms/notice/getAllNotice，这个 url 请求将映射到后台控制器类 NoticeController 中的 getAllNotice 方法；method 属性用于设置请求方式，这里为 post 方式；contentType 属性用于设置发送到服务器的数据编码类型，这里为 application/x-www-form-urlencoded；dataType 属性用于设置服务器返回的数据类型，这里为 json；pagination 属性用于设置是否显示分页，这里为 true，表示允许分页；pageSize 属性用于设置每页的记录行数，这里为 50；pageNumber 属性用于设置首页页码，这里为 1；singleSelect 属性用于设置是否单选，这里为 false，表示可以多选；queryParamsType 属性用于设置参数格式，这里为 undefined；queryParams 属性用于设置查询参数，bootstrap 的 table 控件会将 page 和 rowse 这两个参数传递到控制器类 NoticeController 中的 getAllNotice 方法；cache 属性用于设置是否启用 AJAX 数据缓存，这里为 false，表示禁用；sidePagination 属性用于设置在哪里进行分页，可选值为 client 或者 server，这里为 server，表示由服务端处理分页；columns 用于设置列。

Bootstrap table 控件的数据源来自控制器类 NoticeController 中的 getAllNotice 方法的返回结果，代码如下：

```
// 查询所有notice
@RequestMapping(value = "/getAllNotice", method = { RequestMethod.GET,
RequestMethod.POST })
@ResponseBody
public Map<String, Object> getAllNotice(@ModelAttribute Notice notice,
Integer page, Integer rows,HttpServletRequest req, HttpServletResponse rep)
{
    // 获取request中保存的当前登录用户的编号
    String uid = (String) req.getSession().getAttribute("USERID");
    // 将该用户编号设置到对象notice中
    notice.setUserId(uid);
    // 初始化分页类对象
    Pager pager = new Pager();
    pager.setCurPage(page);
    pager.setPerPageRows(rows);
    // 创建对象params，用于封装查询条件
    Map<String, Object> params = new HashMap<String, Object>();
    params.put("notice", notice);
    // 获取满足条件的消息总数
```

```java
        int totalCount = noticeService.count(params);
        // 根据查询条件获取当前页的消息列表
        List<Notice> notices = noticeService.findNotice(notice, pager);
        // 创建对象result,用于保存返回结果
        Map<String, Object> result = new HashMap<String, Object>(2);
        result.put("total", totalCount);
        result.put("rows", notices);
        return result;
    }
```

getAllNotice 方法有 5 个参数，一个是 Notice 类型的参数 notice，用于封装表单传递来的查询条件；有两个 Integer 类型的参数 page 和 rows，用于接收从前端 Bootstrap table 控件传递来的页码和每页显示的记录数；有一个 HttpServletRequest 类型的参数 req，代表客户端的请求；有一个 HttpServletResponse 类型的参数 rep，代表服务器的响应。

在 getAllNotice 方法中，首先初始化一个分页类对象 pager，给其设置 curPage 和 perPageRows 两个属性值；然后创建 Map<String, Object>类型的对象 params，用于封装查询条件；接着依次调用业务接口 NoticeService 的 count 方法，获取满足条件的消息总数，调用 findNotice 方法，获取满足条件的消息列表；再创建 Map<String, Object>类型的对象 result，保存查询结果数据；最后将返回结果转为 JSON 格式，以字符串的形式发送到前端页面 SendNoticeList.jsp，为 Bootstrap 的 table 控件提供数据源。

在业务接口 NoticeService 中，声明两个方法，代码如下：

```java
// 根据条件查询通知总数
Integer count(Map<String, Object> params);
// 分页显示通知
List<Notice> findNotice(Notice notice, Pager pager);
```

在实现类 NoticeServiceImpl 中，实现 count 方法，代码如下：

```java
@Override
public Integer count(Map<String, Object> params) {
    return noticeDao.count(params);
}
```

在实现类 NoticeServiceImpl 的 count 方法中，直接调用接口 NoticeDao 中的 count 方法，代码如下：

```java
// 根据条件查询通知总数
@SelectProvider(type = NoticeDynaSqlProvider.class, method = "count")
Integer count(Map<String, Object> params);
```

在实现类 NoticeServiceImpl 中，实现 findNotice 方法，代码如下：

```java
@Override
public List<Notice> findNotice(Notice notice, Pager pager) {
    // 创建对象 params
    Map<String, Object> params = new HashMap<String, Object>();
    // 将封装有查询条件的 notice 对象放入 params
    params.put("notice", notice);
    // 根据条件计算消息总数
    int recordCount = noticeDao.count(params);
```

```
    // 给 pager 对象设置 rowCount 属性值(记录总数)
    pager.setRowCount(recordCount);
    if (recordCount > 0) {
        // 将 page 对象放入 params
        params.put("pager", pager);
    }
    // 分页获取消息
    return noticeDao.selectByPage(params);
}
```

findNotice 方法有两个参数，一个是 Notice 类型的参数 notice，用于封装前端页面传递来的查询条件；另一个是 Pager 类型的参数 pager，用于封装从前端页面传递来的页码和每页记录数。在 findNotice 方法中，创建了一个 Map<String, Object>类型的对象 params，用来存放两个对象，一个是 notice 对象，另一个是 pager 对象。对于 pager 对象来说，放入 params 前，还需设置 pager 对象的 rowCount 属性值，这个值是通过调用 NoticeDao 接口的 count 方法获得的。设置好 params 对象后，再调用 NoticeDao 接口的 selectByPage 方法获得当前页的消息列表。

在 NoticeDao 接口中，selectByPage 方法的代码如下：

```
// 分页查询通知
@SelectProvider(type = NoticeDynaSqlProvider.class, method = "selectWithParam")
public List<Notice> selectByPage(Map<String, Object> params);
```

这样，已发送消息列表的显示功能就实现了。接下来，介绍消息查看功能的实现过程。

3. 消息查看

在已发送消息列表中，单击某一行记录中的查看按钮，将执行 JavaScript 函数 check，其代码如下：

```
function check(nid) {
    window.location.href = "/ccms/views/notice/checkNotice.jsp?nid=" + nid;
}
```

此时，页面重定向到 checkNotice.jsp，并将参数 nid 传递过去。checkNotice.jsp 页面位于 src/main/webapp/views/notice 目录下，其效果如图 20-31 所示。

图 20-31　查看消息

在 checkNotice.jsp 页面中，最核心的代码位于$(function() { })代码段中，如下所示：

```javascript
$(function() {
    var nid = GetQueryString('nid');
    $.ajax({
        url : '/ccms/notice/getNoticeById',
        type : 'post',
        async : false,
        cache : false,
        data : {
            id : nid
        },
        dataType : 'json',
        success : function(data) {
            $('#noticeList').html();
            if (data.success) {
                $.each(
                    data.rows,
                    function(key, val) {
                        var content = '<div class="ibox">'
                                + '<div class="ibox-content">'
                                + '<a class="btn-link">'
                                + '<h2>';
                        content += val.title;
                        content += '</h2></a><div class="small m-b-xs"><strong>';
                        content += val.noticebelong;
                        content += '</strong> <span class="text-muted"><i class="fa fa-clock-o"></i>'
                                + val.operatetime
                                + '</span></div><p>';
                        content += val.content;
                        content += '</p><div class="row"><div class="col-md-6"><h5>类型：</h5><button class="btn btn-white btn-xs" type="button">';
                        content += val.typeName;
                        content += '</button></div>';
                        content += '<div class="col-md-6"><div class="small text-right"><h5></h5>';
                        content += '</div></div></div>'
                        $('#noticeList').append(
                                content);
                    });
            } else {
                swal({
                    title : "系统提示",
                    text : "暂无消息",
                    type : "warning"
                }, function() {
                });
            }
```

```javascript
        },
        error : function(aa, ee, rr) {
            swal({
                title : "系统提示",
                text : "请求服务器失败,请稍候再试",
                type : "warning"
            }, function() {
            });
        }
    });
    //视图
    var myChart = echarts.init(document.getElementById('echart'),
            'macarons');
    $.ajax({
        url : '/ccms/notice/getNoticeDetailNum',
        type : 'post',
        async : false,
        cache : false,
        data : {
            nid : nid
        },
        dataType : 'json',
        success : function(data) {
            if (data.success) {
                var read = data.read;
                var vote = data.vote;
                option = {
                    tooltip : {
                        trigger : 'item',
                        formatter : "{a} <br/>{b}: {c} ({d}%)"
                    },
                    legend : {
                        orient : 'vertical',
                        x : 'left',
                        data : [ '赞成', '反对', '已查收', '未查收' ]
                    },
                    series : [ {
                        name : '投票',
                        type : 'pie',
                        selectedMode : 'single',
                        radius : [ 0, '30%' ],
                        label : {
                            normal : {
                                position : 'inner'
                            }
                        },
                        labelLine : {
                            normal : {
                                show : false
                            }
                        },
```

```
                        data : vote
                    }, {
                        name : '查看',
                        type : 'pie',
                        radius : [ '40%', '55%' ],
                        data : read
                    } ]
                };
                // 使用刚指定的配置项和数据显示图表。
                myChart.setOption(option);
            } else {
                swal({
                    title : "系统提示",
                    text : "请求失败,请稍后再试",
                    type : "warning"
                }, function() {
                });
            }
        },
        error : function(aa, ee, rr) {
            swal({
                title : "系统提示",
                text : "请求服务器失败,请稍候再试",
                type : "warning"
            }, function() {
            });
        }
    });
    // 指定图表的配置项和数据
});
```

在$(function() { })代码段中,首先通过 jQuery AJAX 方式向后台服务器发送请求,请求的地址为/ccms/notice/getNoticeById,这个 url 请求将映射到控制器类 NoticeController 中的 getNoticeById 方法;data 用于指定发送到服务器的数据,这里为 id,表示消息编号;success 用于指定请求成功后调用的回调函数,getNoticeById 方法的返回结果将传递到该回调函数的参数 data,然后显示消息的相关信息;error 用于指定请求失败时被调用的函数。

在控制器类 NoticeController 中,getNoticeById 方法的代码如下:

```java
// 根据 id 获取详细信息
@RequestMapping(value = "/getNoticeById", method = { RequestMethod.GET,
        RequestMethod.POST })
@ResponseBody
public Map<String, Object> getNoticeById(Notice notice, HttpServletRequest
req, HttpServletResponse rep) {
    Map<String, Object> result = new HashMap<String, Object>(2);
    List<Notice> list = noticeService.getNoticeById(notice.getId());
    if (list.size() > 0) {
        result.put("success", true);
    } else {
        result.put("success", false);
```

```
    }
    result.put("rows", list);
    return result;
}
```

在 getNoticeById 方法中,调用了业务接口 NoticeService 中的 getNoticeById 方法,根据 id 获取消息。

在 NoticeService 接口中,getNoticeById 方法的声明如下:

```
// 根据id获取notice信息
public List<Notice> getNoticeById(String id);
```

在实现类 NoticeServiceImpl 中,实现 getNoticeById 方法,代码如下:

```
@Override
public List<Notice> getNoticeById(String id) {
    return noticeDao.getNoticeById(id);
}
```

在实现类 NoticeServiceImpl 的 getNoticeById 方法中,直接调用接口 NoticeDao 中的 getNoticeById 方法。

在接口 NoticeDao 中,getNoticeById 方法的代码如下:

```
// 根据id获取通知
@Select("select n.id,n.userId,n.title,n.content,n.operatetime,
n.type,su.name as noticebelong,"
        + "(case type when 1 
then '消息' when 2 then '投票' when 3 then '报修' "
        + "when 4 then '通知' end) as typeName from notice as n "
        + "left join sys_user as su on n.userId=su.userCode where n.id = #{id}
")
public List<Notice> getNoticeById(@Param("id") String id);
```

在$(function() { })代码段中,再次通过 jQuery AJAX 方式向后台服务器发送请求,这次请求的 url 为 /ccms/notice/getNoticeDetailNum,这个 url 请求将映射到控制器类 NoticeController 中的 getNoticeDetailNum 方法;data 用于指定发送到服务器的数据,这里为 nid,表示消息编号;success 用于指定请求成功后调用的回调函数,getNoticeDetailNum 方法返回结果将传递到该回调函数的参数 data,作为图表控件的数据源;error 用于指定请求失败时被调用的函数。

在控制器类 NoticeController 中,getNoticeDetailNum 方法的代码如下:

```
// 根据id获取信息详情
@RequestMapping(value = "/getNoticeDetailNum", method = { RequestMethod.GET,
RequestMethod.POST })
@ResponseBody
public Map<String, Object> getNoticeDetailNum(String nid,
HttpServletRequest req, HttpServletResponse res) {
    Map<String, Object> result = new HashMap<String, Object>(2);
    List<Echarts> elr = noticeService.getNoticeDetailNum(nid);
    List<Echarts> elv = noticeService.getNoticeVoteNum(nid);
    result.put("success", true);
```

```
        result.put("read", elr);
        result.put("vote", elv);
        return result;
    }
```

在 getNoticeDetailNum 方法中，依次调用了业务接口 NoticeService 中的 getNoticeDetailNum 方法，根据消息编号获取已读未读消息数目；调用 getNoticeVoteNum 方法，根据消息编号获取投票信息。

在 NoticeService 接口中，getNoticeDetailNum 和 getNoticeVoteNum 方法的声明如下：

```
// 获取已读未读消息数目
public List<Echarts> getNoticeDetailNum(String nid);
// 获取投票信息
public List<Echarts> getNoticeVoteNum(String nid);
```

在实现类 NoticeServiceImpl 中，实现 getNoticeDetailNum 方法，代码如下：

```
@Override
public List<Echarts> getNoticeDetailNum(String nid) {
    List<Echarts> els = noticeDao.getNoticeReadNum(nid);
    return els;
}
```

在实现类 NoticeServiceImpl 的 getNoticeDetailNum 方法中，直接调用了接口 NoticeDao 中的 getNoticeReadNum 方法，其代码如下：

```
// 获取已读未读消息数目
@Select("select count(*) as value ,(case flag when 0 then '未查收' when 1 then '已查收' end) as name " + "from answer where nid = #{nid} group by flag")
public List<Echarts> getNoticeReadNum(@Param("nid") String nid);
```

在实现类 NoticeServiceImpl 中，实现 getNoticeVoteNum 方法，代码如下：

```
@Override
public List<Echarts> getNoticeVoteNum(String nid) {
    List<Echarts> elv = noticeDao.getNoticeVoteNum(nid);
    return elv;
}
```

在实现类 NoticeServiceImpl 的 getNoticeVoteNum 方法中，直接调用了接口 NoticeDao 中的 getNoticeVoteNum 方法，其代码如下：

```
// 获取投票信息
@Select("select count(*) as value ,(case vote when 1 then '赞成' when 2 then '反对' end) as name "+ "from answer where nid = (select id from notice where type = '2' and id=#{id}) " + "and (vote=1 or vote =2) group by vote")
public List<Echarts> getNoticeVoteNum(@Param("id") String id);
```

至此，发送消息功能就讲解完了。接下来，介绍接收消息功能的实现过程。

20.8.2 接收消息

在后台登录页中，若以用户名学生 AA、密码 123456 登录系统，则以单位用户的身份进入系统首页面 index.jsp。在消息管理栏目下，单击接收消息菜单，打开接收消息列表页 ReveiveNoticeList.jsp，该页面位于 src/main/webapp/views/notice 目录下，页面效果如图 20-32 所示。

图 20-32　接收消息列表页

在 ReveiveNoticeList.jsp 页面中，需要实现接收消息列表显示、消息查收等功能。

1. 接收消息列表显示

为了显示所有发给学生 AA 的消息，在 ReveiveNoticeList.jsp 页面中，首先定义了一个 id 为 noticeList 的<div>标签，如下所示：

```
<div class="row" id="noticeList"></div>
```

然后，在$(function() { })代码块中，使用<div>标签来展示接收到的消息信息，代码如下：

```
$(function() {
    $.ajax({
        url : '/ccms/answer/getNotice',
        type : 'post',
        async : false,
        cache : false,
        data : {},
        dataType : 'json',
        success : function(data) {
            $('#noticeList').html();
            if (data.success) {
                $.each(
                    data.rows,
                    function(key, val) {
                        var content = '<div class="ibox">'
                            + '<div class="ibox-content">'
                            + '<a class="btn-link">'
                            + '<h2>';
                        content += val.title;
                        content += '</h2></a><div class="small m-b-xs"><strong>';
                        content += val.noticebelong;
                        content += '</strong> <span class="text-muted"><i class="fa fa-clock-o"></i>' + val.operatetime + '</span></div><p>';
```

```javascript
                        content += val.content;
                        content += '</p><div class="row"><div class="col-md-6"><h5>类型：</h5><button class="btn btn-white btn-xs" type="button">';
                        content += val.typeName;
                        content += '</button></div>';
                        content += '<div class="col-md-6"><div class="small text-right"><h5>操作：</h5>';
                        if (val.flag == '1') {
                            content += '<button class="btn btn-primary btn-xs" type="button">';
                            content += '已处理</button>';
                        } else {
                            if (val.type == '2') {
                                content += '<button class="btn btn-primary btn-xs flag" type="button" id="vote_true" onclick="changerFlag(\'1\',\'' + val.id + '\')">';
                                content += '赞同</button>';
                                content += '  <button class="btn btn-primary btn-xs flag" type="button" id="vote_true" onclick="javascript:changerFlag(\'2\',\'' + val.id + '\')">';
                                content += '反对</button>';
                            } else {
                                content += '<button class="btn btn-primary btn-xs flag" type="button" id="check" onclick="javascript:changerFlag(\'0\',\'' + val.id + '\')">';
                                content += '点击查收</button>';
                            }
                        }
                        content += '</div></div></div>'
                        $('#noticeList').append(content);
                    });
                } else {
                    swal({
                        title : "系统提示",
                        text : "暂无消息",
                        type : "warning"
                    }, function() {
                    });
                }
            },
            error : function(aa, ee, rr) {
                swal({
                    title : "系统提示",
                    text : "请求服务器失败,请稍候再试",
                    type : "warning"
                }, function() {
                });
            }
        });
    });
```

在$(function() { })代码块中，通过jQuery AJAX方式向后台服务器发送请求，请求的地址为/ccms/answer/getNotice，这个 url 请求将映射到控制器类 AnswerController 中的 getNotice 方法；data 参数用于指定发送到服务器的数据；success 用于指定请求成功后调用的回调函数，getNotice 方法的执行结果将传递给该函数的参数 data，然后将接收到的消息数据展示出来。

在控制器类 AnswerController 中，getNotice 方法的代码如下：

```java
package com.ccms.controller;
……
import com.ccms.pojo.Notice;
import com.ccms.service.AnswerService;
import com.ccms.tools.JsonUtil;
import com.google.gson.JsonObject;
@Controller
@RequestMapping(value = "/answer", method = { RequestMethod.GET, RequestMethod.POST })
public class AnswerController {
    @Autowired
    private AnswerService answerService;
    JsonUtil<Notice> json = new JsonUtil<Notice>();
    // 查询所有 notice
    @RequestMapping(value = "/getNotice", method = { RequestMethod.GET, RequestMethod.POST })
    @ResponseBody
    public Map<String, Object> getNotice(Notice notice, HttpServletRequest req, HttpServletResponse rep) {
        String uid = (String) req.getSession().getAttribute("USERID");
        String unitid = req.getSession().getAttribute("UNITINFOID").toString();
        Map<String, Object> result = new HashMap<String, Object>(2);
        List<Notice> list = answerService.getAllNotice(unitid);
        if (list.size() > 0) {
            result.put("success", true);
        } else {
            result.put("success", false);
        }
        result.put("rows", list);
        return result;
    }
    ……
}
```

在 getNotice 方法中，调用业务接口 AnswerService 中的 getAllNotice 方法，根据单位编号查询所有消息。

在 AnswerService 接口中，getAllNotice 方法的声明如下：

```java
// 根据单位查询所有消息
public List<Notice> getAllNotice(String unitId);
```

在接口 AnswerService 的实现类 AnswerServiceImpl 中，实现 getAllNotice 方法，代码如下：

```java
package com.ccms.service.impl;
……
import com.ccms.dao.AnswerDao;
import com.ccms.pojo.Notice;
import com.ccms.service.AnswerService;
@Service("answerService")
@Transactional(propagation = Propagation.REQUIRED, isolation =
Isolation.DEFAULT)
public class AnswerServiceImpl implements AnswerService {
    @Autowired
    AnswerDao answerDao;
    @Override
    public List<Notice> getAllNotice(String uid) {
        return answerDao.getAllNotice(uid);
    }
    ……
}
```

在实现类 AnswerServiceImpl 的 getAllNotice 方法中，直接调用了接口 AnswerDao 中的 getAllNotice 方法，其代码如下：

```java
// 查询
@Select("select DISTINCT n.id,n.userId,n.title,n.content,n.operatetime,"
    + "n.type,aw.flag,su.name as noticebelong,(case type when 1 then '消息' " + "when 2 then '投票' when 3 then '报修' when 4 then '通知' end) as typeName
" + "from notice as n " + "left join sys_user as su on n.userId=su.userCode
" + "left join answer as aw on n.id = aw.nid " + "where n.id in (select nid from answer where uid=#{uid}) and uid=#{uid} " + " order by operatetime desc")
List<Notice> getAllNotice(@Param("uid") String uid);
```

这样，接收消息列表显示功能就实现了。接下来，介绍消息查收功能的实现过程。

2. 消息查收

在接收消息列表中，如果类型是消息，则单击"查收"按钮查收消息；如果类型是投票，则可单击"赞同"或"反对"按钮查收消息。单击这三个按钮后，都将执行 JavaScript 代码编制的函数 changerFlag。changerFlag 函数有两个参数，第一个是 voet，如果单击"查收"按钮，传递给 voet 的值为 0；如果单击"赞同"按钮，传递给 voet 的值为 1；如果单击"反对"按钮，传递给 voet 的值为 2。另一个参数是 nid，表示消息编号。changerFlag 函数的代码如下：

```javascript
//修改状态位
function changerFlag(vote, nid) {
    $.ajax({
        url : '/ccms/answer/changerFlag',
        type : 'post',
        async : false,
        cache : false,
        data : {
            vote : vote,
```

```js
            nid : nid
        },
        dataType : 'json',
        success : function(data) {
            if (data.success) {
                swal({
                    title : "系统提示",
                    text : "投票成功",
                    type : "success"
                }, function() {
                    location.reload();
                });
            } else {
                swal({
                    title : "系统提示",
                    text : "投票失败,请稍后再试",
                    type : "warning"
                }, function() {
                });
            }
        },
        error : function(aa, ee, rr) {
            swal({
                title : "系统提示",
                text : "请求服务器失败,请稍候再试",
                type : "warning"
            }, function() {
            });
        }
    });
}
```

在函数 changerFlag 中,通过 jQuery AJAX 方式向后台服务器发送请求,请求的地址为 /ccms/answer/changerFlag,这个 url 请求将映射到控制器类 AnswerController 中的 changerFlag 方法;data 用于指定发送到服务器的数据,这里传递 vote 和 nid 两个参数值。

在控制器类 AnswerController 中,changerFlag 方法的代码如下:

```java
// 改变状态位
@RequestMapping(value = "/changerFlag", method = { RequestMethod.GET,
        RequestMethod.POST })
public String changerFlag(String vote, String nid, HttpServletRequest req,
        HttpServletResponse res) {
    String unitid = req.getSession().getAttribute("UNITINFOID").toString();
    int result = answerService.upFlag(vote, nid, unitid);
    JsonObject jObject = new JsonObject();
    jObject.addProperty("success", result);
    try {
        ServletOutputStream jos = res.getOutputStream();
        jos.write(jObject.toString().getBytes("utf-8"));
        jos.flush();
```

```
            jos.close();
        } catch (IOException e) {
            e.printStackTrace();
        }
        return null;
    }
```

在 changerFlag 方法中，调用业务接口 AnswerService 中的 upFlag 方法，修改消息状态标记。

在 AnswerService 接口中，upFlag 方法的声明如下：

```
// 修改状态位
public int upFlag(String vote, String nid, String uid);
```

在实现类 AnswerServiceImpl 中，实现 upFlag 方法，代码如下：

```
@Override
public int upFlag(String vote, String nid, String uid) {
    if (vote == null || vote.trim().length() <= 0) {
        vote = "0";
    }
    return answerDao.upFalg(vote, nid, uid);
}
```

在实现类 AnswerServiceImpl 的 upFlag 方法中，直接调用了接口 AnswerDao 中的 upFalg 方法，其代码如下：

```
// 更改通知标记
@Update("update answer set flag=1,vote=#{vote} where nid = #{nid} and uid =#{uid}")
int upFalg(@Param("vote") String vote, @Param("nid") String nid, @Param("uid") String uid);
```

在图 20-32 所示的接收消息列表页中，先单击"查收"按钮，再单击"赞同"按钮。此时，在接收消息列表中，这两条消息的状态均显示为已处理，如图 20-33 所示。

图 20-33　接收消息后的消息状态

至此，单位用户功能就讲解完了。

20.9 小　　结

本章基于 Spring、Spring MVC 与 MyBatis 整合框架，采用注解方法并结合前端基于 Bootstrap 的 H+框架，详细讲解了校园通讯管理系统的具体实现过程。系统的主要功能包括平台管理员功能(院校管理员管理、院校管理)、院校管理员功能(单位管理、角色管理、用户管理)和单位用户功能(发消息、接收消息)，并按照三层架构开发每个功能模块。

通过本章的学习，希望读者能够进一步熟练掌握 Spring、Spring MVC 与 MyBatis 框架整合开发的基本步骤、方法和技巧。

第 21 章 电商网站

本章将基于 Spring、Spring MVC 与 MyBatis 整合框架,并结合前端 Vue 框架实现一个简单的电商网站。

21.1 需求与系统分析

本章所实现的电商网站功能比较简单,主要包括商品列表显示、商品详情显示、购物车管理、订单提交。

21.2 数据库设计

本章所使用的数据库与第 19 章相同,数据库名为 eshop,电商网站只使用了商品信息表 product_info、订单信息表 order_info 和订单明细表 order_detail 这三张表。

商品信息表 product_info 的字段说明如表 21-1 所示。

表 21-1 商品信息表 product_info

字段名	类型	主外键	说明
id	int(4)	PK	商品 id 标识,主键,自增
code	varchar(16)		商品编号

续表

字段名	类　型	主外键	说　明
name	varchar(255)		商品名称
tid	int(4)	FK	商品类别 id
brand	varchar(20)		品牌
pic	varchar(255)		商品图片
num	int(4)		商品数量
price	decimal(10,0)		商品价格
intro	longtext		商品介绍
status	int(4)		商品状态

订单信息表 order_info 的字段说明如表 21-2 所示。

表 21-2　订单信息表 order_info

字段名	类　型	主外键	说　明
id	int(4)	PK	订单 id 标识，主键，自增
uid	int(4)	FK	客户 id
status	varchar(16)		订单状态
ordertime	datetime		订单下单时间
orderprice	decimal(8,2)		订单价格

订单明细表 order_detail 的字段说明如表 21-3 所示。

表 21-3　订单明细表 order_detail

字段名	类　型	主外键	说　明
id	int(4)	PK	订单明细 id，主键，自增
oid	int(4)	FK	订单 id
pid	int(4)	FK	商品 id
num	int(4)		购买数量

21.3　环境搭建与配置文件

本章实现的电商网站采用了前后台分离技术，可以参照第 17 章的内容，完成电商网站后台框架搭建及相关配置文件的编写。网站后台的目录结构如图 21-1 所示。

网站前端基于 Vue 脚手架 vue-cli 快速构建而成，其目录结构如图 21-2 所示。由于篇幅所限，有关使用 vue-cli 构建项目的知识，读者可以通过本章配置的视频学习。

在网站后台目录结构中，com.eshop.controller 包用于存放控制器类，com.eshop.service 包用于存放业务逻辑层接口，com.eshop.service.impl 包用于存放业务逻辑层接口的实现类，com.eshop.dao 包用于存放数据访问层接口，com.eshop.pojo 包用于存放实体类。

dbconfig.properties 为存储数据库连接信息的属性文件，applicationContext.xml 为 Spring 框架的配置文件，dispatcherServlet-servlet.xml 为 Spring MVC 框架的配置文件。

图 21-1　网站后台的目录结构　　　　图 21-2　网站前端的目录结构

在网站前端目录结构中，views 目录用于存放每个路由页面的.vue 文件；components 目录用于存放公共组件；router 目录用于存放路由配置文件。在网站前端开发过程中，使用了 Vue.js 的路由插件 vue-router 和状态管理插件 Vuex，首先要在 main.js 文件中导入并进行初始化，代码如下：

```
import Vue from 'vue';
import VueRouter from 'vue-router';
import Routers from './router/router';
import Vuex from 'vuex';
import App from './App.vue';
import './style.css';
import axios from 'axios'
Vue.use(VueRouter);
Vue.use(Vuex);
// 路由配置
const RouterConfig = {
    // 使用 HTML5 的 History 路由模式
    mode: 'history',
    routes: Routers
};
const router = new VueRouter(RouterConfig);
router.beforeEach((to, from, next) => {
    window.document.title = to.meta.title;
    next();
```

```
});
router.afterEach((to, from, next) => {
    window.scrollTo(0, 0);
});
const store = new Vuex.Store({
    state: {
        // 后续开发添加
    },
    getters: {
        // 后续开发添加
    },
    mutations: {
        // 后续开发添加
    },
    actions: {
        // 后续开发添加
    }
});
new Vue({
    el: '#app',
    router: router,
    store: store,
    render: h => {
        return h(App)
    }
});
```

其中，router.js 文件用于配置路由，Vuex 默认设置了 state、getters、mutations 和 actions，在后续开发中将会逐步添加。style.css 文件包含全局使用 CSS 样式，可以在 main.js 中直接导入。

21.4 创建实体类

在 com.eshop.pojo 包中，依次创建实体类 ProductInfo、OrderInfo、OrderDetail 和 CartItem。实体类 ProductInfo 用于封装商品信息，代码如下：

```java
package com.eshop.pojo;
public class ProductInfo {
    private int id;
    private String name;
    private String brand;
    private String pic;
    private double price;
    private String intro;
    public String getPic() {
        return pic;
    }
    public void setPic(String pic) {
        this.pic = "http://localhost:8080/eshop/product_images/" + pic;
```

```
        }
        // 省略其他属性的 getter 和 setter 方法
}
```

实体类 OrderInfo 用于封装订单信息，代码如下：

```
package com.eshop.pojo;
public class OrderInfo {
    private Integer id;
    private int uid;
    private String status;
    private String ordertime;
    private double orderprice;
    // 省略上述属性的 getter 和 setter 方法
}
```

实体类 OrderDetail 用于封装订单明细信息，代码如下：

```
package com.eshop.pojo;
public class OrderDetail {
    private int id;
    private int oid;
    private int pid;
    private int num;
// 省略上述属性的 getter 和 setter 方法
}
```

CartItem 类用于封装购物车信息，代码如下：

```
package com.eshop.pojo;
public class CartItem {
    private int id;
    private int count;
    // 省略上述属性的 getter 和 setter 方法
}
```

21.5 创建几个 Dao 接口

在 com.eshop.dao 包中，创建数据访问层接口 ProductInfoDao 和 OrderInfoDao，使用 MyBatis 注解完成数据表的操作。

在接口 ProductInfoDao 中，编写如下代码：

```
package com.eshop.dao;
import java.util.List;
import org.apache.ibatis.annotations.Param;
import org.apache.ibatis.annotations.Select;
import com.eshop.pojo.ProductInfo;
public interface ProductInfoDao {
    // 获取所有商品
    @Select("select * from product_info")
    public List<ProductInfo> selectProductInfo();
```

```
    // 根据商品id获取商品
    @Select("select * from product_info where id = #{id}")
    public ProductInfo selectProductInfoById(@Param("id") int id);
}
```

在接口 ProductInfoDao 中，selectProductInfo 方法用于获取所有商品；selectProductInfoById 方法用于根据商品编号获取商品。

在接口 OrderInfoDao 中，编写如下代码：

```
package com.eshop.dao;
import org.apache.ibatis.annotations.Insert;
import org.apache.ibatis.annotations.Options;
import com.eshop.pojo.OrderDetail;
import com.eshop.pojo.OrderInfo;
public interface OrderInfoDao {
    // 保存订单主表
    @Insert("insert into order_info(uid,status,ordertime,orderprice) " +
"values(#{uid},#{status},#{ordertime},#{orderprice})")
    @Options(useGeneratedKeys = true, keyProperty = "id")
    int saveOrderInfo(OrderInfo oi);
    // 保存订单明细
    @Insert("insert into order_detail(oid,pid,num) values(#{oid},#{pid},#{num})")
    @Options(useGeneratedKeys = true, keyProperty = "id")
    int saveOrderDetail(OrderDetail od);
}
```

在接口 OrderInfoDao 中，saveOrderInfo 方法通过@Insert 注解映射一条 INSERT 语句，保存订单信息；saveOrderDetail 方法通过@Insert 注解映射一条 INSERT 语句，保存订单明细。

21.6 创建 Service 接口及实现类

在 com.eshop.service 包中，创建业务逻辑层接口 ProductInfoService 和 OrderInfoService。在接口 ProductInfoService 中声明两个方法，代码如下：

```
package com.eshop.service;
import java.util.List;
import com.eshop.pojo.ProductInfo;
public interface ProductInfoService {
    // 获取所有商品
    public List<ProductInfo> getProductInfo();
    // 根据商品编号获取商品
    public ProductInfo getProductInfoById(int id);
}
```

在接口 OrderInfoService 中声明两个方法，代码如下：

```
package com.eshop.service;
import com.eshop.pojo.OrderDetail;
```

```java
import com.eshop.pojo.OrderInfo;
public interface OrderInfoService {
    // 添加订单主表
    public int addOrderInfo(OrderInfo oi);
    // 添加订单明细
    public int addOrderDetail(OrderDetail od);
}
```

在 com.eshop.service.impl 包中，创建 ProductInfoService 接口的实现类 ProductInfoServiceImpl，代码如下：

```java
package com.eshop.service.impl;
......
import com.eshop.dao.ProductInfoDao;
import com.eshop.pojo.ProductInfo;
import com.eshop.service.ProductInfoService;
@Service("productInfoService")
@Transactional(propagation = Propagation.REQUIRED, isolation = Isolation.DEFAULT)
public class ProductInfoServiceImpl implements ProductInfoService {
    @Autowired
    ProductInfoDao productInfoDao;
    @Override
    public List<ProductInfo> getProductInfo() {
        return productInfoDao.selectProductInfo();
    }
    @Override
    public ProductInfo getProductInfoById(int id) {
        return productInfoDao.selectProductInfoById(id);
    }
}
```

创建 OrderInfoService 接口的实现类 OrderInfoServiceImpl，代码如下：

```java
package com.eshop.service.impl;
......
 org.springframework.transaction.annotation.Transactional;
import com.eshop.dao.OrderInfoDao;
import com.eshop.pojo.OrderDetail;
import com.eshop.pojo.OrderInfo;
import com.eshop.service.OrderInfoService;
@Service("orderInfoService")
@Transactional(propagation = Propagation.REQUIRED, isolation = Isolation.DEFAULT)
public class OrderInfoServiceImpl implements OrderInfoService {
    @Autowired
    OrderInfoDao orderInfoDao;
    @Override
    public int addOrderInfo(OrderInfo oi) {
        return orderInfoDao.saveOrderInfo(oi);
    }
    @Override
```

```
    public int addOrderDetail(OrderDetail od) {
        return orderInfoDao.saveOrderDetail(od);
    }
}
```

21.7　商品列表页

本章的电商网站实现了前后台分离，由于前端页面发送到后台服务器的 AJAX 请求无法跨域，因此这里使用了 Nginx 进行代理。Nginx (engine x) 是一个高性能的 HTTP 和反向代理服务器，也是一个 IMAP/POP3/SMTP 服务器。由于篇幅所限，Nginx 的配置过程此处不再详述，读者可以通过本章视频教程学习。

配置并启动 Nginx，然后在 Windows 命令窗口中通过 npm run dev 命令，启用前端项目 shopping。在浏览器地址栏中，输入 http://localhost:8888/shopping/index.html 这个 url，就可以看到商品列表页了，效果如图 21-3 所示。

图 21-3　商品列表页

在商品列表页中，可以按照品牌筛选(如 APPLE、ThinkPad)，也可以按照价格在筛选的基础上再进行排序。

商品列表页是通过组件 list.vue 来实现的，该文件位于 views 目录下。在介绍这个组件之前，先来看下单个商品是如何展示的。单个商品展示是通过另一个组件 product.vue 来实现的，该组件位于 components 目录下，其代码如下：

```
<template>
    <div class="product">
        <router-link :to="'/product/' + info.id" class="product-main">
            <img :src="info.pic">
            <h4>{{ info.name }}</h4>
            <div class="product-cost">¥ {{ info.price }}</div>
```

```html
            <div class="product-add-cart" @click.prevent="handleCart">加入购物车</div>
        </router-link>
    </div>
</template>
<script>
    export default {
        // 声明需要从父级接收数据
        props: {
            info: Object
        },
        methods: {
            handleCart () {
                this.$store.commit('addCart', this.info.id);
            }
        }
    };
</script>
<!-- 以下样式只作用于当前的.vue文件 -->
<style scoped>
    .product{
        width: 25%;
        float: left;
    }
<!-- 此处省略了其他样式定义 -->
</style>
```

在product.vue文件中，单个商品展示的样式定义在<style scoped>标记部分。由于商品属性比较多，为了方便父子组件传递参数，在props选项中定义了一个属性info来接收对象格式的数据，父组件可以直接将数据传递过来赋值给info。

有了单个商品展示组件，还需要从后台获取商品数据，由于Vue.js本身没有提供AJAX方法，因此后台数据的获取是通过第三方的HTTP库axios实现的。在命令端口中，首先切换到项目路径，这里为d:/shopping，然后输入如下命令来安装axios插件。

```
npm install --save axios
```

商品列表相关的数据是通过Vuex来维护的，在main.js文件的Vuex中声明与数据列表(商品列表和购物车列表)相关的state、mutations和actions。

```js
const store = new Vuex.Store({
    state: {
// 商品列表
        productList: [],
// 购物车列表
        cartList: []
    },
    mutations: {
        // 添加商品列表
        setProductList(state, data) {
            state.productList = data;
        }
```

```
        },
        actions: {
            // 请求商品列表
            getProductList(context) {
                // 通过 ajax 获取
                axios.get('/eshop/product/getProduct').
then(function(result) {
                    context.commit('setProductList', result.data);
                })
            }
        }
});
```

在 Vuex 的 actions 选项中，定义了一个函数 getProductList，该函数在后面将要实现的组件 list.vue 内可以通过 this.$store.dispatch 触发。在 getProductList 函数中，首先通过 axios 向后台服务器发送/eshop/product/getProduct 这个 url 请求，服务器返回的数据绑定到参数 result 中；然后通过调用 mutations 选项中定义的 setProductList 方法，将数据设置到商品列表 productList 中。

eshop/product/getProduct 这个 url 请求将映射到后台项目 eshop 中的控制器类 ProductInfoController 的 getProductInfo 方法，其代码如下：

```
package com.eshop.controller;
......
import com.eshop.pojo.ProductInfo;
import com.eshop.service.ProductInfoService;
@Controller
@RequestMapping("/product")
public class ProductInfoController {
    @Autowired
    ProductInfoService productInfoService;
    @RequestMapping(value = "/getProduct", method = {
RequestMethod.GET, RequestMethod.POST })
    @ResponseBody
    public List<ProductInfo> getProductInfo() {
        List<ProductInfo> pList = productInfoService.getProductInfo();
        return pList;
    }
}
```

在 getProductInfo 方法中，调用业务接口 ProductInfoService 中的 getProductInfo 方法获取所有商品数据，并将其转换为 JSON 格式，再发送到前端页面。

前面曾提到过，商品列表页是通过组件 list.vue 来实现的。现在可以看下这个文件了，与商品列表显示相关的代码如下：

```
<template>
    <div v-show="list.length">
        <!-- 遍历商品列表，构建多个 components/product.vue 组件-->
        <Product v-for="item in list" :info="item" :key="item.id">
</Product>
        <div class="product-not-found" v-show="!list.length">暂无相关商品
</div>
```

```
        </div>
    </template>
    <script>
        // 导入单个商品展示组件
        import Product from '../components/product.vue';
        export default {
            components: { Product },
            computed: {
                list () {
                    // 从 Vuex 获取商品列表数据
                    return this.$store.state.productList;
                }
            },
            mounted () {
                // 初始化时，通过 Vuex 的 action 请求数据
                this.$store.dispatch('getProductList');
            }
        }
    </script>
    <style scoped>
        // 此处省略了 list.vue 组件的样式
    </style>
```

在组件 list.vue 初始化时，首先调用 Vuex 的 action 中的 getProductList 方法请求商品列表数据，然后遍历商品列表，构建多个 components/product.vue 组件，每个 components/product.vue 组件用于单个商品展示。

打开浏览器，输入地址 http://localhost:8888/shopping/index.html，就可以看到商品列表了，如图 21-4 所示。

图 21-4　商品列表

接下来，介绍如何实现对商品按品牌筛选和按价格排序的功能。

1．按商品价格排序

为了实现按照价格排序，在遍历商品列表时就不能直接使用数据 list 了，也不能直接对 list 进行操作，否则会破坏原有数据。因此，这里使用计算属性来动态返回过滤后的数据。在 list.vue 组件中，与排序相关的代码如下：

```html
<template>
    <div v-show="list.length">
        <div class="list-control">
            <!-- 价格排序按钮 -->
            <div class="list-control-order">
                <span>排序: </span>
                <span
                    class="list-control-order-item"
                    :class="{on: order === ''}"
                    @click="handleOrderDefault">默认</span>
                <span
                    class="list-control-order-item"
                    :class="{on: order.indexOf('price') > -1}"
                    @click="handleOrderCost">价格
                    <template v-if="order === 'price-asc'">↑</template>
                    <template v-if="order === 'price-desc'">↓</template>
                </span>
            </div>
        </div>
        <!-- 遍历商品列表,构建多个 components/product.vue 组件-->
        <Product v-for="item in orderedAndFilteredList" :info="item" :key="item.id"></Product>
        <div class="product-not-found" v-show="!orderedAndFilteredList.length">暂无相关商品</div>
    </div>
</template>
<script>
    // 导入单个商品展示组件
    import Product from '../components/product.vue';
    export default {
        components: { Product },
        computed: {
            list () {
                // 从 Vuex 获取商品列表数据
                return this.$store.state.productList;
            },
            orderedAndFilteredList () {
                // 复制原始数据
                let list = [...this.list];
                // 按品牌过滤,此功能稍后实现
                // 按价格排序
                if (this.order !== '') {
                    if (this.order === 'price-desc') {
                        list = list.sort((a, b) => b.price - a.price);
                    } else if (this.order === 'price-asc') {
                        list = list.sort((a, b) => a.price - b.price);
                    }
                }
                return list;
            }
        },
        data () {
```

```
            return {
                // 排序依据，取值price-desc：价格降序，price-asc：价格升序
                order: ''
            }
        },
        methods: {
            // 处理默认排序
            handleOrderDefault () {
                this.order = '';
            },
            // 处理按价格排序
            handleOrderCost () {
                if (this.order === 'price-desc') {
                    this.order = 'price-asc';
                } else {
                    this.order = 'price-desc';
                }
            }
        },
        mounted () {
            // 初始化时，通过Vuex的action请求数据
            this.$store.dispatch('getProductList');
        }
    }
</script>
```

在遍历商品列表时，使用计算属性orderedAndFilteredList，会返回排序后的数据，排序依据通过 data:order 指定，默认为空(即 price-desc)。price-asc 时按照价格升序，price-desc 时按照价格降序。排序时通过 JavaScript 数组的 sort 方法，对数组中前后两个元素进行比较。在对<Product>循环时，将 list 改为 orderedAndFilteredList，显示的就是过滤后的数据了。此外，在<template>模板里还添加了价格排序按钮。单击价格按钮，通过执行 handleOrderCost 方法，实现升序和降序两种状态的切换。此时浏览网站，单击价格按钮，就能看到商品按照价格排列了，如图21-5 所示。

图 21-5　按照价格降序排列

2. 按品牌筛选

为了实现按品牌筛选，首先需要获取品牌数据。在 main.js 文件中，品牌数据可以作为 getters 从 Vuex 的 productList 中遍历获取，相关代码如下：

```
// 数组去重
function getFilterArray(array) {
    const res = [];
    const json = {};
    for(let i = 0; i < array.length; i++) {
        const _self = array[i];
        if(!json[_self]) {
            res.push(_self);
            json[_self] = 1;
        }
    }
    return res;
}
const store = new Vuex.Store({
    state: {
        // 商品列表
        productList: [],
        // 购物车列表
        cartList: []
    },
    getters: {
        brands: state => {
            const brands = state.productList.map(item => item.brand);
            return getFilterArray(brands);
        }
    }
});
```

getters 中的 brands 依赖于 productList，通过 map 方法将 productList 中的 brand 数据过滤出来，再利用自定义函数 getFilterArray 将数组中重复的品牌去除。

然后，在 list.vue 组件中将 Vuex 中的品牌数据引入进来，并完成商品列表过滤。此外，还需添加品牌筛选按钮，相关代码如下：

```
<template>
    <div v-show="list.length">
        <div class="list-control">
            <!-- 品牌筛选按钮 -->
            <div class="list-control-filter">
                <span>品牌：</span>
                <span
                    class="list-control-filter-item"
                    :class="{on: item === filterBrand}"
                    v-for="item in brands"
                    @click="handleFilterBrand(item)">{{ item }}</span>
            </div>
            <!-- 价格排序按钮 -->
```

```html
        <!-- 遍历商品列表，构建多个 components/product.vue 组件-->
        <Product v-for="item in orderedAndFilteredList" :info="item" :key="item.id"></Product>
        <div class="product-not-found" v-show="!orderedAndFilteredList.length">暂无相关商品</div>
    </div>
</template>
```
```javascript
<script>
    // 导入单个商品展示组件
    import Product from '../components/product.vue';
    export default {
        components: { Product },
        computed: {
            list () {
                // 从 Vuex 获取商品列表数据
                return this.$store.state.productList;
            },
            brands () {
                return this.$store.getters.brands;
            },
            orderedAndFilteredList () {
                // 复制原始数据
                let list = [...this.list];
                // 按品牌过滤
                if (this.filterBrand !== '') {
                    list = list.filter(item => item.brand === this.filterBrand);
                }
                // 排序……
                // 此功能前面已实现，由于篇幅所限，此处省略相关代码
                return list;
            }
        },
        data () {
            return {
                // 过滤依据
                filterBrand: '',
                // 排序依据
                order: ''
            }
        },
        methods: {
            // 处理品牌筛选
            handleFilterBrand (brand) {
                if (this.filterBrand === brand) {
                    this.filterBrand = '';
                } else {
                    this.filterBrand = brand;
                }
            },
            // 处理默认排序……
            // 处理按价格排序……
        },
```

```
    mounted () {
      // 初始化时，通过 Vuex 的 action 请求数据
        this.$store.dispatch('getProductList');
    }
  }
</script>
```

通过对 brands 遍历，生成多个品牌按钮。第一次单击品牌按钮时，会执行 handleFilterBrand 函数，设置要筛选的品牌名称。再次单击该按钮时，会取消设置。

此时浏览网站，先单击一个品牌按钮，再单击价格按钮，就能看到商品先按照品牌筛选，再按照价格排序的效果了，如图 21-6 所示。

至此，商品列表页的功能就实现了。接下来，介绍商品详情页的实现过程。

图 21-6　按品牌筛选和按价格排序

21.8　商品详情页

商品详情页为 views 目录下的 product.vue，页面效果如图 21-7 所示。

图 21-7　商品详情页

首先，需要在 router.js 文件中为商品详情页 product.vue 配置路由，代码如下：

```js
// 引入组件
import product from "../views/product.vue";
const routers = [{
    {
        path: '/product/:id',
        meta: {
            title: '商品详情'
        },
        component: product
    }
];
export default routers;
```

在商品列表页中，单击某个商品跳转到商品详情页时，需要将商品的 id 传递过去。因此商品详情的路由接收一个参数 id，这个参数就是商品的 id。

在商品详情页中，通过$route 可以获取当前路由的参数，也就是商品编号；然后通过 AJAX 向后台服务器发送请求，获取这个商品的信息，相关代码如下：

```html
<script>
    import axios from 'axios'
    export default {
        data() {
            return {
                // 获取路由中的参数
                id: parseInt(this.$route.params.id),
                product: null
            }
        },
        methods: {
            getProduct() {
                var self=this;
                // 通过 ajax，根据商品 id 向后台服务器获取商品信息
                axios.get('/eshop/product/getProductById/' + this.id).then(function(result) {
                    self.product = result.data;
                })
            }
        },
        mounted() {
            // 初始化时，请求数据
            this.getProduct();
        }
    }
</script>
```

eshop/product/getProductById 这个 url 请求将映射到后台项目 eshop 中的控制器类 ProductInfoController 的 getProductInfoById 方法，并将商品 id 传递过去，其代码如下：

```java
@RequestMapping("/getProductById/{id}")
```

```
@ResponseBody
public ProductInfo getProductInfoById(@PathVariable("id") Integer id){
    ProductInfo product=productInfoService.getProductInfoById(id);
    return product;
}
```

从后台服务器获取数据后，就可以将数据写入商品详情页的模板<template>中，相关代码如下：

```
<template>
    <div v-if="product">
        <div class="product">
            <div class="product-image">
                <img :src="product.pic">
            </div>
            <div class="product-info">
                <h1 class="product-name">{{ product.name }}</h1>
                <div class="product-cost">¥ {{ product.price }}</div>
                <div class="product-add-cart" @click="handleAddToCart">加入购物车</div>
            </div>
        </div>
        <div class="product-desc">
            <h2>产品介绍</h2>
            {{ product.intro }}
        </div>
    </div>
</template>
```

此时浏览网站，就可以看到图 21-7 所示的商品详情页了。

21.9 购物车页

购物车主要用于暂存客户购买的商品，客户还可以删除购物车的商品、清空购物车和修改购物车中商品数量，购物车页面效果如图 21-8 所示。

图 21-8 购物车页

将商品放入购物车是通过 Vuex 来完成的，在 main.js 的 Vuex 中，首先配置 state 和 mutations 两个选项，相关代码如下：

```
const store = new Vuex.Store({
    state: {
        // 商品列表
        productList: [],
        // 购物车列表
        cartList: []
    },
    ……
    mutations: {
        ……
        // 添加到购物车
        addCart(state, id) {
            // 先判断购物车是否已有，如果有，则数量+1
            const isAdded = state.cartList.find(item => item.id === id);
            if(isAdded) {
                isAdded.count++;
            } else {
                state.cartList.push({
                    id: id,
                    count: 1
                })
            }
        },
        // 修改购物车中的商品数量
        editCartCount(state, payload) {
            const product = state.cartList.find(item => item.id === payload.id);
            product.count += payload.count;
        },
        // 删除购物车中的商品
        deleteCart(state, id) {
            const index = state.cartList.findIndex(item => item.id === id);
            state.cartList.splice(index, 1);
        },
        // 清空购物车
        emptyCart(state) {
            state.cartList = [];
        }
    }
    ……
});
```

在 Vuex 的 state 选项中，cartList 用于保存购物车记录，其格式是数组。数组 cartList 中的每一个元素都是一个对象，包含商品 id 和购买数量这两个数据。

在 mutations 选项中，依次定义了 addCart、editCartCount、deleteCart 和 emptyCart 四个方法。

addCart 方法用于将商品添加到购物车，其接收的参数是商品 id。添加前，先判断 cartList

中是否已存在该商品，如果已存在，则数量加 1，否则将商品添加到 cartList。

editCartCount 方法用于修改购物车中的商品数量；deleteCart 方法用于删除购物车中指定 id 的商品；emptyCart 方法用于清空购物车。

有了购物车数据，就可以将这些数据显示出来了。在 app.vue 文件中，首先定义购物车入口，相关代码如下：

```
<template>
    <div>
        <div class="header">
            <router-link to="/list" class="header-title">电商网站</router-link>
            <div class="header-menu">
                <router-link to="/cart" class="header-menu-cart">
                    购物车
                    <span v-if="cartList.length">{{ cartList.length }}</span>
                </router-link>
            </div>
        </div>
        <router-view></router-view>
    </div>
</template>
```

购物车页为 cart.vue，该文件位于 views 目录下，其代码如下：

```
<template>
    <div class="cart">
        <div class="cart-header">
            <div class="cart-header-title">购物清单</div>
            <div class="cart-header-main">
                <div class="cart-info">商品信息</div>
                <div class="cart-price">单价</div>
                <div class="cart-count">数量</div>
                <div class="cart-cost">小计</div>
                <div class="cart-delete">删除</div>
            </div>
        </div>
        <div class="cart-content">
            <div class="cart-content-main" v-for="(item, index) in cartList">
                <div class="cart-info">
                    <img :src="productDictList[item.id].pic">
                    <span>{{ productDictList[item.id].name }}</span>
                </div>
                <div class="cart-price">¥ {{ productDictList[item.id].price }}</div>
                <div class="cart-count">
                    <span class="cart-control-minus" @click="handleCount(index, -1)">-</span> {{ item.count }}
                    <span class="cart-control-add" @click="handleCount(index, 1)">+</span>
```

```html
                    </div>
                    <div class="cart-cost">¥ {{ productDictList[item.id].price * item.count }}</div>
                    <div class="cart-delete">
                        <span class="cart-control-delete" @click="handleDelete(index)">删除</span>
                    </div>
                </div>
                <div class="cart-empty" v-if="!cartList.length">购物车为空</div>
            </div>
            <div class="cart-footer" v-show="cartList.length">
                <div class="cart-footer-desc">
                    共计 <span>{{ countAll }}</span> 件商品
                </div>
                <div class="cart-footer-desc">
                    应付总额 <span>¥ {{ costAll }}</span>
                </div>
                <div class="cart-footer-desc">
                    <div class="cart-control-order" @click="handleOrder">下单</div>
                </div>
            </div>
        </div>
    </div>
</template>
<script>
    import axios from 'axios'
    export default {
        computed: {
            cartList() {
                return this.$store.state.cartList;
            },
            productList() {
                return this.$store.state.productList;
            },
            productDictList() {
                const dict = {};
                this.productList.forEach(item => {
                    dict[item.id] = item;
                });
                return dict;
            },
            countAll() {
                let count = 0;
                this.cartList.forEach(item => {
                    count += item.count;
                });
                return count;
            },
            costAll() {
                let cost = 0;
                this.cartList.forEach(item => {
```

```
                    cost += this.productDictList[item.id].price *
item.count;
                });
                return cost;
            }
        },
        methods: {
            handleCount(index, count) {
                if(count < 0 && this.cartList[index].count === 1) return;
                this.$store.commit('editCartCount', {
                    id: this.cartList[index].id,
                    count: count
                });
            },
            handleDelete(index) {
                this.$store.commit('deleteCart',
this.cartList[index].id);
            }
        }
    }
</script>
<style scoped>
    // 此处省略了 cart.vue 的样式
</style>
```

在购物车页 cart.vue 的<script>中，首先定义了 cartList、productList、productDictList、countAll 和 costAll 这 5 个计算属性。其中，计算属性 cartList 用于从 Vuex 中获取购物车数据 cartList；计算属性 productList 用于从 Vuex 中获取商品列表数据 productList；计算属性 productDictList 用于将商品列表数据 productList 转换为字典，方便快速读取。productDictList 是对象，key 是商品 id，value 是商品信息；计算属性 countAll 用于计算商品总数量；计算属性 costAll 用于计算商品总价格。

然后定义了 handleCount 和 handleDelete 两个方法，handleCount 方法用于修改 cartList 中指定商品的购买数量，handleDelete 方法用于删除 cartList 中指定的商品。这两个方法都接收参数 index，index 是遍历 cartList 时的索引。此外，这两个方法交给了 Vuex 中的 mutations 来操作数据。在 main.js 中，Vuex 中的 mutations 的代码如下：

```
const store = new Vuex.Store({
    state: {
        // 购物车列表
        cartList: []
    },
    mutations: {
        ……
        // 修改购物车中的商品数量
        editCartCount(state, payload) {
            const product = state.cartList.find(item => item.id ===
payload.id);
            product.count += payload.count;
        },
```

```
            // 删除购物车中的商品
            deleteCart(state, id) {
                const index = state.cartList.findIndex(item => item.id === id);
                state.cartList.splice(index, 1);
            }
        }
        ......
});
```

21.10 订单提交

在购物车页面中,单击"下单"按钮,执行函数 handleOrder。这个函数定义在 list.vue 文件的<script>中的 methods 选项内,相关代码如下:

```
<script>
    import axios from 'axios'
    export default {
        ......
        methods: {
            ......
            // 处理下单
            handleOrder() {
                let param = new URLSearchParams();
                let cart = JSON.stringify(this.$store.state.cartList) ;
                param.append('cart', cart);
                axios({
                    method:
                        'post',
                    url:
                        '/eshop/order/handlerOrder',
                    data: param
                }) ;
                this.$store.commit('emptyCart');
                window.alert('下单成功');
            }
        }
    }
</script>
```

在 handleOrder 方法中,首先通过 axios 向后台服务器发送请求,请求的 url 为 /eshop/order/handlerOrder,这个请求将映射到控制器类 OrderInfoController 中的 handlerOrder 方法;data 用于指定要传递的参数,这里将传递的参数封装在 URLSearchParams 类型的对象 param 中,param 里存放的是购物车数据。然后,调用 Vuex 的 emptyCart 方法清空购物车。

在控制器类 OrderInfoController 中,handlerOrder 方法的代码如下:

```
package com.eshop.controller;
......
```

第21章 电商网站

```java
import com.fasterxml.jackson.core.type.TypeReference;
import com.fasterxml.jackson.databind.JsonMappingException;
import com.fasterxml.jackson.databind.ObjectMapper;
@Controller
@RequestMapping("/order")
public class OrderInfoController {
    @Autowired
    OrderInfoService orderInfoService;
    @Autowired
    ProductInfoService productInfoService;
    @RequestMapping(value = "/handlerOrder", method = {
RequestMethod.GET, RequestMethod.POST })
    @ResponseBody
    public String handlerOrder(String cart) throws JsonParseException,
JsonMappingException, IOException {
        // 创建ObjectMapper对象,实现JavaBean和JSON的转换
        ObjectMapper mapper = new ObjectMapper();
        // 将JSON字符串转换成List<CartItem>集合
        List<CartItem> ciList = mapper.readValue(cart, new
TypeReference<ArrayList<CartItem>>() {
        });
        OrderInfo oi = new OrderInfo();
        oi.setUid(1);
        oi.setStatus("未付款");
        oi.setOrdertime(new SimpleDateFormat("yyyy-MM-dd
HH:mm:ss").format(new Date()));
        double orderPrice = 0;
        for (CartItem ci : ciList) {
            ProductInfo pi =
productInfoService.getProductInfoById(ci.getId());
            orderPrice += ci.getCount() * pi.getPrice();
        }
        oi.setOrderprice(orderPrice);
        // 保存订单信息
        orderInfoService.addOrderInfo(oi);
        for (CartItem ci : ciList) {
            OrderDetail od=new OrderDetail();
            od.setOid(oi.getId());
            od.setPid(ci.getId());
            od.setNum(ci.getCount());
            // 保存订单明细
            orderInfoService.addOrderDetail(od);
        }
        return "sucess";
    }
}
```

在 OrderInfoController 类的 handlerOrder 方法中，首先将前端页面传递来的 JSON 格式的购物车数据转换成 List<CartItem>集合；然后调用业务接口 OrderInfoService 中的 addOrderInfo 保存订单信息到数据表 order_info；再循环调用接口 OrderInfoService 中的 addOrderDetail 方法保存订单明细。

21.11 小　　结

本章基于 Spring、Spring MVC 与 MyBatis 整合框架，采用注解方法并结合前端 Vue 框架，详细讲解了一个简单的电商网站的具体实现过程。电商网站实现的功能包括商品列表页显示、商品详情页显示、购物车页显示和订单提交，并按照三层架构开发每个功能模块。

通过本章的学习，希望读者能够熟练掌握 Spring、Spring MVC 与 MyBatis 框架整合开发的基本步骤、方法和技巧。